Advances in
Neurochemistry

Volume 2

Advances in Neurochemistry

ADVISORY EDITORS

A Continuation Order Plan is available for this series. A continuation order will bring delivery of each new volume immediately upon publication. Volumes are billed only upon actual shipment. For further information please contact the publisher.

Advances in

Neurochemistry

Volume 2

Edited by

B. W. Agranoff

Mental Health Research Institute and
Department of Biological Chemistry
University of Michigan
Ann Arbor, Michigan

and

M. H. Aprison

Institute of Psychiatric Research and
Department of Psychiatry and Biochemistry
Indiana University School of Medicine
Indianapolis, Indiana

PLENUM PRESS • NEW YORK AND LONDON

The Library of Congress cataloged the first volume of this series as follows:

Advances in neurochemistry/edited by B. W. Agranoff and M. H. Aprison.
– New York: Plenum Press, [1975-
v.   : ill.; 24 cm.
Includes bibliographies and index.

1. Neurochemistry.   I. Agranoff, Bernard W., 1926-   II. Aprison,
M. H., 1923-
[DNLM: 1. Neurochemistry–Period.   W1 AD684E]
QP356.3.A37   612'.822   75-8710

Library of Congress Card Catalog Number 75-8710

ISBN-13: 978-1-4615-8239-7   e-ISBN-13: 978-1-4615-8237-3
DOI: 10.1007/978-1-4615-8237-3

© 1977 Plenum Press, New York
Softcover reprint of the hardcover 1st edition 1977

A Division of Plenum Publishing Corporation
227 West 17th Street, New York, N.Y. 10011

CONTRIBUTORS

MANFRED L. KARNOVSKY • *Departments of Biological Chemistry and Psychiatry, Harvard Medical School, Boston, Massachusetts*

SEYMOUR KAUFMAN • *Laboratory of Neurochemistry, National Institute of Mental Health, Bethesda, Maryland*

DANIEL E. KOSHLAND, JR. • *Department of Biochemistry, University of California, Berkeley, California*

MASANORI OTSUKA • *Department of Pharmacology, Faculty of Medicine, Tokyo Medical and Dental University, Tokyo, Japan*

PETER REICH • *Departments of Biological Chemistry and Psychiatry, Harvard Medical School, Boston, Massachusetts*

T. L. SOURKES • *Laboratory of Neurochemistry, Department of Psychiatry, McGill University, Montreal, Quebec, Canada*

S. N. YOUNG • *Laboratory of Neurochemistry, Department of Psychiatry, McGill University, Montreal, Quebec, Canada*

PREFACE

In the Preface to Volume 1, we stated:

> This series recognizes that investigators who have entered neurochemistry from the biochemical tradition have a rather specialized view of the brain. Too often, interdisciplinary offerings are initially attractive but turn out to recite basic biochemical considerations. We have come to believe that there are now sufficiently large numbers of neurochemists to support a specialized venture such as the present one. We have begun with consideration of traditional areas of neurochemistry which show considerable scientific activity. We hope they will serve the neurochemist both for general reading and for specialized information. The reader will also have the opportunity to reflect on the unbridled speculation that results from the disinhibiting effects on the author who has been invited to write a chapter.
>
> We plan occasionally also to offer reviews of areas not completely in the domain of neurochemistry which we nevertheless feel to be sufficiently timely to be called to the attention of all who use chemical principles and tools in an effort to better understand the brain.

The contributions to the present volume pursue these goals. We believe the series has set high standards and has continued to uphold them. In accordance with the principle stated in the last paragraph of the Preface Volume 1, we include in this volume Koshland's "Sensory Response in Bacteria" (Chapter 5). We draw attention not so much to the fact that a bacterium has a "sense organ," but to its use of a temporal mechanism for chemoreception which may, like other bacterial mechanisms, have its counterpart in eukaryotes and in particular, in neurons.

B. W. Agranoff
M. H. Aprison

CONTENTS

CHAPTER 3

SUBSTANCE P AND SENSORY TRANSMITTER

MASANORI OTSUKA

CHAPTER 4

BIOCHEMISTRY OF SLEEP

MANFRED L. KARNOVSKY AND PETER REICH

CHAPTER 5
SENSORY RESPONSE IN BACTERIA
DANIEL E. KOSHLAND, JR.

PHENYLKETONURIA:
Biochemical Mechanisms

SEYMOUR KAUFMAN

Laboratory of Neurochemistry
National Institute of Mental Health
Bethesda, Maryland

1. INTRODUCTION

As implied by the title, this review of phenylketonuria (PKU) will be neither comprehensive nor encyclopedic. Rather, it will be limited to those aspects of the disease where sufficient biochemical knowledge is available to support meaningful discussion—admittedly, often speculative—about underlying mechanisms. Recent comprehensive reviews are available (Hsia, 1970; Knox, 1972).

To those interested in the normal development and functioning of the brain, PKU—especially its pathophysiology—has held the promise of providing unique insights. The reason for this hope can probably be traced to the remarkable specificity of tissue damage that is characteristic of the disease. Thus, while mental retardation is a consequence of many untreated inborn errors of metabolism, with most of them the developing brain is but one of the organs that is adversely affected by the disease. In galactosemia, for example, mental retardation is one of the characteristic symptoms, but there is also evidence of damage to the liver, kidneys, eyes, etc. By contrast, in PKU, the damage to the brain is quite specific.

This disease also provides us with one of nature's invaluable paradigms for studying and, ultimately, understanding aspects of the dynamic

1

metabolic interaction between different organs; for although the brain appears to be uniquely damaged in PKU, that organ does not even possess the enzyme system that is affected in the disease—the phenylalanine hydroxylating system, which is located exclusively in the liver and kidneys. The damage to the brain in PKU, therefore, is clearly secondary to the primary defect in these distant organs.

2. HISTORY OF THE DISEASE AND ITS GENERAL CHARACTERISTICS

Surprisingly, the first biochemically signficant observation on PKU was made not by a scientist but by the mother of two mentally retarded children. When, in 1934, she brought the children to A. Fölling, a young Norwegian biochemist and physician, she mentioned to him that since early infancy, a peculiar "mousy" odor had clung to the children. Apparently suspecting that the odor might be due to the urinary excretion of acetoacetate, Fölling added ferric chloride to samples of the urine. Instead of turning the red-brown color that is characteristic of acetoacetate, the urine turned green. In a period of less than 2 months, Fölling isolated from the urine the compound responsible for the green color and identified it as phenylpyruvate (Fölling, 1934a) (Later it was shown that the compound responsible for the mousy odor was phenylacetate). Within a short time, Fölling discovered eight more mentally retarded patients who also excreted phenylpyruvate. In 1938, Fölling made his second important observation on the biochemical characteristics of the new disease when he reported that these patients had elevated levels of phenylalanine in their blood and urine (Fölling, 1938). He correctly postulated that the new disease was caused by an inherited defect in phenylalanine metabolism (Fölling, 1934a), calling it *imbecilitas phenylpyruvica*. Subsequently, it was referred to as *phenylpyruvic amentia, phenylpyruvic oligophrenia*, Fölling's disease, and phenylketonuria. The last name, introduced by Penrose and Quastel (1937), appears to have been generally adopted.

Individuals with PKU are relatively healthy in all aspects but one—they are mentally retarded. The vast majority are idiots, i.e., their I.Q.s are less than 50 (Jervis, 1954; Paine, 1957; Hsia *et al.*, 1958; Partington, 1962). They also tend to show a deficiency in pigmentation and have blond hair and blue eyes (Paine, 1957). About one-third of the patients have some degree of eczema or other skin conditions (Knox, 1972).

In addition to mental retardation, the patients show other signs of central nervous system (CNS) pathology. Most have abnormal EEG patterns, and about one-fourth have a history of convulsive seizures, usually starting between 6 and 18 months of age. Other characteristic neurological manifestations of the disease are increased muscle tone, hyperactive tendon reflexes, tremors, and adventitious hyperkinesis (Knox, 1972). When it is not masked by severe mental retardation, disturbed behavior appears to be another salient feature of the disease. Screaming, sudden outbursts of violent activity, and increased irritability, are common. A significant number of the patients show behavior that has been interpreted as psychotic (see Cowie, 1971, for review).

Of all the symptoms of classical PKU, it is primarily the mental retardation, presumably a reflection of structural damage to the brain, that quickly becomes irreversible unless therapy (restriction of phenylalanine intake) is instituted early. Almost all the other symptoms can still be reversed by the dietary therapy long after the brain damage becomes permanent. As will be discussed, the etiology of these two kinds of symptoms (the reversible and the irreversible) are probably different.

Considering the overt signs of CNS involvement in PKU, there was a surprisingly long lag period before any pathological correlates of the disease in the CNS were established. It is now generally agreed that there is a defect in myelinization in the brains of PKU patients, which is accompanied by myelin breakdown products (Poser and Van Bogaert, 1959; Crome and Pare, 1960; Malamud, 1966; Menkes, 1968). Possible biochemical mechanisms for this defect will be discussed later.

Since effective treatment of the disease depends on early diagnosis (as will be shown), it is fortunate that there are biochemical changes that are manifested many months before the first signs of mental retardation or neurological deterioration. Nevertheless, not until it was realized that despite a lack of a functional phenylalanine hydroxylase system, the newborn PKU infant is not clinically abnormal [although it is somewhat underweight (Saugstad, 1972)], and all the urinary and blood biochemical abnormalities that characterize the disease develop postnatally, was accurate early diagnosis of PKU possible. An additional complication that compounded the first attempts at early diagnosis was ignorance of the fact that not all the biochemical abnormalities develop simultaneously.

Thus, although the elevation in blood phenylalanine in PKU infants is the earliest chemical change, and is therefore used as the basis for the diagnosis of the disease in the newborn infant, even this alteration is minimal at birth. This developmental characteristic of the disease, so critical for its reliable diagnosis in the newborn infant, was first reported by Armstrong and Binkley (1956). They showed that the concentration of phenylalanine in the blood of a PKU baby was normal at birth and that the level rose steadily thereafter, reaching a peak at 24 days. More recent studies have confirmed this finding and have shown that the maximum phenylalanine levels are reached, on the average, by the 6th or 7th day of postnatal life

(Holtzman *et al.*, 1974*a*). After this peak has been reached, there are indications that the levels fall again; i.e., the phenylalanine levels in older PKU children are lower than those of children younger than 3 years of age (Armstrong and Low, 1957). Because of this age-dependent rise in phenylalanine concentration, the earlier a PKU infant is screened, the greater the chance that the blood phenylalanine concentration will not be elevated (Holtzman *et al.*, 1974*b*) and the higher the probability that the diagnosis of PKU will be missed. Furthermore, it has been found that the time course of the increase in blood phenylalanine levels during the first week of life is different for males and females, with females showing a lag of 2–3 days before the rise (Holtzman *et al.*, 1974*b*). This sex difference, coupled with the drive for earlier and earlier screening (which, in turn, stems from the desire for the early initiation of therapy), probably accounts for the recently discovered predominance of male PKU infants over females that are detected in screening programs of newborn infants (Hsia and Dobson, 1970). It should be noted that surveys of retarded children (in contrast to surveys of all newborn infants) yield equal numbers of males and females with PKU (Jervis, 1954).

Beyond the tricky period immediately after birth, an increased phenylalanine level in the blood is a reliable diagnostic criterion for PKU or one of its variants. The fasting plasma level of phenylalanine, determined most accurately by ion-exchange column chromatography, is elevated twenty- to thirtyfold from a normal value of 0.051–0.054 mM (0.84 to 0.89 mg%) (Stein and Moore, 1954; Stein *et al.*, 1954; Evered, 1956; Linneweh and Ehrlich, 1962). Serum values are slightly higher (Perry *et al.*, 1967*a*). In a useful attempt to introduce a common language into this complex field, classical PKU patients have been defined as those who show persistent plasma phenylalanine levels of over 25 mg% (>1.5 mM) (Hsia, 1970).

As already mentioned, Fölling's original discovery of PKU stemmed not from his finding of elevated levels of phenylalanine in the blood of these patients but, rather, from his identification in the urine of an unusual product, phenylpyruvate. Since then, a long and ever-growing list of phenylalanine-derived metabolites, such as *o*-hydroxyphenylacetate (Armstrong *et al.*, 1955), phenylacetate (Woolf, 1951), phenylacetylglutamine (Woolf, 1951), and most recently, mandelate (Blau, 1970) have been found in elevated amounts in the urine and blood of PKU patients.

Subsequently, the list was shown to include metabolites of tryptophan, such as indoleacetic acid (Armstrong and Robinson, 1954) and indican (Bessman and Tada, 1960), as well as metabolites of tyrosine, such as *p*-hydroxyphenyllactic acid (Bickel *et al.*, 1955; Chalmers and Watts, 1974) and *p*-hydroxyphenylpyruvic acid (Chalmers and Watts, 1974; Hoffman and Gooding, 1969). Although it has been claimed that the *p*-hydroxyphenyl compounds that are excreted by phenylketonuric patients are derived in part from phenylalanine (Chalmers and Watts, 1974), a claim that would imply the presence of quite high residual phenylalanine hydroxylase activity in these patients, recent experiments with deuterated phenylalanine have shown that these parahydroxy derivatives are not derived from phenylalanine, but only from tyrosine (Curtius *et al.*, 1972*b*). As will be discussed, the evidence indicates that these disturbances in tryptophan and tyrosine metabolism are secondary to the primary disturbance in phenylalanine metabolism.

The earliest methods used to diagnose PKU followed Fölling's lead and were based on the measurement with $FeCl_3$ of phenylpyruvate in the urine. While this measurement is still useful in the diagnosis of the disease in children, it has limited value in newborns. Its limitations were pointed out as long ago as 1956, when it was shown that a PKU baby did not excrete detectable amounts of phenylpyruvate until it was 34 days old (Armstrong and Binkley, 1956). It was also shown that phenylpyruvate is no longer detectable in the urine when the serum phenylalanine levels fall below 15–20 mg/100 ml (Armstrong and Low, 1957). Both the relatively late onset of phenylpyruvate excretion and its relationship to phenylalanine blood levels have been confirmed (Rey *et al.*, 1974) and extended to the characteristics of the excretion of *o*-hydroxyphenylacetate (Rey *et al.*, 1974, Zelnicek and Slama, 1971). Despite these disadvantages and the added drawback of the nonspecificity of the $FeCl_3$ test for phenylpyruvate [*p*-hydroxyphenylpyruvate and xanthurenic acid also give a green color with $FeCl_3$ (Gibbs and Woolf, 1959)], there are still advocates of the use of measurements of urinary phenylpyruvate excretion in the diagnosis of PKU in the newborn (Allen *et al.*, 1964; Allen and Wilson, 1964).

PKU is transmitted as a recessive, autosomal trait (Jervis, 1954, 1939). The incidence of the disease is about 1 in 20,000 (Jervis, 1954; Berman *et al.*, 1969), although there is considerable geographic variation [e.g., the incidence in Ireland is only 1 in 4000 (Cahalane, 1968)]. On the basis of an incidence for the homozygous condition of 1 in 20,000, it can be calculated that the number of heterozygotes in the general population would be approximately 1 in 70.

Although heterozygotes for PKU have no signs of the disease, they can be detected by a phenylalanine tolerance test: they show a more prominent and persistent rise in plasma phenylalanine (Hsia *et al.*, 1956; Berry *et al.*, 1957) and a smaller rise in tyrosine (Jervis, 1960) than do normals. Heterozygotes for PKU can also be detected under the basal conditions; they show elevated plasma phenylalanine levels (Hsia *et al.*, 1956; Knox and Messinger, 1958) and elevated ratios of plasma phenylalanine to plasma tyrosine (Perry *et al.*, 1967b); both these deviations from the norm have been used. These methods of detection, however, suffer from an excessive overlap of the values for the heterozygotes and the normal individuals: about 20–30% of the heterozygotes cannot be classified with certainty. The discrimination can be improved by relating the plasma phenylalanine-to-tyrosine ratio to the plasma phenylalanine concentration (Rosenblatt and Scriver, 1968). With this method, 42 of 43 presumed heterozygotes were correctly identified.

As is often the case in medicine, an effective therapy for PKU was developed long before any clear picture of the pathogenesis of the disease was available. Bickel *et al.* (1954), extrapolating from the known beneficial effects of galactose restriction in the treatment of galactosemia, treated a 2-year-old PKU child with a low phenylalanine diet for a period of 2½ years. The treatment led not only to a fall in the level of phenylalanine in the blood and urine, but also to "an appreciable improvement in the patient's mental status" (Bickel *et al.*, 1954).

This first report was quickly followed by others that described the results of treatment of an additional 11 PKU patients of varying ages with low phenylalanine diets (Armstrong and Tyler, 1955; Bickel *et al.*, 1954; Woolf *et al.*, 1955). All these early studies reported that the dietary treatment led to normalization of the clinical and biochemical symptoms. Although it was clear that the beneficial effects of the diet included relief of most of the neurological symptoms, such as seizures, the improvement in I.Q. was less dramatic. There was agreement, however, that this therapy would probably be most effective if the diet were instituted at the earliest possible age.

Subsequent experience with the diet has confirmed this early impression and has shown that there is an inverse relationship between the ultimate I.Q. of the patient and the age at which the diet had been started; early treatment leads to I.Q.s that are close to 100 (Hsia, 1970). There is evidence that beyond the age of 3 or 4 years, the diet is ineffective, which is an indication that the damage to the brain after this period of time is irreversible (Knox, 1972). Consistent with this conclusion is the evidence that the low phenylalanine diet can be discontinued after the age of about 3 years without further deterioration in mental development (Kang *et al.*, 1970*b*; Hackney *et al.*, 1968; Horner *et al.* 1962; Hudson, 1967; Vandeman, 1963). The precise date for termination of the diet is still under study.

Until 8 years ago, the effectiveness of the dietary treatment of PKU was not uniformly accepted (Birch and Tizard, 1967; Bessman, 1966). More recent results, particularly from those studies in which the I.Q. of treated and untreated (or treated late) PKU siblings were compared, show that early treatment with the diet is effective; i.e., in the vast majority of the cases, the treated sibling had a higher I.Q. (Hsia, 1970; Smith and Woolf, 1974).

3.   ELUCIDATION OF THE METABOLIC BLOCK IN PKU

It was 13 years after Fölling's discovery of PKU in 1934 before the position of the metabolic block in the disease could be described with precision. During this long interval, it was not possible to go beyond the vague statement that the disease was caused by "a disturbance in phenylalanine metabolism."

The reason for this delay was that both before and during this period, the picture of normal mammalian phenylalanine catabolism was confusing and incomplete. In a sense, the field of intermediary metabolism of the aromatic amino acids had to catch up with the biochemical observations that were being made on PKU patients. As it turned out, it was knowledge gained from this disease, as well as that from another inborn error of metabolism, alcaptonuria, that helped complete the modern picture of normal phenylalanine and tyrosine metabolism.

With an armamentarium that did not yet include the ultimate weapon—isotopic tracers—chemists at the turn of the century attacked problems in intermediary metabolism with a variety of surprisingly potent techniques. They could overwhelm the whole animal or an isolated organ (by perfusion) with the substance under study and look for the accumulation of metabolites; they could administer derivatives of the substance with the hope that the blocking group would not be removed *in vivo,* and that it would prevent the total combustion of the substance and therefore allow the accumulation of the corresponding derivative of the normal metabolite; or, they could take advantage of the few human metabolic diseases that had been described at the time and determine whether a substance, when given to the patient, could increase the excretion of a metabolic product proximal to the metabolic block. For this area of intermediary metabolism, it was fortunate that the disease, alcaptonuria, had already been described. It was known that patients with this disease could carry out what was thought to be the normal conversion of tyrosine to homogentisic acid, but that the further metabolism of homogentisic acid was blocked, leading to the excretion of the latter compound. This mutant was brilliantly exploited in the study of aromatic amino acid catabolism.

Starting with the earlier knowledge that both phenylalanine and tyrosine could be completely combusted in animals to CO_2, H_2O, ammonia, and urea, Neubauer, adding results obtained mainly from studies with alcaptonurics, had formulated, in 1909, a scheme for the metabolism of both amino acids (Figure 1). With minor modifications (omission of the intermediates between p-hydroxyphenylpyruvate and homogentisic acid, insertion of fumarylacetoacetate between homogentisic acid and "acetone bodies," and elimination of the direct hydroxylation of phenylpyruvate), this scheme

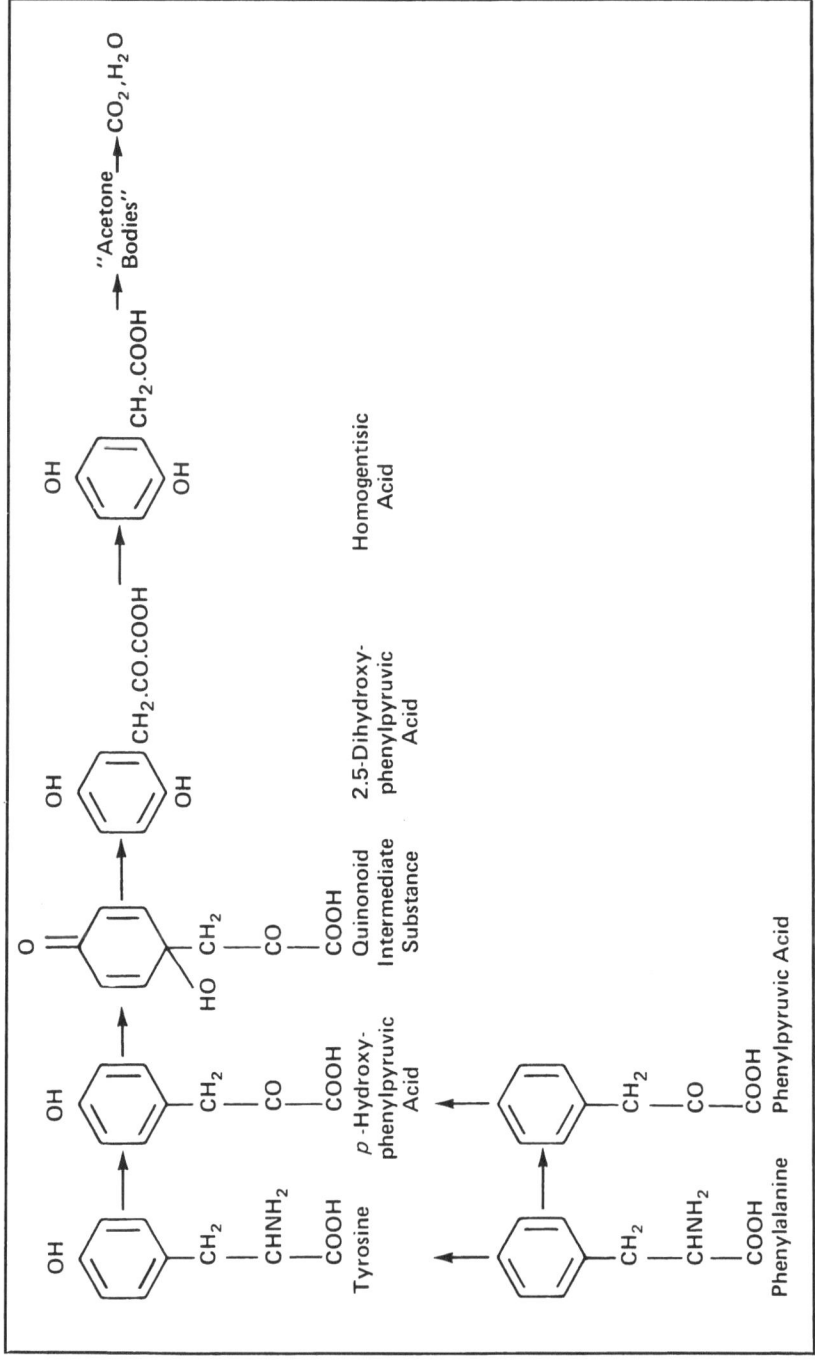

FIGURE 1. Neubauer's scheme for phenylalanine and tyrosine metabolism.

is in accord with the modern formulation of this catabolic pathway (Figure 2).

The proposed pathway was supported by an impressive weight of evidence, much of which can be summarized by one of Neubauer's generalizations (1904): all aromatic acids that can be easily oxidized in the whole organism will give rise to homogentisic acid in alcaptonurics. Not only phenylalanine and tyrosine, but also phenylpyruvate, phenyllactate, and *p*-hydroxyphenylpyruvate, were compounds that fit the generalization.

Embden *et al.* (1906) extended this generalization one step further by adding a third important characteristic: any aromatic compound that can be completely oxidized in animals and that gives homogentisic acid in alcaptonurics will also give rise to extra acetoacetate in a perfused liver preparation.

The only aspect of Neubauer's scheme that remained ambiguous was that part showing how phenylalanine enters the pathway. As can be seen, phenylalanine's entry is depicted as a loop with two alternative pathways: (a) direct conversion to tyrosine and (b) conversion to phenylpyruvate. It was precisely this issue of how phenylalanine enters the catabolic pathway that would continue to plague the field for almost half a century and that would hamper the early attempts to accurately describe the metabolic block in PKU.

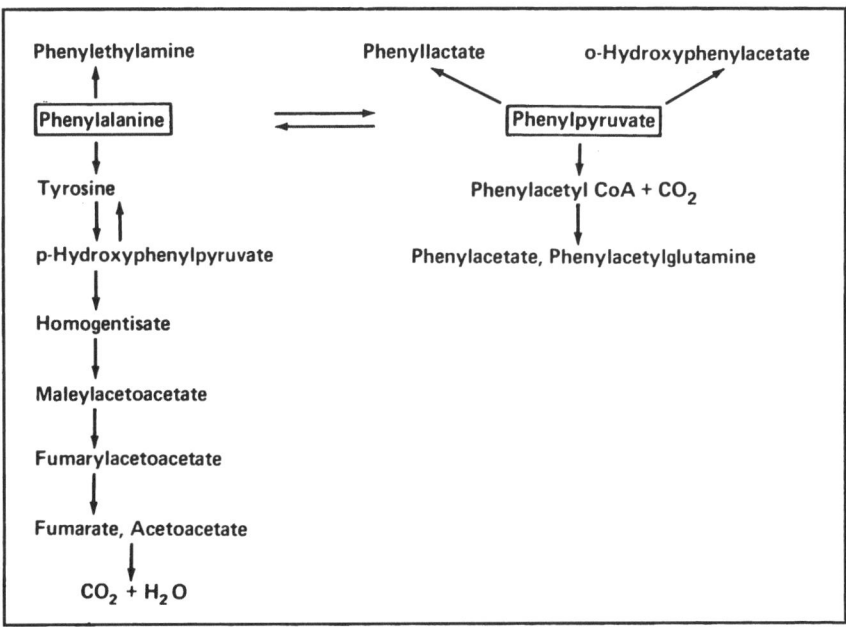

FIGURE 2.   Modern (1976) formulation of the metabolism of phenylalanine and tyrosine.

For, in spite of the evidence in favor of the Neubauer scheme, there were also results that it could not accommodate. Dakin, utilizing mainly the second approach to problems in intermediary metabolism mentioned previously, i.e., administration of derivatives of normal metabolites, had prepared both p-methylphenylalanine and p-methoxyphenylalanine and had come to the conclusion that both compounds "undergo practically complete oxidation in the normal organism in *precisely the same fashion as phenylalanine and tyrosine do*" (Dakin, 1911*a*). In subsequent experiments, he obtained results with these compounds that appeared to violate Neubauer's generalization; i.e., he found that these methylated derivatives gave rise to acetoacetate when perfused through the surviving liver of the dog (Dakin, 1911*a*). Furthermore, when fed to an alcaptonuric, they did not lead to extra excretion of homogentisic acid (Dakin, 1911*b*). Based on these results, Dakin concluded that the complete oxidation of phenylalanine proceeded through a pathway in which neither tyrosine nor homogentisic acid were normal intermediates (1911*a*). Rather, he postulated that phenylalanine was oxidized to CO_2 and H_2O by first being converted to phenylpyruvate in a pathway that can be abbreviated as: phenylalanine → phenylpyruvate → open chain form of phenylpyruvate → acetoacetate → CO_2 and water (Dakin, 1911*a*).

In 1913, Embden and Baldes reported the first convincing evidence for the direct conversion of phenylalanine to tyrosine when they showed that perfusion of the isolated dog liver with phenylalanine led to the formation of extra tyrosine. They also provided strong evidence against phenylpyruvate being a normal intermediate in the oxidation of phenylalanine by showing not only that phenylpyruvate did not give rise to acetoacetate in the perfused liver, but that it actually inhibited the formation of the latter compound from phenylalanine and tyrosine (Embden and Baldes, 1913).

Although Embden's results appeared to provide irrefutable evidence in favor of a simplified version of the Neubauer scheme (i.e., one without the loop involving entry of phenylpyruvate), the field could neither digest nor refute Dakin's contradictory results obtained with the methyl and methoxy derivatives of phenylalanine. Indeed, it was only recently that a partial explanation of Dakin's results was forthcoming. In 1957, Pirrung *et al.* showed that most, if not all, of the extra acetoacetate that is formed when the isolated dog liver is perfused with DL p-methoxyphenylalanine is due to deamination of the D-isomer of the methoxy derivative; the L-isomer does not give rise to extra acetoacetate. This ketogenic effect of the D-isomer is apparently due to the ability of ammonia to stimulate the formation in the liver of acetoacetate (Edson, 1935), a fact that, of course, was not known when Dakin did his work. It is also now clear that the methyl and methoxy derivatives of phenylalanine are not completely oxidized in animals but are excreted in part unchanged and in part as the p-methyl- and p-methoxy-phenylaceturic acid derivatives (Umezawa *et al.*, 1959).

In view of the confusing status of the field at the time of the discovery of PKU, it is understandable that Fölling's first attempt to explain the nature of the defect was incorrect. He suggested that in PKU there were two biochemical lesions, the first being an abnormal deamination of phenylalanine to phenylpyruvate, and the second being an inability to metabolize the latter compound (Fölling, 1934a).

For a time, these proposals seemed to receive some support. Thus, in 1937, Penrose and Quastel reported that PKU patients excrete more of an administered dose of phenylpyruvate than do controls. As had already been proposed by Fölling, these workers suggested that the metabolic disturbance in PKU was due "largely to a diminished rate of oxidation or rupture of the benzene ring of phenylpyruvate" (Penrose and Quastel, 1937).

The issue remained unresolved for another decade. In 1947, Jervis reported the results of a series of incisive and relatively straightforward experiments. He showed that the administration of phenylalanine to animals and normal humans led to a prompt rise in tyrosine in the blood. When the same experiment was carried out on PKU patients, there was no increase in tyrosine in the blood (Jervis, 1947). On the basis of these results, he postulated that the metabolic error in PKU is an inability to convert phenylalanine to tyrosine; i.e., the defect was in phenylalanine hydroxylase.

In 1953, Jervis demonstrated the defect in phenylalanine hydroxylase directly by showing that postmortem liver samples from three control patients could catalyze the conversion of phenylalanine to tyrosine *in vitro*, whereas liver samples from two PKU patients could not (Jervis, 1953).

It is now known from experiments with deuterium-labeled phenylalanine that phenylalanine is converted to tyrosine and that the reaction takes place even in the presence of dietary tyrosine (Moss and Schoenheimer, 1940). It is also known, mainly from metabolic studies on PKU patients, that the conversion of phenylalanine to tyrosine is an obligatory step in the complete oxidation of phenylalanine to CO_2 and water: there are no alternate pathways by which the benzene ring of phenylalanine can be completely combusted in animals. The alanine side chain of the amino acid can be modified extensively, e.g., by deamination, decarboxylation, and oxidation, but the benzene ring remains intact. When the major catabolic pathway is blocked, as in PKU, most of the phenylalanine is metabolized by these pathways that lead to modification of the alanine side chain, and these phenylalanine derivatives, all with their benzene rings intact, are excreted in the urine. A summary of the normal catabolic pathway that leads to metabolism of the whole molecule and of the pathway that leads only to metabolism of the side chain (mainly *via* phenylpyruvate) is shown in Figure 2 (compare with Neubauer's scheme in Figure 1).

As shown in the scheme, phenylalanine and phenylpyruvate are interconvertible. It is known, for example, that phenylpyruvate can satisfy the nutritional requirement for phenylalanine in most animals (Wood and Cooley, 1954), presumably because it can be converted to phenylalanine by transamination.

In light of these facts, it is worth reexamining the results obtained by Penrose and Quastel (1937). As already mentioned, they found that phenylketonurics have a diminished capacity to metabolize a dose of phenylpyruvate. Since part of the administered phenylpyruvate would normally be metabolized via phenylalanine (and then oxidized to CO_2 and H_2O), it is not surprising that phenylketonurics, with a block in phenylalanine hydroxylase, would, as a consequence, have a decreased capacity to metabolize phenylpyruvate. Part of their conclusion, therefore, is correct: phenylketonurics do have at least a partial block in phenylpyruvate oxidation. What these early workers did not realize, because it had not yet been clearly established that phenylalanine must first be converted to tyrosine before it can be completely oxidized, is that the block in phenylpyruvate oxidation in PKU is secondary to the primary block in the conversion of phenylalanine to tyrosine.

Jervis' work appeared to close the chapter on this aspect of PKU. Subsequent work has provided no reason to challenge his conclusion that the block in PKU lay between phenylalanine and tyrosine. It was only at the molecular level that his conclusion (i.e., that phenylalanine hydroxylase is defective in PKU) proved to be based on inadequate data. Another 20 years of basic work on the nature of the enzyme system that catalyzes the conversion of phenylalanine to tyrosine was needed before the affected component in PKU could be identified with certainty.

Just as lack of knowledge of the normal metabolism of phenylalanine and tyrosine had hampered attempts to identify the metabolic lesion in PKU, lack of knowledge about the phenylalanine hydroxylating system delayed efforts to accurately describe the block at the molecular level. The ability to accomplish this next step in the analysis of the defect in PKU had to wait on a detailed analysis of the phenylalanine hydroxylase system.

In 1953, when Jervis demonstrated that PKU liver samples were unable to convert phenylalanine to tyrosine, essentially nothing was known about the enzyme system that catalyzes the conversion. Indeed, at that time the reaction could not even be written as a balanced equation without invoking the participation of the imaginary "½ O_2" as one of the reactants (Figure 3). Since there was no basis for believing that the catalytic system consisted of anything more than one component (the enzyme), Jervis had equated "no measurable hydroxylase activity" with "no hydroxylase."

A start toward analyzing the enzymology of the conversion of phenylalanine to tyrosine was made in 1952 when it was reported that a soluble extract from rat liver could catalyze the hydroxylation of phenylalanine (Udenfriend and Cooper, 1952). It was also found that DPN$^+$ and a "cosubstrate" (an alcohol or aldehyde) were required.

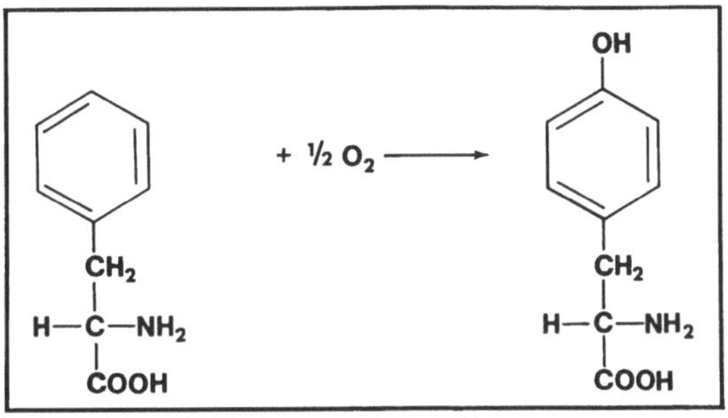

FIGURE 3. Early formulation of the enzymatic conversion of phenylalanine to tyrosine.

The significance of the peculiar requirement for DPN^+ and a cosubstrate was clarified when it was shown that the cofactor in the hydroxylation reaction was actually the reduced form of the pyridine nucleotide (Mitoma, 1956; Kaufman, 1957). In addition, two proteins from rat liver, one relatively stable and one quite labile, were shown to be involved (Mitoma, 1956). The two proteins were partially purified from rat (the labile one) and from sheep (the stabile one) liver extracts (Kaufman, 1957). Although DPNH was reported to be more active than TPNH when the two enzymes were prepared from rat liver (Mitoma, 1956), with the mixed sheep and rat liver system, TPNH was more active than DPNH (Kaufman, 1957).

The stoichiometry of the hydroxylation reaction with the latter system was shown to be in accord with Eq. 1 (Kaufman, 1957).

$$TPNH + H^+ + phenylalanine + O_2 \rightarrow TPN^+ + tyrosine + H_2O \quad (1)$$

This dual requirement for a biological reducing agent, TPNH, and molecular oxygen indicated that the hydroxylase was a mixed-function oxidase (i.e., an enzyme that catalyzes an oxidative reaction in which one atom of molecular oxygen is reduced to H_2O while the other is incorporated into the substrate) (Kaufman, 1962). This indication was verified when it was shown that the oxygen in the tyrosine formed from phenylalanine is derived from molecular oxygen and not from water (Kaufman et al., 1962).

The next advance in our knowledge of the phenylalanine hydroxylase system came with the discovery that in addition to the two enzymes and one coenzyme (i.e., reduced pyridine nucleotide), still another coenzyme was an essential component of the system (Kaufman, 1958). This nonprotein factor was isolated from rat liver extracts and identified in 1963 as an unconjugated pterin, the 7,8-dihydro derivative of biopterin (Kaufman, 1963), the structure of which is shown in Figure 4.

Since there was a 5-year period (from 1958 to 1963) during which the detailed structure of the naturally occurring cofactor was not known and the compound was not readily available, it was fortunate that synthetic

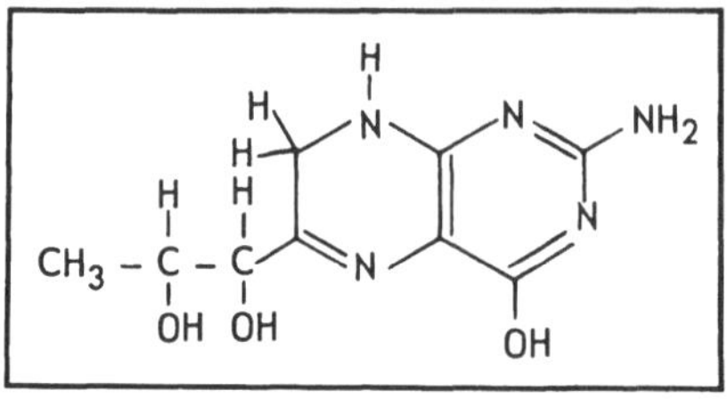

FIGURE 4. Structure of 7,8-dihydro-biopterin.

model pterin compounds with high hydroxylation cofactor activity were known. It was through the use of two of these compounds, 2-amino-4-hydroxy-6,7-dimethyl-5,6,7,8-tetrahydropteridine (DMPH₄) and the corresponding 6-monomethyl derivative (Kaufman and Levenberg, 1959), that the roles in the hydroxylation reaction of the two enzymes and the two coenzymes were elucidated.

The modern formulation of the enzymatic conversion of phenylalanine to tyrosine is shown in Figure 5 (Kaufman, 1959, 1964, 1971). The rat liver enzyme is phenylalanine hydroxylase. It catalyzes a coupled oxidation of phenylalanine and the tetrahydropterin (pterin is the trivial name for a 2-amino-4-hydroxypteridine) to tyrosine and the quinonoid dihydropterin, respectively; molecular oxygen is the electron acceptor. These are the minimum requirements for phenylalanine hydroxylation—the hydroxylase; the two substrates, O₂ and phenylalanine; and an active tetrahydropterin cofactor. With these components, however, one cannot get more tyrosine than the amount of tetrahydropterin present, for as the scheme shows, 1 mol tyrosine is formed for each mole of tetrahydropterin added (for certain pterins, such as the 7-methyl compound, the hydroxylation reaction is partially "uncoupled" and there is only about 0.3 mol tyrosine formed for each mole of tetrahydropterin oxidized (Storm and Kaufman, 1968).

Since the amount of phenylalanine that is metabolized per day in the liver exceeds by many orders of magnitude the amount of biopterin in the liver [6 μg/g in rat liver (Rembold, 1964)], nature provided a mechanism that allows the pterin cofactor to function catalytically. The sheep liver enzyme plays this role. It catalyzes a reaction that regenerates the tetrahydro form of the cofactor from the dihydro form, utilizing reduced pyridine nucleotide as the reductant. Once its role had been established, this enyme could be named for the reaction that it catalyzed, rather than for its tissue of origin—quinonoid dihydropteridine reductase or, simply, dihydropteridine reductase (Kaufman, 1971). Recently, the reductase from a variety of tissues has been shown to be more active with DPNH than with TPNH (Nielsen, 1969; Scrimgeour and Cheema, 1971; Craine et al., 1972).

There are two important characteristics of the quinonoid dihydropterin: the first is that it can be reduced nonenzymatically to the tetrahydro level by a variety of electron donors, such as ascorbate and mercaptans (Kaufman, 1959). Because of this property, the role of dihydropteridine reductase can be filled, *in vitro,* with certain reducing agents. The second property is that the quinonoid compound is extremely unstable, undergoing rapid tautomerization to the more stable 7,8-dihydro derivative (Kaufman, 1959). The latter compound, in contrast to the quinonoid dihydropterin, cannot be reduced at a significant rate by either dihydropteridine reductase or, nonenzymatically, by ascorbate or mercaptans (at 25°C, at neutral pH) (Kaufman, 1959).

These are the four essential components of the phenylalanine hydroxylase system: the two enzymes, phenylalanine hydroxylase and dihydropteridine reductase, and the two coenzymes, a tetrahydropterin and reduced pyridine nucleotide.

The instability of the quinonoid dihydropterin intermediate and its rearrangement to the 7,8-dihydro compound led to the discovery of the fifth component of the system.

As has already been mentioned, the hydroxylation cofactor was originally isolated from liver in the 7,8-dihydro form. Since the cofactor in this form is active in the phenylalanine hydroxylase system (it was isolated on the basis of this activity) and since 7,8-dihydropterins are essentially inactive as substrates for dihydropteridine reductase, it was evident that there must be a mechanism for reduction of 7,8-dihydrobiopterin to the active tetrahydro form and that this reaction was taking place in the presence of the partially purified phenylalanine hydroxylase and reductase fractions that were being used at that time.

These inferences proved to be correct. It was shown that impure preparations of dihydropteridine reductase from sheep liver are contaminated with another enzyme, dihydrofolate reductase. Although previously it had been thought that this enzyme is specific for conjugated pterins like 7,8-dihydrofolate, it was shown that it is also capable of catalyzing the reduction of 7,8-dihydrobiopterin to tetrahydrobiopterin according to Eq. 2 (Kaufman, 1967*a*).

TPNH + H$^+$ + 7,8-dihydrobiopterin →
$$\text{TPN}^+ + 5,6,7,8\text{- tetrahydrobiopterin} \quad (2)$$

The scheme for the hydroxylation of phenylalanine starting with 7,8,-dihydrobiopterin is shown in Figure 5.

In contrast to dihydropteridine reductase, dihydrofolate reductase is not active with 7,8-dihydro derivatives of the model pterin cofactors, such as the 6,7-dimethyl pterin (Kaufman, 1959; Morales and Greenberg, 1964). Because of this specificity, 7,8-dihydro-6,7-dimethylpterin is not active in the phenylalanine hydroxylase system, whereas 7,8-dihydrobiopterin is active.

Whereas dihydropteridine reductase, as mentioned previously, is more active with DPNH than with TPNH, the reverse is true for dihydrofolate reductase. The activity of the latter enzyme from rat liver is 15 times higher

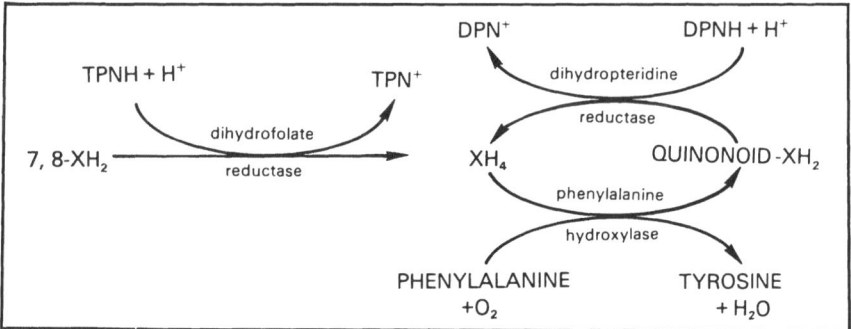

FIGURE 5. Modern formulation of the enzymatic conversion of phenylalanine to tyrosine (XH_2 is dihydroptenin; XH_4 is tetrahydropterin).

and that from sheep liver is 6 times higher with TPNH than with DPNH (Craine *et al.*, 1972).

The different pyridine nucleotide specificity of these two reductases indicates that the nucleotide specificity of the phenylalanine hydroxylase system *in vitro* in any crude tissue preparation will be determined by two variables: (a) the state of oxidation of biopterin and (b) the relative activities of dihydropteridine and dihydrofolate reductases.

As can be seen in Figure 5, if, under any condition, 7-8-dihydrobiopterin is the predominant form of the cofactor, the nucleotide specificity of the hydroxylation reaction will be determined by the reductase reaction that is rate-limiting. If the dihydropteridine reductase reaction is limiting, the initial rate of the hydroxylation reaction will be faster with DPNH than with TPNH; if the dihydrofolate-reductase-catalyzed reaction is limiting, the hydroxylation reaction will be faster with TPNH than with DPNH.

If, on the other hand, tetrahydrobiopterin is the predominant form of the cofactor, the dihydrofolate-reductase-catalyzed reaction will play only a minor role in the conversion of phenylalanine to tyrosine, and the nucleotide specificity of the system will be determined by the specificity of dihydropteridine reductase; i.e., the hydroxylation reaction will be relatively specific for DPNH.

These considerations can probably account for the earlier, apparently conflicting, observations on the nucleotide specificity of the phenylalanine hydroxylase reaction in crude liver preparations (Kaufman, 1957; Mitoma, 1956).

Although dihydrofolate reductase is an essential component of the hydroxylation system in the presence of 7,8-dihydrobiopterin, it is not known with certainty whether it plays any role in the hydroxylation reaction *in vivo*. There are two reasons for suspecting that it does: (a) in all probability, the *de novo* synthesis of biopterin proceeds through the 7,8-dihydro derivative (Kaufman, 1967*b*) so that dihydrofolate reductase would be obligatorily required to convert this compound to the tetrahydro level; (b) dihydrofolate reductase probably serves to salvage some of the 7,8-dihydrobiopterin that might be formed from the quinonoid compound in any situation where the dihydropteridine-reductase-catalyzed reaction is not fast enough to keep pace with the hydroxylase (see Figure 5).

Finally, there are two other components that may play a role in the phenylalanine hydroxylase system, but only under special conditions, and only in the presence of the naturally occurring cofactor, tetrahydrobiopterin.

The first of these is a protein, called *phenylalanine hydroxylase stimulator* (PHS), that has been isolated from rat liver extracts in pure form (Kaufman, 1970; Huang *et al.*, 1973). There is evidence to support the idea that PHS is an enzyme that catalyzes the breakdown to products of an intermediate in the hydroxylation reaction (Huang and Kaufman, 1973). PHS is believed not to be required in the presence of cofactor analogues, such as $DMPH_4$, because the corresponding intermediate formed from this pterin breaks down rapidly (to products) even in the absence of PHS (Huang and Kaufman, 1973).

Another factor that markedly stimulates (thirty- to fiftyfold) the hydroxylation reaction in the presence of tetrahydrobiopterin, but not in the presence of $DMPH_4$, is lysolecithin (and related phospholipids) (Fisher and Kaufman, 1972*b*, 1973*b*). While both PHS and lysolecithin can be regarded as components of the phenylalanine hydroxylase system in the test tube, it is not known whether they participate in the hydroxylation reaction *in vivo*.

In summary, there are four components that are directly involved in the hydroxylation reaction: phenylalanine hydroxylase, dihydropteridine reductase, tetrahydrobiopterin, and reduced pyridine nucleotide. The fifth component, dihydrofolate reductase, is probably involved *in vivo* in at least an ancillary manner. The sixth and seventh factors, PHS and phospholipids, although not essential, can markedly stimulate the reaction *in vitro*.

Phenylalanine hydroxylase is present in the soluble fraction when liver tissue is extracted with isotonic solutions (Udenfriend and Cooper, 1952). Although it was originally reported to be present exclusively in liver (Udenfriend and Cooper, 1952), it was subsequently detected in kidney and pancreas (Tourian *et al.*, 1969). Its occurrence in the latter tissue, however, has not been confirmed (McGee *et al.*, 1972).

The enzyme has been obtained from rat liver in essentially pure form (Kaufman and Fisher, 1970). It is a multimeric protein composed of two electrophoretically distinguishable subunits (mol.wt. 51,000–55,000) (Kaufman and Fisher, 1970). In addition to the monomers, the enzyme exists as dimers (mol.wt. 110,000) and tetramers (mol.wt. 210,000) (Kaufman and Fisher, 1970). A minor amount of a third species of the tetrameric form of the enzyme, which accounts for 5–10% of the total hydroxylase activity, has recently been detected in rat liver extracts (Barranger *et al.*, 1972). It was also reported that fetal rat liver contains only two of the three isozymes seen in adult liver (Barranger *et al.*, 1972). No differences in kinetic properties of the isozymes have been reported (Barranger *et al.*, 1972).

Although the three different forms of the enzyme may indeed be isozymes, the possibility has not been ruled out that within the liver cell there exists only a single type of monomer (and hence only one species of dimer and one species of tetramer) and that the "isozymes" that have been detected are formed from this single species (by partial proteolysis or by some other kind of modification) during the isolation of the enzyme. Because of this possibility of *in vitro* artifact, and because of the failure to detect functional differences between the different forms of the enzyme, the physiological significance of isozymes of phenylalanine hydroxylase is not known. Despite these qualifications, the detection of "isozymes" has generated considerable speculation about their relevance to PKU (Barranger *et al.*, 1972).

Phenylalanine hydroxylase is a metalloprotein. It contains about 1 mol Fe per monomer (Fisher *et al.*, 1972). The Fe is essential for hydroxylase activity, but the mechanism of its participation in the reaction is not known (Fisher *et al.*, 1972).

4. IDENTIFICATION OF THE AFFECTED COMPONENT OF THE PHENYLALANINE HYDROXYLASE SYSTEM IN CLASSICAL PKU

As progress was made in the analysis of the complex phenylalanine hydroxylase system, parallel progress was made in identifying the affected component in PKU.

Jervis' work had shown that the system that catalyzes the conversion of phenylalanine to tyrosine is inactive in PKU liver. The first step beyond this point was taken after it had been established that at least two proteins were somehow involved in the hydroxylation reaction, but before their roles in the system had been elucidated. In 1957, it was reported that only one of these two proteins, the more stable one, was active in PKU liver samples, whereas the "labile enzyme" was missing (Mitoma *et al.*, 1957; Wallace *et al.*, 1957).

The next step was taken after it was shown that a pterin cofactor is an essential component of the phenylalanine hydroxylase system (Kaufman, 1958*a*). Since, under many conditions, this cofactor is by far the most labile component of the system (Kaufman, 1958*a*), the earlier work had not even established that an enzyme is missing in PKU liver. In 1958, it was reported that the pterin cofactor is present in PKU liver samples in amounts comparable to those present in control samples (Kaufman, 1958b). It was also shown that the "rat liver enzyme" activity was not detectable and that the "sheep liver enzyme" activity was present in the PKU liver samples. The demonstration that the cofactor is present provided the first strong evidence that an enzyme is missing in PKU; the subsequent identification of the rat and sheep liver enzymes as phenylalanine hydroxylase and dihydropteridine reductase, respectively (Kaufman, 1959), provided the first sound basis for the conclusion that the affected enzyme in PKU is phenylalanine hydroxylase.

Until recently, the evidence in favor of this conclusion was strong but not conclusive. The reason for the small degree of uncertainty that still existed can be traced to the type of evidence that was available. That evidence was based on experiments in which it had been shown that liver extracts from PKU patients are inactive in converting phenylalanine to tyrosine under conditions where extracts from control patients were active. To identify the inactive component of a multicomponent system, the next step would be to supplement the inactive extract with the individual components of the system (ideally, one at a time) until the hydroxylating activity is restored. What was done in practice in all of the earlier experiments was to restore activity with the addition of fairly crude preparations of phenylalanine hydroxylase. It is evident that if subsequent work had led to a resolution of that crude hydroxylase preparation into two or more components, some uncertainty would have been introduced into the original conclusion. There is, therefore, a relationship between the firmness of the conclusion that phenylalanine hydroxylase is the affected component and the purity of phenylalanine hydroxylase (or any other component of the system) that is used in supplementation experiments.

After phenylalanine hydroxylase from rat liver was obtained in essentially pure form (Kaufman and Fisher, 1970), this last element of uncertainty was dispelled. It has been shown that the phenylalanine hydroxylating activity can be restored to an inactive PKU liver sample by the addition of essentially pure rat liver phenylalanine hydroxylase. The addition of none of the other components of the system could substitute for the hydroxylase. Furthermore, direct assays showed that PHS, the most recently identified protein component of the system, is present in normal amounts in a PKU liver sample (Huang et al., 1973). From these results, it can be concluded that phenylalanine hydroxylase is firmly established as the only affected component of the hydroxylase system in the classical form of PKU.

This conclusion has been further strengthened by results obtained with a specific antiserum to essentially pure rat liver phenylalanine hydroxylase. The antiserum, elicited in sheep, cross-reacts with normal human phenylalanine hydroxylase (Friedman *et al.*, 1972c). Under conditions where 5–10% of the normal human liver enzyme could have been detected with the double immunodiffusion technique, no precipitin line was seen with an extract from a PKU liver sample (Friedman *et al.*, 1972*b*).

5. THE NATURE OF THE GENETIC DEFECT IN CLASSICAL PKU

The results of the immunological studies just discussed ruled out only a single possibility for the molecular basis of the phenylalanine hydroxylase deficiency in PKU: normal amounts of a mutant form of the hydroxylase with all its antigenic determinants intact, but devoid of hydroxylase activity, were not present in PKU liver. These results left open the possibility that PKU was caused by a deletion mutation of the gene that codes for the hydroxylase.

To discriminate between this possibility and the remaining possibilities for explaining the basis of the phenylalanine hydroxylase deficiency, a PKU liver sample was examined with an assay for the hydroxylase that was considerably more sensitive than those that had been used in any of the earlier studies on PKU liver. In particular, the assay tried to exploit two characteristics of the hydroxylase: (1) the dramatically lower K_m of the enzyme for phenylalanine in the presence of tetrahydrobiopterin as compared with its K_m in the presence of, e.g., the 6,7-dimethylterahydropterin (Kaufman, 1970), and (2) the stimulation of the hydroxylase by lysolecithin in the presence of tetrahydrobiopterin (Fisher and Kaufman, 1973*b*).

With the more sensitive assay conditions that have just been outlined, a small amount of enzyme-dependent conversion of ^{14}C-labeled phenylalanine to ^{14}C-labeled tyrosine was detected (Friedman *et al.*, 1973). The hydroxylase activity (per milligram of protein) was equal to about 0.27% of that present in a liver extract from a non-PKU control. This value is well below (by a factor of 20–40) the limits of detection of the immunological tests that we had used previously. The hydroxylated product was unequivocally identified as tyrosine by chromatographic and enzymatic methods (Friedman *et al.*, 1973).

Because of the minute amounts of activity that were detected and because there is some cross-specificity among the various pterin-dependent aromatic amino acid hydroxylases—e.g., adrenal and brain tyrosine hydroxylase and pineal and brain tryptophan hydroxylase can convert phenylalanine to tyrosine (Kaufman and Fisher, 1974; Tong and Kaufman, 1975)—the hydroxylating enzyme in the PKU liver sample was characterized with the use of rather specific inhibitors of the three pterin-dependent hydroxylases. The results of the inhibitor analysis are shown in Table 1. First, it can be seen that tyrosine formation was completely dependent on added tetrahydrobiopterin and was stimulated more than twofold by lysolecithin. Since liver has detectable amounts of tyrosine hydroxylase, presumably localized to the nerve endings in that organ, it was possible that this enzyme was responsible for the phenylalanine hydroxylating activity in the

TABLE 1. Characteristics of the Enzyme from the
Liver of a Patient with PKU that Converts
[^{14}C]Phenylalanine to [^{14}C]Tyrosine

Additions (+) or omissions (−)	Activity (pmol tyrosine/hr per mg protein)
Experiment A[a]	
Control	27
−Lysolecithin	12.5
−Tetrahydrobiopterin	0.15
Experiment B[b]	
Control	30
+0.1 mM 3-iodotyrosine	26
Experiment C[c]	
Control	29
+1 mM p-chlorophenylalanine	0.0
+ 1 mM 6-fluorotryptophan	27
+Adrenal tyrosine hydroxylase	13,600
+Adrenal tyrosine hydroxylase +0.1 mM 3-iodotyrosine	250

[a] Total volume of reaction mixture was 0.05 ml. The extract was treated with ammonium sulfate to 75% of saturation, and the precipitate was dissolved in 0.12 M KCl buffered with 0.01 M Tris-HCl (pH 7.0); 400 μg of this ammonium sulfate fraction was added to each assay. Phenylalanine concentration was 0.05 mM (300,000 cpm/tube), and tetrahydrobiopterin concentration was 0.05 mM. Each value has been corrected for any radioactivity in the tyrosine area (equivalent to 24 pmol tyrosine/hr per mg protein) after incubation with a boiled enzyme control.

[b] Total volume was 0.1 ml. Extract, 0.05 ml, containing 1 mg protein, was added to each tube. Phenylalanine content was 0.11 mM (750,000 cpm), and tetrahydrobiopterin content was 0.025 mM. The radioactivity in the tyrosine area after incubation with a boiled enzyme control was equivalent to 31 pmol tyrosine/hr per mg protein.

[c] Total volume was 0.05 ml. Extract, 0.015 ml (0.3 mg protein), was added to each tube. Phenylalanine content was 0.036 mM (375,000 cpm), and tetrahydrobiopterin content was 0.03 mM. The radioactivity in the tyrosine area after incubation with a boiled enzyme control was equivalent to 14 pmol tyrosine/hr per mg protein. The tyrosine hydroxylase used had been purified through the substituted Sepharose step from bovine adrenal glands; 30 μg of this fraction was added to each reaction tube assayed. The tyrosine content was 0.05 mM (400,000 cpm/tube), and the tetrahydrobiopterin content was 0.1 mM.

PKU liver sample. The results of Experiment B (Table 1) ruled out this possibility. The potent and specific tyrosine hydroxylase inhibitor, 3-iodo-tyrosine, inhibited tyrosine formation by only 14%. Since rat liver phenylalanine hydroxylase is not inhibited at all by this concentration of iodotyrosine, it was concluded that no more than 14% of the hydroxylation that was observed could have been due to tyrosine hydroxylase. It can also be calculated (from the reported activity of tyrosine hydroxylase in liver) that the observed specific activity of phenylalanine hydroxylase in the PKU liver sample is about 50 times greater than the reported value for tyrosine hydroxylase. The results of Experiment C (Table 1) show that the hydroxylase activity was sensitive to the nonspecific amino acid hydroxylase inhibitor, p-chlorophenylalanine, whereas it was insensitive to the rather specific inhibitor of tryptophan hydroxylase, 6-fluorotryptophan. The results of these studies strongly indicate that at least 86% of phenylalanine hydroxylating activity in this sample was due to a form of phenylalanine hydroxylase.

The detection of some phenylalanine hydroxylase in PKU liver obviously ruled out the possibility that there is a deletion of the structural gene for the hydroxylase as the genetic basis for this disease. The results were compatible with the idea that the hydroxylase activity in classical PKU is due to the presence of either low amounts of normal hydroxylase, or a structurally altered hydroxylase with low catalytic activity.

In an attempt to distinguish between these two alternatives, the kinetic properties of the hydroxylase in normal and in PKU liver were compared. The first difference is in the degree of stimulation by lysolecithin. As already mentioned (Experiment A, Table 1), the enzyme in PKU liver was stimulated about twofold by lysolecithin. It has previously been shown that the normal human liver enzyme is stimulated three- to fourfold (Friedman *et al.*, 1973). It is possible, but not likely, that this represents a difference in endogenous phospholipid content between the two livers, rather than a difference in properties between the two enzymes.

A more convincing difference in properties was uncovered when the apparent K_m value for phenylalanine was determined. Although the K_m values themselves may not be significantly different, it was found that the enzyme from PKU liver is not inhibited by excess phenylalanine up to 0.2 mM. By contrast, the normal human enzyme is 50% inhibited at 0.14 mM phenylalanine (Friedman *et al.*, 1973).

Finally, the capacity of the antiserum to rat liver hydroxylase to inhibit the normal and PKU liver hydroxylases was studied. It was found that the maximum inhibition that could be observed was 64% with the normal and 18% with the PKU liver hydroxylases (Friedman *et al.*, 1973).

All these results, which are summarized in Table 2, indicate that in this PKU patient there is a structurally altered hydroxylase with very low activity rather than a very low amount of normal enzyme; i.e., this PKU

TABLE 2. Properties of a PKU and a Normal Human
Liver Phenylalanine Hydroxylase

Property	Normal	PKU
K_m for phenylalanine (nM)	28–39	37
Substrate inhibition above 0.1 mM phenylalanine	Present	None
Specific activity[a] (nmol tyrosine/hr per mg protein)	56	0.150
Stimulation by lysolecithin(%)	300–400	100
Inhibition by antiserum(%)	63	18

[a] V_{max} with tetrahydrobiopterin and lysolecithin.

patient has a mutation in the gene coding for the structure of phenylalanine hydroxylase.

To explain why no cross-reacting protein could be detected in PKU liver with the specific antiserum to phenylalanine hydroxylase, several alternative models can be proposed: (a) The mutation in phenylalanine hydroxylase affects not only the enzyme's catalytic properties, but also its antigenic determinants, in such a way that it no longer forms a precipitating complex with the antibodies. According to this model, PKU liver may contain a normal amount of the structurally altered hydroxylase, but the method used cannot detect it. (b) The altered hydroxylase is synthesized at a normal rate but is degraded more rapidly than is the normal enzyme, perhaps as a consequence of its altered structure. According to this model, PKU liver contains less than the normal amount of the structurally altered hydroxylase. However, other models could also be devised that could account for our results.

Since there is no reason to suspect that this PKU patient is anything but typical, it seems likely that the above conclusion will apply to most, if not all, patients with classical PKU. Just how general the conclusion is can be determined only by future studies on the hydroxylase in other PKU patients.

The detection of some conversion of phenylalanine to tyrosine in the liver of a PKU patient and the characterization of the responsible enzyme as phenylalanine hydroxylase is the first direct evidence that the hydroxylase is not completely missing or inactive in classical PKU.

It will be noted from Table 2 that the comparison between the phenylalanine hydroxylase activity in the PKU and the normal liver samples refers to maximum velocities. Since the concentrations of phenylalanine in the blood (and tissues) of untreated PKU patients is elevated, the hydroxylase in the PKU patient will be operating closer to V_{max} than will the enzyme in normals. For this reason, the *in vivo* phenylalanine hydroxylase activity in PKU will likely be somewhat higher relative to control values than would be estimated from the *in vitro* assays; i.e., the 0.27% of normal activity determined *in vitro* for the PKU patient could be increased to 0.5–0.6% of the normal activity *in vivo*.

It had previously been shown that PKU patients have measurable ability to convert [^{14}C]phenylalanine to [^{14}C]tyrosine *in vivo* (Udenfriend and Bessman, 1953). The ratio of the specific radioactivities of the tyrosine and phenylalanine isolated from plasma proteins was about 10% that from control patients. Although this ratio was considered to be a "rough index of the extent of conversion of phenylalanine to tyrosine" (Udenfriend and Bessman, 1953), there is no obvious way of translating these data into a quantitative measure of phenylalanine hydroxylase activity. It is certainly too rough an index of phenylalanine hydroxylase activity to sustain the conclusion that PKU patients have 10% of the normal ability to convert phenylalanine to tyrosine (Knox and Hsia, 1957). Another barrier to attempts to correlate *in vivo* and *in vitro* results is the possible contribution to the measured hydroxylation reaction *in vivo* of bacterial enzymes in the gut and that of other pterin-dependent aromatic amino acid hydroxylases (Kaufman and Fisher, 1974) in the adrenal medulla and in the central and peripheral nervous system.

Recently, the metabolism of deuterated phenylalanine was studied in a normal child, a phenylketonuric, and a hyperphenylalaninemic patient. In contrast to the normal child, neither of the patients formed detectable amounts of deuterated tyrosine or metabolites derived from tyrosine (Curtius *et al.*, 1972*b*). Thus, this method of assessing phenylalanine hydroxylation *in vivo*, unlike the one used earlier (Udenfriend and Bessman, 1953), indicated that the block in phenylalanine hydroxylation is essentially complete. Presumably, the deuterated phenylalanine method is not sensitive enough to detect the approximately 5% of normal phenylalanine hydroxylase activity that would be expected in the case with hyperphenylalaninemia (see section 6). It is not surprising, therefore, that it failed to detect the much lower activity expected for the classical PKU case.

In addition to trying to evaluate reported differences in the *in vivo* and *in vitro* estimates of phenylalanine hydroxylase activity in PKU patients, it is necessary to examine critically reports that claim to have detected much higher residual phenylalanine hydroxylase activities in the liver samples from PKU patients as well as those that indicate a bewildering range of values in variants of PKU.

The reliable assay of any enzyme activity in crude tissue preparations is difficult; e.g., other enzymes may be present that can degrade the substrate, an essential cofactor, or the product, and inhibitors or activators may influence the reproducibility of results. These difficulties are compounded when the enzyme being studied is part of a complex system, as is the case with phenylalanine hydroxylase. In addition, as will be discussed, it is now clear that PKU is not a single entity, but rather is a heterogeneous group of diseases.

In the face of this complexity, of both biochemical and genetic origin, the field has been ill-served by the gratuitous element of chaos that is introduced with the use of *in vitro* assays for phenylalanine hydroxylase, which may not be reliable.

As has already been discussed (Kaufman and Max, 1971), one of the practices that can limit the value of *in vitro* measurements of phenylalanine hydroxylase activity is the failure to use an assay mixture that contains nonlimiting amounts of all the known components of the system. Clearly, it is only when all these components are present that the measurement of the conversion of phenylalanine to tyrosine becomes a measure of phenylalanine hydroxylase activity.

An assay that does contain all these components, which is suitable for measuring phenylalanine hydroxylase activity in needle biopsy liver samples from patients, has been described (Kaufman, 1969). The components include: phenylalanine, dihydropteridine reductase, reduced pyridine nucleotide (DPNH, or TPNH, or both), a reduced pyridine-nucleotide-generating system, catalase, and a tetrahydropterin. Catalase has been shown to protect the hydroxylating system from H_2O_2-mediated inactivation (the autooxidation of tetrahydropterins generates H_2O_2) (Kaufman, 1971).

In the desire to simplify this complex system, some investigators have omitted catalase, dihydropteridine reductase, and a pyridine-nucleotide-regenerating system (Justice *et al.*, 1967; LaDu and Zannoni, 1967). While it is likely that most "normal" human liver extracts will not be limiting in any of these components [e.g., it has been shown that extracts from normal human liver have sufficient dihydropteridine reductase so that added amounts of this enzyme do not stimulate the conversion of phenylalanine to tyrosine in these extracts (Kaufman, 1969)], there is no assurance that this will be true in every case. Unless it is shown that these components are not limiting in the individual tissue sample being examined, any hydroxylase assay carried out in their absence may not be a valid measure of phenylalanine hydroxylase activity. Since tetrahydropterins are very easily oxidized [the half-life for $DMPH_4$ is about 6–7 min at 20°C (Kaufman, 1964)] and are probably even less stable in the presence of crude tissue extracts, the use of large amounts of tetrahydropterins is unlikely to eliminate the need for adding a tetrahydropterin-regenerating system. Furthermore, there are distinct disadvantages to the use of large amounts of $DMPH_4$: they can lead to significant rates of nonenzymatic hydroxylation of phenylalanine and, in the absence of an efficient tetrahydropterin-regenerating system, the 7,8-dihydropterin that is formed may inhibit the hydroxylase.*

For these reasons, the report that a patient with classical PKU as well as one with "mild" PKU (plasma phenylalanine of 17 mg/100 ml) both had undetectable levels of phenylalanine hydroxylase in their livers cannot be rigorously interpreted. The assay mixture used to measure phenylalanine hydroxylase activity included only phenylalanine, $DMPH_4$ and DPNH, and nicotinamide (Justice *et al.*, 1967); all the other essential factors were assumed to be present in the extracts in nonlimiting amounts. The failure to detect phenylalanine hydroxylase activity under these conditions, especially in the case with "mild" PKU, could have been due to a partial deficiency of phenylalanine hydroxylase that appeared to be more severe than it really was because of an additional deficiency of one of the compo-

---

* The use of oversimplified assays for the hydroxylase has probably generated other problems that are related to PKU.

    With the use of one of these simplified assays, LaDu and Zannoni (1967) and Zannoni and Moraru (1969) have obtained kinetic results for the hydroxylase that are at variance with those reported by others (Kaufman, 1971; Kaufman and Fisher, 1974; Treiman and Tourian, 1973). Specifically, with their assay, LaDu and his colleagues found that the apparent K_m for $DMPH_4$ was not a constant but, rather, increased as the concentration of phenylalanine increased; i.e., "at high concentrations of phenylalanine, higher concentrations of reduced pteridine are necessary to obtain the maximum velocity of hydroxylation" (LaDu and Zannoni, 1967).

    The precise reason the simplified assay conditions used by LaDu and his co-workers have yielded results that have not been reproduced under other assay conditions is not known. Whatever the reason, the exceptional results observed with this assay have generated serious misconceptions about certain aspects of the biochemistry of PKU.

    Thus, although the question of whether the K_m for the pterin cofactor is a constant or whether it varies with the concentration of phenylalanine may seem like an esoteric one, this question touches many others. First, if the K_m for $DMPH_4$ were not a constant that is a recognizable property of the nomal hydroxylase, we would be denied an important criterion by which a putative mutant form of the enzyme could be characterized (see Section 6 for examples of the use of this criterion).

    Second, based on the unique results obtained with the simplified assay for phenylalanine hydroxylase, LaDu has proposed that an unusual type of hyperphenylalaninemia would result from the presence of a variant form of phenylalanine hydroxylase that had a reduced affinity for the pterin cofactor. The characteristics of this unusual form of hyperphenylalaninemia were predicted to be as follows: relatively normal phenylalanine metabolism if the phenylalanine concentration were low or normal, but if elevated concentrations of phenylalanine were produced by dietary intake or a loading dose, the rate of metabolism would decrease and a sustained high blood level of phenylalanine would result as if the patient had a marked deficiency of phenylalanine hydroxylase (LaDu, 1967). Regrettably, this proposal is based on a property of phenylalanine hydroxylase—a variation of the K_m for $DMPH_4$ with phenylalanine concentration—that, as mentioned above, has not been confirmed with other assays. The proposal is therefore unsupported by reproducible data.

nents that was not added to the system: catalase, dihydropteridine reductase, or even a DPNH-regenerating system.

The use of essentially the same assay system may have contributed to the recent finding of widely scattered (covering a tenfold range) phenylalanine hydroxylase levels in non-PKU human liver samples (McLean *et al.*, 1973). That something was amiss in this study is indicated by the low mean specific hydroxylase activity (28 μmol tyrosine per hr per gram protein at 37°C) that was found. This value is only one-third that obtained (74 μmol tyrosine per hour per gram protein; range, less than twofold) with an assay system in which all the essential components had been added (Kaufman and Max, 1971). Since the latter value was obtained at 25°C, it can be estimated (by assuming that the hydroxylation reaction has a Q10 of 2.0, i.e., that the reaction will go twice as fast at 35°C as at 25°C) that this value of 74 at 25°C would be about 160 at 37°C. Thus, under the assay conditions used by McLean and co-workers, the hydroxylase activity was only about one-sixth of what it might have been if all the required components had been added. Although the detection of only a fraction of the hydroxylase activity may be a trivial consequence of the use of this type of ill-defined assay, the uncertainty about what was actually being measured in each sample limits the value of what was potentially an extremely useful study: the determination of phenylalanine hydroxylase levels in male and female fetuses of various ages. As it is, there is no way of knowing whether the widely scattered differences in hydroxylase activity that were found were due to differences in phenylalanine hydroxylase levels or to variations in the levels of one of the other ancillary components of the system.

An even more serious flaw in methodology probably marred a recent study that claimed to have detected relatively large amounts of phenylalanine hydroxylase activity in liver samples from patients with classical PKU. Using tyrosine formation, determined fluorometrically, as a measure of phenylalanine hydroxylase activity in liver biopsy samples, Grimm *et al.* (1975) reported that 14 cases of classical PKU had an average phenylalanine hydroxylase level equal to 2.7% that of a group of control patients. In every case but one, the addition of $DMPH_4$ to the PKU liver samples led to a decrease in apparent hydroxylase activity. By contrast, added $DMPH_4$ stimulated the phenylalanine hydroxylase activity in controls (thirty- to fortyfold), in one heterozygote (6.5-fold), and in two patients with hyperphenylalaninemia (two- to fivefold). From their results, these workers concluded that the effect of added cofactor on the hydroxylase activity could serve as an important criterion for the classification of various forms of PKU.

The finding of relatively high phenylalanine hydroxylase levels in classical PKU liver samples is at variance with the findings from other studies, which either failed to detect any activity (Jervis, 1953; Justice *et al.*, 1967) or which, with the use of highly sensitive techniques, detected 0.27% of the normal activity (Friedman *et al.*, 1973). As we have previously proposed (Friedman *et al.*, 1973), the formation of tyrosine that was detected by Grimm and co-workers in classical PKU was probably not due to the phenylalanine-hydroxylase-catalyzed conversion of phenylalanine to tyrosine, but to the proteolytic liberation of tyrosine from proteins that are present in liver extracts. The fact that the tyrosine formation that was detected was not pterin-dependent supports this interpretation, for even the minute amount of [^{14}C]tyrosine that is formed from [^{14}C]phenylalanine in a classical PKU patient is completely dependent on added tetrahydrobiopterin (Friedman *et al.*, 1973). [Although the explanation for the inhibition by DMPH$_4$ of tyrosine formation observed by Grimm *et al.* is not known with certainty, it should be emphasized that tetrahydropterins are not inert substances; they are potent metal chelators and will, on oxidation, generate hydrogen peroxide and superoxide ions (Fisher and Kaufman, 1973*a*); either of these properties could account for inhibition of proteolytic activity.]

Our proposed explanation for the high rates of tyrosine formation reported by Grimm and co-workers for classical PKU patients as being due to proteolytic liberation of tyrosine, rather than to phenylalanine hydroxylase activity, can also account for their observation that the extent of stimulation of tyrosine formation by added DMPH$_4$ appeared to be characteristic of each type of patient studied by them. In a situation where the total amount of tyrosine formed is due to the sum of two processes—a rather constant amount of non-pterin-dependent proteolysis and a variable amount of pterin-dependent hydroxylation—the extent of stimulation of tyrosine formation by added pterin will obviously depend on the relative contribution to the total amount of tyrosine formed made by the two processes. The greater the amount of hydroxylation, the greater will be the observed stimulation by added pterin. This situation can account for the rank order of pterin stimulation that they observed: normals are greater than hyperphenylalaninemics, which are greater than classical PKU.

Beyond the complications due to proteolytic liberation of tyrosine, the extent of stimulation of phenylalanine hydroxylase in tissue extracts by added pterins will depend mainly on the extent of dilution of the tissue extract being assayed and, hence, ultimately, on the sensitivity of the phenylalanine hydroxylase assay that is being used (more sensitive assays allow greater dilution of extracts). Another determining factor is the care with which the tissue is handled and the relative activities of the endogenous phenylalanine hydroxylase and dihydropteridine reductase. Even if the tissues or extracts are allowed to warm up, an excess of the reductase (and the endogenous levels of DPNH and TPNH) over the hydroxylase will tend to keep the biopterin in the active tetrahydro form; in the reverse situation, some tetrahydrobiopterin will be oxidized and a greater stimulation of phenylalanine hydroxylase by added tetrahydropterin will be seen. Again, the effect of these variables on the results will be minimized if a complete phenylalanine hydroxylase assay system is used, since the *added* excess reductase and reduced pyridine nucleotides will convert the dihydrobiopterin to the tetrahydro form and keep it in that form.

The potential hazards in the assay of phenylalanine hydroxylase that have just been discussed have been encountered, ironically, because phenylalanine hydroxylase activity is deceptively easy to measure in normal human and some animal liver extracts. The error introduced by, for example, reliance on measurements of tyrosine formation (determined either fluorometrically or colorimetrically), rather than on measurements of radioactive tyrosine formed from radioactive phenylalanine, is insignificant when phenylalanine hydroxylase activity is determined in normal liver extracts, because the rate of hydroxylase-catalyzed conversion of phenylalanine to tyrosine is much faster than the protease-dependent formation of tyrosine. The same is true for the nonenzymatic hydroxylation of phenylalanine that is seen in the presence of high concentration of $DMPH_4$ (Abita *et al.*, 1974). Many types of assay, therefore, pass muster with normal liver, which contains relatively high phenylalanine hydroxylase levels. These same assay conditions may prove to be hazardous, however, when tissues with little or no phenylalanine hydroxylase activity are assayed.

6. VARIANTS OF PKU

As long as PKU was diagnosed exclusively by means of urinary tests for phenylpyruvate, the disease looked like a single entity with fairly uniform characteristics, which, if untreated, led to mental retardation. This detector was simply opaque to those disturbances in phenylalanine metabolism that do not lead to excretion of phenylpyruvate.

As already mentioned, the first promising results of early dietary treatment of PKU stimulated a drive for early screening of the disease. When this drive led to the finding that elevations in blood levels of phenylalanine could precede, by weeks, the urinary excretion of phenylpyruvate, the stage was set for a switch in diagnostic methods from those that rely on detection of phenylpyruvate in the urine to those that measure phenylalanine in the blood. The switch was facilitated with the development of an inexpensive, quick, semiquantitative test for phenylalanine in small samples of blood (Guthrie, 1961). As others have noted, this change in diagnostic procedures converted a simple disease into a complex one, since it led to the discovery of PKU variants. As will be discussed, these conditions can have many causes, but, in contrast to classical PKU, they may not have the serious consequences of untreated classical PKU.

The discovery of the complexities of these abnormalities in phenylalanine metabolism has been described as the opening of a Pandora's box (Carpenter *et al.*, 1968). Although the metaphor may have been used pejoratively—the discovery certainly has led to temporary chaos (see below)—it should be recalled the Pandora's box contained, after all, not only evils but also hope. There is a chance, therefore, that in this sense the metaphor is apt, and that the present state of confusion is a prelude to deeper understanding of the underlying causes of disturbed phenylalanine metabolism.

That the field has been in a state of semantic turmoil ever since PKU variants were discovered is evidenced by the plethora of terms that have been used to describe these different conditions (Hsia, 1970). The terms include hyperphenylalaninemia without phenylketonuria, persistent hyperphenylalaninemia, atypical phenylketonuria, neonatal phenylalaninemia, mild phenylketonuria, normal phenylketonuria, transient hyperphenylalaninemia, and phenylalaninemia.

Part of the semantic problem is that "PKU" is an imprecise term that has a dual meaning; it describes not only a disease with certain characteristics, but also some of the symptoms of the disease, e.g., urinary excretion of phenylketones, i.e., phenylpyruvate. The confusion caused by this dual meaning of "PKU" can be seen in the extreme case with the term "mild phenylketonuria," which really means phenylketonuria without phenylketonuria, the first use of the term referring to the disease and the second to the symptom.

A more recent term that has been introduced to describe variants, "phenylalaninemia," is an even more unfortunate choice than PKU. Since phenylalanine is a normal blood constituent, every living animal suffers from the disease of phenylalaninemia; indeed, it is a disease that if cured would be fatal.

Ideally, diseases that are caused by inborn errors of metabolism should be named according to the enzyme deficiency that leads to the metabolic abnormality. With this idea in mind, the following classification of PKU variants can be proposed. Since it is now clear that the symptom, phenylketonuria, is secondary to—and not an invariant consequence of—the symptom, hyperphenylalaninemia (i.e., hyperphenylalaninemia is closer to the enzyme deficiency than is phenylketonuria), the proposed classification is based on the possible causes of hyperphenylalaninemia.

I. Hyperphenylalaninemia due to defects in the functioning of the phenylalanine hydroxylase system.
   A. Primary defects that directly affect one of the components of the system—"phenylalanine hydroxylase system deficiencies."
      1. Phenylalanine hydroxylase deficiency
      2. Dihydropteridine reductase deficiency
      3. "Tetrahydrobiopterin synthase" deficiency (lack of biopterin)
      4. Dihydrofolate reductase deficiency
      5. PHS deficiency
   B. Primary defects in another metabolic pathway that *indirectly* affect the activity of the phenylalanine hydroxylase system.
II. Hyperphenylalaninemia due to defects in a pathway of phenylalanine metabolism other than hydroxylation, e.g., phenylalanine transaminase deficiency. As will be discussed later, there are questions about how realistic this last category is as a possibility.

If enzyme assays were available on tissues from all patients with hyperphenylalaninemia, precise diagnosis of the enzyme deficiency would obviously be no problem. The challenge that the field faces today is to try to identify the enzyme deficiency from the data that are available. These data are limited to (a) blood levels of phenylalanine and its metabolites on unrestricted diets; (b) urinary excretion of phenylalanine and its metabolites, mainly phenylpyruvate, and those products derived from it; (c) nutritional tolerance data, e.g., amounts of dietary phenylalanine that can be tolerated without causing hyperphenylalaninemia; and (d) acute tolerance data, where a challenging dose of phenylalanine (usually 0.1 g/kg) is administered and the change in blood levels of phenylalanine (and tyrosine), as well as excretion of phenylpyruvate, are monitored.

It has long been recognized that the acute tolerance test, where kinetics of the change in phenylalanine levels can be measured, has great potential for differential diagnosis. Indeed, based on this test, some attempts have been made not only to identify the affected enzyme but even to infer some of its altered properties (Woolf *et al.,* 1968). As we will see, these attempts to detect altered properties in the absence of an analysis of the factors that can influence phenylalanine tolerance results were probably premature.

In order to more fully exploit the data that can be readily obtained from any patient, such as results from acute tolerance tests, it is necessary to examine normal phenylalanine metabolism in greater detail than has been done up to this point.

In healthy individuals, the metabolism of phenylalanine is partitioned between three major pathways: hydroxylation, transamination, and protein synthesis.* Since we are interested only in those reactions that lead to a net disposition of phenylalanine, with protein synthesis we can neglect "turnover" and restrict our attention to net protein synthesis. For adults, therefore, phenylalanine metabolism is less complex than it is in growing children because it is partitioned between only two major pathways: hydroxylation and transamination.

The fraction of phenylalanine that is disposed of via each pathway will be a function of the relative affinities of the three enzyme systems and their velocities. If these parameters differ for each of the three enzymes, the fraction of phenylalanine handled by the different pathways will obviously vary with the tissue level of phenylalanine.

In order to attempt to quantitate the relative contribution of each pathway to phenylalanine metabolism, therefore, one must know something about the kinetic properties of the enzymes involved, and, to do this, it is evident that one must be able to identify the responsible enzymes.

It is only with the transamination reaction that there is some ambiguity in this identification, for it is now clear that there are at least two, and perhaps three, enzymes that can catalyze the transamination of phenylalanine to phenylpyruvate. In 1964, it was reported that tyrosine aminotransferase (TAT), a cytoplasmic enzyme in liver, which is specific for α-ketoglutarate as the amino acceptor, can utilize phenylalanine as well as other aromatic amino acids (Jacoby and LaDu, 1964). The reported results indicated that tyrosine aminotransferase could account for most of phenylalanine transamination under the assay conditions used.

In 1971, however, it was shown that hepatic mitochondrial aspartate aminotransferase (which can utilize oxaloacetate and α-ketoglutarate, but not pyruvate, as the amino group acceptor) could also utilize a wide variety of aromatic amino acids, including phenylalanine (Miller and Litwack,

---

*Minor reactions, such as decarboxylation, will be neglected. That this pathway is in fact minor is supported by the properties of the decarboxylase, e.g., the K_m for phenylalanine of the enzyme from hog kidney is 42 mM and its relatively low rate with phenylalanine as a substrate (Christenson *et al.,* 1970). Although these properties indicate that a larger fraction of phenylalanine would be disposed of via decarboxylation as phenylalanine concentrations go up, as in PKU, even in this condition the fraction of total phenylalanine that is metabolized by this pathway would be expected to be small. This expectation is consistent with findings with PKU patients. When these patients were given an inhibitor of monoamine oxidase to block the further metabolism of the decarboxylation product, phenylethylamine, they did excrete much more of the amine than did similarly treated controls. Nonetheless, the amount of phenylethylamine excreted, 5–20 mg/day (Oates *et al.,* 1963), is minor compared with the grams of total transaminated products, such as phenylpyruvate and *o*-hydroxyphenylacetate, that are excreted per day by PKU patients (Knox, 1972). The conclusion that decarboxylation of phenylalanine becomes a "major route" for phenylalanine metabolism when tissue levels of phenylalanine are elevated (David *et al.,* 1974) does not appear to be valid, at least for humans.

1971). A problem that must be resolved before one can start to analyze phenylalanine metabolism quantitatively is whether both of these transaminases handle phenylalanine in the organism and, if not, which one does.

Fortunately, data are available, at least for the enzymes from rat liver, that strongly indicate that the cytoplasmic enzyme normally functions as a tyrosine transaminase, whereas the mitochondrial enzyme normally handles phenylalanine transamination *in vivo*.

The relevant properties of the two transaminases that support these conclusions, together with the kinetic properties of phenylalanine hydroxylase and the protein-synthesizing system, are shown in Table 3. The most important values are those in the last line showing the velocities at serum values of phenylalanine and tyrosine. It can be seen that for tyrosine, the activity of the mitochondrial transaminase is insignificant compared with that of the cytoplasmic enzyme, whereas for phenylalanine, the reverse is true. These data lead to the important conclusion that *in vivo*, different enzymes are responsible for most of the transamination of these two amino acids.*

---

* A cytoplasmic, anionic isozyme of aspartate aminotransferase also occurs in most tissues. This isozyme, which, like the mitochondrial enzyme, presumably can utilize phenylalanine as a substrate, may account, together with tyrosine aminotransferase, for part of the measurable phenylalanine α-ketoglutarate transaminase that has been observed in liver extracts (Lin *et al.*, 1958, Auerbach and Waisman, 1959; Fuller *et al.*, 1972). Since the activity of the anionic isozyme in liver appears to be much lower than that of the mitochondrial enzyme, and since its K_m for aspartate (and presumably for phenylalanine) is markedly higher—in all tissues examined—than that of the mitochondrial enzyme (Kopelovich *et al.*, 1971; Nisselbaum and Bodansky, 1964), the contribution to phenylalanine metabolism of the anionic isozyme will not be considered in our analysis.

In addition to the phenylalanine transaminase activity of aspartate aminotransferase, a phenylalanine-pyruvate transaminase has also been detected in extracts of liver (Auerbach and Waisman, 1959, Lin *et al.*, 1958; Lin and Knox, 1958; Fuller *et al.*, 1972). Although it has been suggested (Civen *et al.*, 1967) that the activity may be due to alanine aminotransferase (but see Fuller *et al.*, 1972), recent evidence indicates that the enzyme in rat liver is identical with histidine-pyruvate aminotransferase (Okuno *et al.*, 1975). The enzyme occurs in both the supernatant and mitochondrial fractions. It is not known, however, whether the properties of the soluble and mitochondrial enzymes are the same. Although its activity *in vitro* is only slightly less than that of phenylalanine α-ketoglutarate transaminase in liver supernatant fractions (Fuller *et al.*, 1972), there are reasons for believing that the cytoplasmic phenylalanine-pyruvate transaminase does not play a major role in phenylalanine transamination in the whole organism. The reason for doubting the quantitative significance of its contribution to phenylalanine metabolism is the fact that the enzyme is present in the cytosol of liver together with a much more active transaminase (Fuller *et al.*, 1972) that also utilizes pyruvate as a substrate, alanine-glutamate transaminase. In addition to the higher activity of the latter enzyme, phenylalanine-pyruvate transaminase has a much poorer Km for pyruvate—7.4 mM (Lin *et al.*, 1958)—than does the alanine-glutamate transaminase—0.9 mM (Hopper and Segal, 1962). Both the lower activity and the poorer K_m for pyruvate would indicate that little of the potential activity of phenylalanine-pyruvate transaminase would be expressed in the presence of an active alanine-glutamate transaminase. (*cont.*, p. 34)

This conclusion is consistent with other observations. Thus, it is well known that cytoplasmic tyrosine aminotransferase is inducible by glucocorticoids, leading to a fivefold increase in activity of the enzyme (Lin and Knox, 1957). By contrast, the activity of the mitochondrial transaminase for tyrosine (and presumably for phenylalanine) is only weakly enhanced by steroids (Fuller *et al.*, 1972). The steroid induction of cytoplasmic tyrosine aminotransferase is thought to account, in part, for the substantial decrease in tyrosine concentration in both plasma and liver that is seen in rats after administration of cortisone (Betheil *et al.*, 1965). If cytoplasmic tyrosine aminotransferase were playing a major role in phenylalanine transamination, one would expect that the concentrations of phenylalanine in plasma and liver would also decrease in response to steroids. In fact, in rats treated with cortisone, the hepatic content of phenylalanine is not significantly changed and the plasma concentration actually goes up (Betheil *et al.*, 1965). These observations indicate that most of the phenylalanine transamination *in vivo* is mediated by an enzyme that is not induced by cortisone; i.e., cytoplasmic tyrosine aminotransferase cannot be the enzyme responsible for most of phenylalanine transamination. As will be shown later, if all phenylalanine transamination were mediated by an enzyme the activity of which was elevated fivefold by cortisone, transamination would then account for a sufficient fraction of phenylalanine metabolism so that changes in phenylalanine content should be detectable after steroid administration. Additional evidence against cytoplasmic tyrosine aminotransferase playing a major role in phenylalanine transamination is the reported failure of steroid administration to alter the phenylalanine tolerance curves for normals and for PKU heterozygotes (Blau *et al.*, 1973).

The data in Table 3 also show the hierarchy of K_m values for phenylalanine of the three metabolic pathways under consideration. It can be seen

---

The mitochondrial phenylalanine-pyruvate transaminase probably does make a contribution to total phenylalanine transamination. Because its K_m for pyruvate is high (5.0 mM) (Okuno *et al.*, 1975) relative to the concentrations of pyruvate in liver (about 0.1 mM) (Spydervold *et al.*, 1974), however, it can be calculated that only a small fraction of its potential activity will be expressed *in vivo*. These calculations indicate that at plasma phenylalanine concentrations of 1.0 mM and above, the rate of phenylalanine transamination catalyzed by aspartate aminotransferase (shown in Figure 6) is much greater than the rate catalyzed by phenylalanine-pyruvate transaminase. Since this is the phenylalanine concentration range of greatest relevance to clinical phenylalanine tolerance data, the activity of this latter enzyme has been omitted from our analysis of phenylalanine metabolism. Because of this omission, the transamination rates shown in Figure 6 at less than 1.0 mM phenylalanine have probably been underestimated.

TABLE 3. Properties of Enzymes Involved in Phenylalanine Metabolism

| | Phenylalanine hydroxylase[a] | Protein synthesis[b] | Transaminase | | | |
| | | | Mitochondrial[c] | | Cytoplasmic[d] | |
			Phenylalanine	Tyrosine	Phenylalanine	Tyrosine
K_m (mM)	0.05	0.02	12	12	80	1.5
V_{max} (μmol/min per g tissue)	0.04	0.0052	0.130	0.08	0.04	0.34
Velocity at 0.08 mM phenylalanine (μmol/min per g)	0.0095	0.0042	0.0009	0.0005	0.00004	0.017

[a] Human liver data. The V_{max} is based on in vitro assays in the presence of tetrahydrobiopterin and lysolecithin at 25°C (Friedman et al., 1973). These data have been used to estimate the V_{max} value at 37°C and in the absence of lysolecithin. [The hydroxylase in rat liver does not appear to be activated by phospholipids in vivo (Milstien and Kaufman, 1975a).] The V_{max} value is in reasonable agreement with in vivo results obtained with humans with the D₂O-liberation assay (Milstien, Puschel, and Kaufman, unpublished results). The K_m for phenylalanine for the human liver enzyme is taken from the paper by Friedman and Kaufman (1973). The value for the velocity of phenylalanine hydroxylase at 0.08 mM phenylalanine has been corrected for the less-than-saturating content of tetrahydrobiopterin that is found in liver. At saturating contents of tetrahydrobiopterin, the velocity at 0.08 mM phenylalanine would be 0.0265 μmol per minute per gram of tissue.

[b] The K_m for protein synthesis was estimated from data for the rat liver system at 37°C (Zamecnik and Keller, 1954). A similar value has been reported for phenylalanine with Ehrlich ascites cells (Rabinowitz et al., 1954). The V_{max} value is from the reported weight gains for infants between the period of birth and 3 months (Nelson, 1959). This rate declines to half this value between 9 and 12 months of life (Nelson, 1959).

[c] The values for the mitochondrial transaminase are for the rat liver enzyme and were determined at 37°C (Miller and Litwack, 1971). Since no value is available for the K_m for phenylalanine, it was assumed to be the same as for tyrosine. That this assumption is reasonable is indicated by the reported K_m value of 13 mM for phenylalanine for rat liver phenylalanine-pyruvate transaminase at 25°C (Lin et al., 1958).

[d] The values for the cytoplasmic enzyme are for the noninduced rat liver tyrosine aminotransferase and were determined at 37°C (Jacoby and LaDu, 1964).

that the values cover a 600-fold range of phenylalanine concentration, with the K_m for protein synthesis being the lowest, followed by the K_m for the hydroxylase, and the transaminase having the highest value. This arrangement makes obvious sense for survival of the organism. The most vital process, presumably protein synthesis, has the highest affinity for phenylalanine (equating low K_m with high "affinity" for the present discussion), whereas the transaminase pathway, which probably functions primarily as a disposal system for excess phenylalanine, has the poorest affinity for the amino acid.

At serum values of phenylalanine, both hydroxylation and protein synthesis would appear to proceed at comparable rates, the rate of transamination being relatively insignificant. This arrangement would favor the most vital process, protein synthesis, should phenylalanine levels fall. By contrast, should phenylalanine levels rise, protein synthesis would quickly become saturated (and therefore proceed at its maximum velocity), followed by phenylalanine hydroxylation, followed by transamination; i.e., an increasing percent of phenylalanine would be disposed of via transamination as phenylalanine levels increased.

These relationships are shown graphically in Figure 6, in which the rates of the three reactions [calculated from the properties of the individual enzyme systems (Table 3)] are depicted as a function of serum phenylalanine concentration. Before using this picture of normal phenylalanine metabolism in the interpretation of phenylalanine tolerance data, it would be useful to try to assess its accuracy. This can be done if the metabolic picture can be further simplified. Thus, it is known that the rate of net protein synthesis declines rapidly after birth so that between 9 and 12 months of age, the rate declines to one-half the value shown in the figure. For adults, of course, the rate of net protein synthesis is zero and phenylalanine metabolism simplifies to only two major reactions: hydroxylation and transamination. For the adult patient with classical PKU, the hydroxylation reaction is essentially zero so that phenylalanine metabolism is simplified still further; i.e., the bulk of phenylalanine in such patients is metabolized through the transamination pathway (some of the excess phenylalanine is excreted unchanged in the urine). Excretion of total transamination products in such patients, therefore, should provide a check on the transamination rates that were calculated from the properties of the isolated enzyme.

Chalmers and Watts have recently reported the results of quantitative measurements of urinary products excreted by a large group of PKU patients (Chalmers and Watts, 1974). From these results, it can be calculated that the average excretion of all products derived from phenylalanine by the transamination pathway is 18.9 mmol/24 hr. By comparison, the amount calculated from the transamination velocities shown in figure 6 is 22.8 mmol/24 hr. Considering the uncertainties involved in the use of the clinical data (e.g., lack of information about patients' weights, their lean body

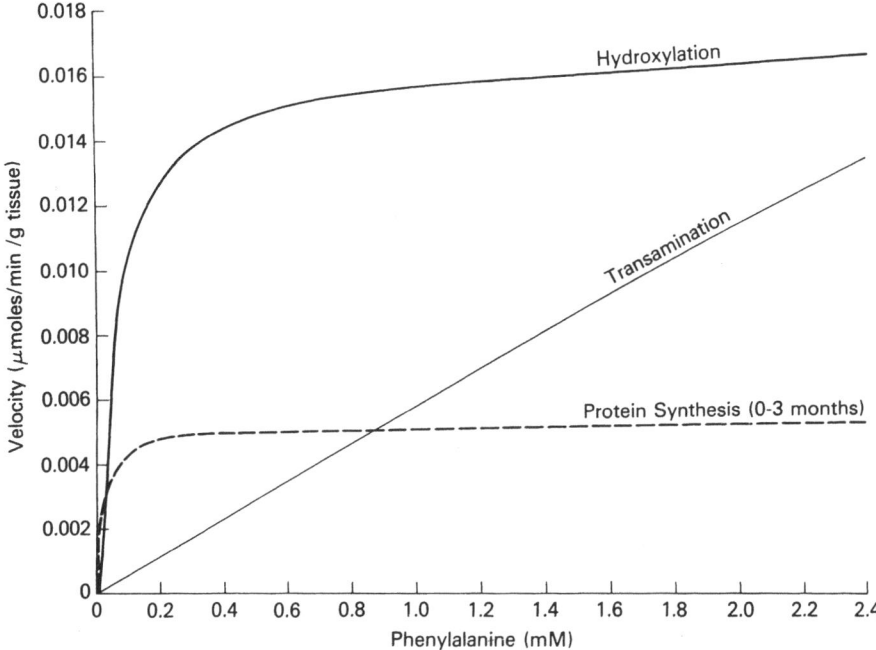

FIGURE 6. Theoretical analysis of phenylalanine metabolism as a function of serum phenyl-alanine concentration. Hypothetical rates, as a function of plasma phenylalanine concentration, for the three major pathways that lead to the net disposal of phenylalanine. The velocities for phenylalanine hydroxylase were corrected for the less-than-saturating contents of tetrahydro-biopterin that are present in human liver. The average value for four different samples of fresh human liver is 13 nmol tetrahydrobiopterin per gram of fresh tissue (Milstien, S. and Kaufman, S., unpublished data). The K_m for tetrahydrobiopterin was taken as 0.025 mM, the value found for the hydroxylase in rat liver slices (Milstien and Kaufman, 1975a). Kidney tissue has a small amount of phenylalanine hydroxylase activity (Tourian et al., 1969). From the results of Ayling et al. (1975) it can be calculated that in the human, total kidney phenylalanine hydroxylase activity (measured in vitro) is less than 5% that of liver. Since in rats the biopterin content in kidney tissue is only 20% that in liver tissue (Rembold, 1964), it is doubtful that even this small kidney activity is fully expressed in the whole organism. For these reasons, the contribution of the kidney hydroxylase to the total has been omitted from the calculations. The velocities for the mitochondrial transaminase in liver and muscle were corrected for the contents of aspartate and phenylalanine reported for adult human liver (Ryan and Carver, 1966) and human muscle (Zachmann et al., 1966) with the use of the standard rate equation for a single enzyme utilizing two substrates simultaneously. The velocities were also corrected for the less-than-saturating content of α-ketoglutarate reported to be present in rat liver (Williamson and Corkey, 1969) with the use of the rate equation for a "Ping-Pong Bi Bi" enzyme-catalyzed reaction (Henson and Cleland, 1964). It was assumed that the content of the keto acid is the same in human liver and muscle as it is in rat liver. The K_m for α-ketoglutarate of 0.39 mM (Kopelovich et al., 1971) of the rat liver mitochondrial aspartate transaminase was used in making the above correc-tions. To calculate the total body transaminase activity, the relative transaminase activity reported for different human tissues was used (Fellman et al., 1969). For convenience, however, this total activity has been expressed per gram of liver tissue. Total net protein synthesis has also been expressed in the same way, i.e., per gram of liver tissue.

masses, the frequency of their eating, etc.), this agreement is reasonable.* The agreement indicates that the enzyme in the whole organism appears to be functioning in the way that can be predicted from its properties that were determined *in vitro*.

Another way of checking the limits of error of the enzyme rate data shown in Figure 6 is to use these velocities to calculate the expected rate of decline in serum phenylalanine concentrations after a challenging load of phenylalanine, i.e., to attempt to construct a theoretical phenylalanine tolerance curve. Tolerance data from a large group of normal adults have been published by Bremer and Neumann (1966*a,b*). Their average $t_{1/2}$ values (i.e., the time after an intravenous injection of phenylalanine at which the phenylalanine concentration is reduced to one-half the extrapolated zero time value) is 88.7 ± 19.4 min (mean value ± standard error). The theoretical $t_{1/2}$ value calculated from the rate data is 93 min. † This agreement between the experimental and theoretical values indicates that

---

* The calculation of the total transaminated products excreted per 24 hr was based on the reported amounts of phenylpyruvate, phenylacetate, and *o*-hydroxyphenylacetate excreted (Chalmers and Watts, 1974). Since phenylacetylglutamine, a major phenylalanine metabolite, is probably derived at least in part from phenylpyruvate, the values reported by Chalmers and Watts were corrected by the addition of the amount of phenylacetyl glutamine that might have been expected using data obtained with other PKU patients (Woolf, 1951).

The quantity of transaminated products expected from the rate curve shown in Figure 6 was calculated from the initial velocity of the transaminase reaction (0.01 μmol per minute per gram of tissue) at the mean serum phenylalanine level (30 mg/100 ml) reported for the patients in the Chalmers and Watts study. Since at this phenylalanine concentration, the transamination reaction follows first-order kinetics, the quantity of products formed per 8-hr period could be estimated from the first-order rate equation (average body weight was taken as 70 kg). The value obtained, 7.6 mmol, was multiplied by 3, to get the amount formed during a 24-hr period, 22.8 mmol, on the assumption that the patients were replenishing their phenylalanine levels by eating three meals in the 24-hr period.

† The greatest uncertainty in making this kind of calculation is lack of knowledge about the rate and extent of equilibration of phenylalanine in the various water compartments in the body. As soon as a challenging dose of phenylalanine is given, two questions have to be answered. First, is the dose of phenylalanine evenly distributed in all compartments? Second, as the plasma phenylalanine falls with time, after a challenging dose of phenylalanine, does the concentration of phenylalanine fall equally in all compartments? There are indications that the answer to both questions is no. In the calculations of theoretical phenylalanine tolerance curves, it was assumed that during the short time of the tolerance test (2–4 hr), phenylalanine in the plasma is in equilibrium with phenylalanine in the total interstitial water but not with that in the intracellular water. The following observations support this assumption with respect to skeletal muscle, by far the largest intracellular water compartment in the body:

1. The concentration of many amino acids, including phenylalanine, is higher in human skeletal muscle than in plasma (Zachmann *et al.*, 1966). In the basal state, therefore, phenylalanine in muscle cells is not equilibrated with the phenylalanine in the plasma water.

2. Despite the higher concentration of phenylalanine in muscle than in plasma, the muscle can sustain still higher concentrations of phenylalanine (relative to phenylalanine in

the sum of the phenylalanine hydroxylase plus the transaminase velocities shown in Figure 6 is probably not in great error. Since we have already reached the same conclusion about the transaminase velocities, we can conclude that the phenylalanine hydroxylase data are also reasonably accurate.

Returning to the picture of normal phenylalanine metabolism that is depicted by the curves shown in Figure 6, it can be seen that at serum levels of phenylalanine of 0.05–0.08 mM, the normal adult metabolizes phenylalanine almost exclusively through the hydroxylation pathway. If the blood level of phenylalanine is elevated, however, as it is after an administration of a challenging dose of phenylalanine, the transamination pathway can account for a significant fraction of phenylalanine metabolism.

Before using the model for phenylalanine metabolism shown* in Figure 6 as an aid in the analysis of PKU variants, there is an important

---

plasma) when a phenylalanine load is given. This conclusion follows from the observation that for most individuals the phenylalanine space is greater than the water space, the latter determined with phenazone (Woolf *et al.,* 1967); i.e., after a challenging dose of phenylalanine, the amino acid is significantly concentrated in some tissues. It would be difficult to account for this result unless the phenylalanine was concentrated in the major tissue mass of the body—skeletal muscle.

3. It can be calculated that during the early point of the fall in plasma phenylalanine, after the administration of a challenging dose of phenylalanine, the phenylalanine in the plasma cannot be in equilibrium with the phenylalanine in total body water. For example, in the study by Bremer and Neumann (1966a), several of the normal adults would have disposed of 135–140% of the administered dose of phenylalanine if, during the time interval of the study (150 min), phenylalanine was coming from total body water.

Using the assumption stated—that phenylalanine in plasma and interstitial water does not equilibrate rapidly with phenylalanine in intracellular water—the theoretical rate of decline of plasma phenylalanine concentration was calculated from the sum of the rates of phenylalanine hydroxylase and the transaminase shown in Figure 6. The calculations were made for successive 30-min periods, starting with the zero time plasma phenylalanine concentration of about 10 mg/100 ml that was reported for normal adults after an intravenous dose of 50 mg phenylalanine per kilogram of body weight (Bremer and Neumann, 1966b). Excretion of phenylalanine in the urine was neglected. At plasma phenylalanine levels of 10 mg/100 ml or less, the urinary loss of phenylalanine is relatively small (Wadman *et al.,* 1971).

* From the analysis of phenylalanine metabolism shown in Figure 6, it seems likely that the dietary requirements of phenylalanine for maximum growth would be a function of the rate of phenylalanine intake. When phenylalanine is consumed in three boluses per day, as is customary for most humans beyond early infancy, more phenylalanine is ingested per unit time than can be utilized for protein synthesis. As a result, blood phenylalanine levels rise and the excess phenylalanine is disposed of via hydroxylation and transamination. At a lower rate of phenylalanine ingestion, it should be possible to use a larger fraction of the ingested phenylalanine for protein synthesis. Thus, the minimum daily requirement of phenylalanine (for growth) might be less if the rate of ingestion of phenylalanine were decreased. In theory (but obviously not in practice), the minimum requirement for growth would likely be achieved by a constant infusion of phenylalanine.

characteristic of phenylalanine metabolism that should be noted. It can be seen that throughout the range of phenylalanine concentrations covered by most phenylalanine tolerance tests, i.e., from 2.4 to 0.24 mM (40–4 mg/100 ml), phenylalanine metabolism in normal adults is composed of a reaction that follows essentially zero-order kinetics (hydroxylation) and a reaction that follows first-order kinetics (transamination), where the former term is the dominant one. The rate of fall of phenylalanine levels in the blood, therefore, would normally be expected to follow an order between zero and first. It is also evident that in any individual who has less than the normal amount of phenylalanine hydroxylase (and therefore less of the zero-order component), such as a PKU heterozygote, the kinetics of the fall in blood levels of phenylalanine would be expected to be closer to first order than that seen in a normal individual.

In apparent disagreement with this prediction, the fall in phenylalanine blood levels during a tolerance test has usually been described as following first-order kinetics (Bremer and Neumann, 1966a,b; Woolf et al., 1967; 1968) for both normals and PKU heterozygotes. A close examination of the data, however, indicates that the disagreement is more apparent than real. Thus, in almost every one of the 20 normals studied by Bremer and Neumann (1966a), the decline in phenylalanine blood levels did not follow strict order kinetics but, rather, followed some composite of first and zero order (i.e., the decrease in phenylalanine concentration as a function of time did not fall off as fast as demanded by first-order kinetics, nor did the rate of decline remain as constant as demanded by zero-order kinetics). Similar results were obtained for more than half the controls and PKU heterozygotes that were studied by Rampini et al. (1969). Thus, the prediction based on the theoretical model for phenylalanine metabolism—that the kinetics of decline in phenylalanine blood levels should be between zero and first order—appears to have been fulfilled in the great majority of the 60 cases that were examined in these two studies. That the decline in blood phenylalanine levels does not follow strict first-order kinetics was probably missed by these investigators because the method of plotting the results that is commonly used (semilog plots of phenylalanine concentrations as a function of time) is, for short time intervals, quite insensitive to significant deviations from first-order kinetics.

Returning to the proposed classification outlined previously, as has been discussed (Kaufman, 1967c), hyperphenylalaninemia could be caused by the lack of any of the essential coenzymes or enzymes of the phenylalanine hydroxylase system. As far as phenylalanine metabolism is concerned, the lack of the hydroxylase, dihydropteridine reductase, or biopterin (i.e., lack of the enzyme system involved in its synthesis) would be indistinguishable; i.e., all these conditions would be expected to show the same abnormalities in phenylalanine tolerance and the same pattern of excretion of phenylalanine-derived metabolites. However, because dihydropteridine reductase and reduced biopterin are also essential components of the tyrosine (Shiman *et al.*, 1971) and tryptophan hydroxylating systems (Friedman *et al.*, 1972*a*) and, hence, are involved in the biosynthesis of dopamine, norepinephrine, epinephrine, and serotonin, individuals lacking either of these components would be expected to show abnormalities in neurotransmitter metabolism and function much more severe than those seen in patients who lack phenylalanine hydroxylase. As will be discussed later, the first patient lacking dihydropteridine reductase was recently described, and he does appear to fulfill these expectations.

Because of its vital role in other areas of metabolism, the total lack of dihydrofolate reductase would probably be a lethal mutation and is therefore unlikely ever to be encountered. A partial lack of this reductase would probably lead to a mild defect in phenylalanine hydroxylation and, perhaps, in catecholamine and serotonin metabolism. This condition should be distinguishable from those just discussed, however, because of the additional disturbances in one-carbon metabolism.

As far as the phenylalanine hydroxylating stimulating protein is concerned, it is not an absolute requirement for phenylalanine hydroxylation *in vitro*. Its complete lack, therefore, would be expected to lead to only partial defects in phenylalanine metabolism and, hence, to only mild hyperphenylalaninemia.

Even based on this brief discussion, an important conclusion can be reached: hyperphenylalaninemia due to a primary defect in any of the components of the phenylalanine hydroxylase system other than the hydroxylase itself should be characterized by symptoms in addition to those seen in classical PKU. Furthermore, these additional symptoms would not be expected to be relieved by dietary restriction of phenylalanine. It is likely, therefore, that these variants can be distinguished from hyperphenylalaninemia due to phenylalanine hydroxylase deficiency without resorting to enzyme assays. What appears to be more difficult to accomplish without enzyme assays is to distinguish between various forms of hyperphenylalaninemia that are caused by different degrees of phenylalanine hydroxylase deficiency.

Despite the plethora of terms that have been used to describe the variants of PKU (as we have already discussed), there is general agreement that these conditions all share one common characteristic—the biochemical abnormalities in all of them are milder than those seen in classical PKU. As mentioned earlier, one of the useful working definitions that has been adopted is that most patients with persistent plasma phenylalanine levels over 25 mg/100 ml have the classical form of PKU, whereas those with levels persisting below 15 mg/100 ml have one of the variants of PKU.

Patients have also been classified on the basis of nutritional tolerances to phenylalanine: those who can tolerate more than 75 mg phenylalanine per kilogram of body weight per day have a variant form of PKU. Even without enzyme assays, both these criteria imply that the underlying enzyme defect must be less severe in these variants than those that lead to classical PKU.

In 1966–1967, these indications were confirmed when the first direct assays for phenylalanine hydroxylase were carried out on liver samples from patients with atypical PKU. The results obtained with the first two patients were reported in 1966 (Kaufman, 1967c). With a radioactive assay in which all the known components of the phenylalanine hydroxylase system were added, and with DMPH₄ used as the pterin cofactor (Kaufman, 1969), low phenylalanine hydroxylase activity, equal to between 3 and 6% of the average for the normals, was detected. The activity was clearly above that of classical PKU patients. (Indeed, it should be noted that with the assay conditions used, no phenylalanine hydroxylase activity has ever been detected in liver samples from classical PKU patients.)

In 1967, hepatic phenylalanine hydroxylase assays were reported for two additional atypical PKU patients, together with results from two with classical PKU (one of these two had a "mild" form of PKU) (Justice et al., 1967). As discussed earlier, interpretation of these results is complicated by the fact that the assay used in this study did not include all the components of the phenylalanine hydroxylase system. As pointed out before, the use of this incomplete assay system could account for the failure to detect any phenylalanine hydroxylase activity in one of the classical PKU patients in whom the disease seemed to be of a milder variety. With the two patients who clearly had atypical PKU (hyperphenylalaninemia without excretion of phenylpyruvate), phenylalanine hydroxylase activities equal to 10 and 50% of control levels were found (Justice et al., 1967). No further characterization of the hydroxylase in these patients was attempted.

Later, we reported the details of our study of the two atypical PKU patients referred to, together with results from a third atypical PKU patient (Kang et al., 1970a; Kaufman and Max, 1971). The average phenylalanine hydroxylase activity for the three patients with a fully supplemented assay system was equal to $3.7 \pm 0.7 \mu$mol tyrosine formed per gram of protein per hour as compared with an average value for five controls of $74 \pm 15 \mu$mol tyrosine formed per gram of protein per hour. Although the phenylalanine hydroxylase activity is only about 5% of that seen in control patients, it has been estimated that it is approximately 30,000 times higher than could be accounted for by the reported phenylalanine hydroxylating activity of tyrosine hydroxylase in liver (Kaufman, 1967c).

In an attempt to determine whether the low phenylalanine hydroxylase activity that was seen in the livers of these patients was due to a low level of normal hydroxylase or to an unknown level of an altered hydroxylase, some of the kinetic properties of the enzyme were studied (Kaufman and Max, 1971). It was found that the K_m for DMPH₄ was indistinguishable from that of control human hydroxylase ($K_m = 0.057$ mM). In addition, it was found that the enzyme in the patient's liver showed the same enhancement in hydroxylase activity (2.5–2.6-fold) when DMPH$_4$ was replaced by the more active 6-methyltetrahydropterin. The only hint of a change in catalytic properties was the finding of a K_m for phenylalanine (in the presence of DMPH$_4$) of 0.88 mM with liver extract from one of the patients, compared with a normal human liver value of 1.25 mM. Although this last difference was suggestive, it was too small to be certain that it was real.

In a subsequent study, the K_m values for phenylalanine of the hydroxylase in liver samples from two additional atypical PKU patients were determined, together with the value for another normal control (Friedman *et al.*, 1972*b*). These results confirmed the first suggestive evidence for a reproducible difference in K_m values: the K_m values for phenylalanine (in the presence of DMPH$_4$) of the three atypical liver samples were 0.88 mM, 0.67 mM, and 0.62 mM, whereas the values for the two control liver samples were 1.20 mM and 1.25 mM. Thus, the low phenylalanine hydroxylase activity in these three patients with atypical PKU appears to be due to the presence of an altered hydroxylase molecule.

Although the slightly lower K_m value for phenylalanine is obviously not responsible for the lower hydroxylase activity, it is an indication that the hydroxylase that is present in these patients is not normal but, rather, is structurally altered with the result that it has altered properties.

At this point, the *in vitro* enzyme studies that have been carried out on classical and atypical PKU patients can be summarized. In classical PKU, the residual activity is equal to 0.27% of normal. By contrast, in atypical PKU, 5% of normal activity has been found. With both types, there is evidence that the residual hydroxylase activity is due not to small amounts of normal enzyme, but to unknown amounts of altered enzyme. With the enzyme from a classical PKU patient, one of the altered properties that has been detected is a decreased sensitivity toward inhibition by excess phenylalanine (in the presence of lysolecithin and tetrahydrobiopterin). The enzyme from atypical PKU patients appears to have a slightly lower K_m for phenylalanine (in the presence of DMPH$_4$) as compared with the enzyme from controls. Enzymes from both types of patients show less stimulation by lysolecithin than do enzymes from non-PKU patients. It should be emphasized that there is no evidence as yet that the kinetic properties of the hydroxylase in classical and atypical PKU are different. So far, they differ only in that the activity in atypical PKU is about 5% of normal, whereas that in the classical PKU is 0.27% of normal.

There have been reports claiming the existence of still another mutant form of phenylalanine hydroxylase, which is distinct from the altered enzyme in either classical or atypical PKU (Woolf *et al.*, 1967, 1968). The existence of this altered hydroxylase has been postulated not on the basis of a study of its properties *in vitro* but, rather, from results of phenylalanine tolerance tests.

The putative new form of phenylalanine hydroxylase was detected during the course of studies of two unrelated children with mild PKU. In each case, the mother of the child seemed to metabolize a dose of phenylalanine in a manner that differed from the way in which either normals or most heterozygotes do; i.e., in these two mothers, phenylalanine concentrations in the blood declined at a constant rate (i.e., the decline followed zero-order kinetics) rather than at a rate that decreases with time. (Strangely, the authors claimed erroneously that these two mothers metabolized phenylalanine "at a rate that decreased as the concentration was raised.") The suggestion was made that "the two unusual parents and their affected offspring possessed a variant form of phenylalanine hydroxylase strongly inhibited by excess phenylalanine and coded for by a third allele at the phenylalanine hydroxylase locus" (Woolf *et al.*, 1968).

There are several reasons why it is difficult to accept this suggestion. First, there is no evidence that the affected children and their mothers share a common disturbance in phenylalanine metabolism. Thus, it was concluded that the affected children suffered from a temporary deficiency of phenylalanine hydroxylase, which, on the basis of tolerance tests, seemed to disappear as the children got older (Stephenson and McBean, 1967). If, as these results indicate, the properties of the altered hydroxylase are not expressed beyond infancy, why should one expect to see them expressed in the parents of these children? Until this question is answered, there is no certainty that the defect in phenylalanine metabolism that is being measured in these two parents, whatever its enzymatic basis, has any relationship to the temporary phenylalanine hydroxylase deficiency manifested by the children.

A second reason for doubting the explanation put forth by Woolf and co-workers is that it cannot explain their observations. A form of phenylalanine hydroxylase that was strongly inhibited by excess phenylalanine could not metabolize phenylalanine at a constant rate, as did the two parents under consideration. Instead, in the range of phenylalanine concentrations that inhibit the hydroxylase, the rate of hydroxylation would increase as phenylalanine concentrations declined.

A final reason for questioning the suggestion made by Woolf and co-workers is that in their analysis of phenylalanine metabolism, they ignored the contribution made by transamination and assumed that phenylalanine is metabolized exclusively by hydroxylation. As shown in Figure 6, however, at high phenylalanine concentrations, transamination cannot be ignored. Furthermore, as pointed out earlier, normal phenylalanine metabolism would not be expected to follow strict first-order kinetics, as claimed by Woolf and co-workers, but rather to follow some composite between zero and first order. Since, as can be seen in Figure 6, it is the transaminase reaction that is responsible for almost all the first-order component (above 0.6 mM phenylalanine), it can be predicted that a deficiency in phenylalanine transamination would lead to phenylalanine metabolism that follows essentially zero-order kinetics.

To see whether this prediction is in accord with the actual results, the first-order rate "constants" for successive time intervals that were reported for these two parents were compared with the "constants" that were calculated from the phenylalanine hydroxylase rate data shown in Figure 6. As can be seen from Table 4, the calculated values are in remarkably close agreement with the values that were actually found. On the basis of these considerations, the proposal can be made that the two unusual parents studied by Woolf and co-workers did not have an altered form of phenylalanine hydroxylase but, instead, had a relative deficiency of phenylalanine transaminase, not necessarily of genetic origin. As implied by this proposal, the evidence in favor of a "third allele" at the phenylalanine hydroxylase locus is extremely weak.

Since it can be estimated from the data published for the human enzyme (Fellman *et al.*, 1969) that the transaminase in skeletal muscle makes up almost 50% of the total, and since it is known that females have less musculature than do males, it is of interest that both of the unusual parents in the families studied by Woolf and co-workers were the mothers. As mentioned earlier, it is not certain that the disturbance in phenylalanine metabolism in the two mothers is related to the temporary phenylalanine hydroxylase abnormality that was detected in their children. If the two metabolic peculiarities are unrelated, one might expect to find the same signs of altered phenylalanine metabolism as those detected in these two women in a group of controls (as opposed to a group of parents of affected children). In this regard, it is significant that one of 26 controls (C6, a male) (Rampini *et al.*, 1969) appeared to have the same kind of disturbance in phenylalanine metabolism as that described by Woolf and co-workers; i.e.,

TABLE 4. A Comparison of Calculated and Actual Rate Constants for Phenylalanine Metabolism in Two Unusual Heterozygotes

Interval (min)	Mother of case 1		Mother of case 2	
	k_1 (found)[a]	k_1 (calc.)[b]	k_1 (found)[a]	k_1 (calc.)[b]
0–50	0.0034	0.0034	0.0028	0.0028
50–100	0.0039	0.0038	0.0032	0.0032
100–150	0.0056	0.0045	0.0039	0.0035

[a] The values for k_1 (found) were taken from the paper by Woolf *et al.* (1968).
[b] The values for k_1 (calc.) were calculated from the data in Figure 6 on the assumption that the transaminase was not active.

following the administration of a dose of phenylalanine, his blood levels of phenylalanine declined at a constant rate. This finding suggests that this unusual pattern of decrease of blood levels of phenylalanine may, in fact, be unrelated to hyperphenylalaninemia.

Before going on to discuss hyperphenylalaninemia that might be caused by the lack of components of the phenylalanine hydroxylase system other than the hydroxylase itself, it is worth considering at this point why classical and atypical PKU have proved to be so difficult to distinguish on the basis of phenylalanine tolerance data. With at least a twentyfold difference in the activity of hepatic phenylalanine hydroxylase in the two conditions, one might have thought that differentiation would present little problem.

In part, the difficulty is due to the fact that during a phenylalanine tolerance test, it is the rate of decline in blood levels of phenylalanine that is measured, rather than the rate of increase of tyrosine concentrations. If the tyrosine that was formed were not further metabolized, its rate of accumulation in the blood would be a far more discriminating measure of phenylalanine hydroxylase activity than is the rate of phenylalanine decrease. Depending on how rapidly the tyrosine is metabolized, however, the rate of tyrosine formation would probably underestimate phenylalanine hydroxylase activity.

Accepting the use of the less desirable measurement of phenylalanine hydroxylase activity—the rate of decline of blood phenylalanine concentrations—what are the predicted limits of discrimination of this test? With the rate data shown in Figure 6, theoretical tolerance curves can be constructed for any hypothetical condition in which the activity of phenylalanine hydroxylase and the transaminase are affected. Figure 7 shows such curves calculated for normal individuals, for those with 50% and 5%, and for a complete lack of phenylalanine hydroxylase, as well as for the hypothetical condition in which transaminase is missing. As already mentioned, for normal adults, a $t_{1/2}$ of 93 min (see Figure 7) was calculated, a value that is in excellent agreement with actual results. In classical PKU with, essentially, a complete lack of phenylalanine hydroxylase activity (and with normal transaminase activity), a $t_{1/2}$ value of 600 min can be calculated (using the first-order rate constant estimated from the slope shown in the figure). This value is considerably smaller than those that have actually been measured. Thus, from the data reported for six PKU patients (Bremer and Neumann, 1966b), an average $t_{1/2}$ value of about 1500 min can be calculated. It has also been reported that it takes classical PKU patients 14 days to eliminate a load of phenylalanine of 0.1 g/kg body weight (Güttler and Wamberg, 1972). These results indicate that there may be a serious discrepancy between the calculated $t_{1/2}$ values and those that have actually been observed for classical PKU patients. Since the calculated $t_{1/2}$ value is

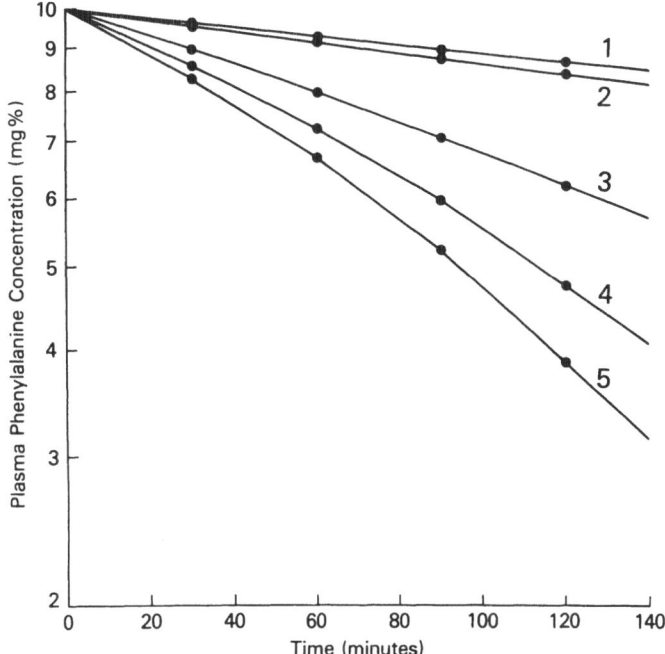

FIGURE 7. Theoretical phenylalanine tolerance curves: curve 1, hydroxylase, zero, trans-minase, 100%; curve 2, hydroxylase, 5%, transaminase, 100%; curve 3, hydroxylase, 50%, transminase, 100%; curve 4, hydroxylase, 100%, transaminase, zero; curve 5, hydroxylase, 100%, transaminase, 100%.

based only on the rates of transamination and since, as already shown, these rates are entirely consistent with the measured excretion of transamination products in PKU patients, the discrepancy is puzzling. A likely explanation of the discrepancy is that it reflects a slow equilibration of phenylalanine in different compartments in the body. As already discussed, there are indications that a loading dose of phenylalanine is concentrated in tissues. Furthermore, from the magnitude of this effect, it is almost certain that muscle must be one of the tissues where the phenylalanine is concentrated. Thus, it seems reasonable to expect that as plasma phenylalanine levels fall, there would be a slow passage of phenylalanine from the intracellular muscle pool into the plasma compartment. As a result of this influx, the apparent rate of phenylalanine disposition would appear to be slower than it actually is (i.e., the $t_{1/2}$ would appear to be larger than it actually is).

A different biochemical process could also lead to a similar result. Thus, if net protein catabolism were occurring in muscle during the period of the tolerance test, which is a likely event considering the fasting state of the patient during the test, this process would provide an additional source of plasma phenylalanine that would also lead to $t_{1/2}$ values that are larger than would be predicted from the enzyme velocity data. (This effect of protein catabolism in muscle on phenylalanine tolerance test results could probably be minimized by performing the tolerance test on patients with adequate carbohydrate intake, rather than on fasting patients.)

Some support for the ideas discussed above comes from the measurements of the early changes in phenylalanine blood levels during a phenylalanine tolerance test conducted on two PKU infants (Hjalmarsson *et al.,* 1971). With both children, the phenylalanine levels declined for the first 2–3 hr after the peak level had been reached. Subsequently, the phenylalanine level actually increased. This pattern suggests that phenylalanine in the plasma compartment was being replenished by an influx of phenylalanine from another compartment.

Finally, it should be noted that if the rate of transfer of phenylalanine from the muscle pool to the plasma pool were slow relative to the total rate of phenylalanine catabolism in normal individuals, this influx of phenylalanine into the plasma compartment might be too slow to affect the measured $t_{1/2}$ values for normal individuals but could lead to serious errors in the $t_{1/2}$ values that are measured for classical and atypical PKU patients. (From the discrepancy between calculated and actual $t_{1/2}$ values for PKU patients the rate of influx of phenylalanine into the plasma would be of the order of 0.3 to 0.4 mg%/hr.)

Curve 2 in Figure 7 shows the expected rate of fall in phenylalanine plasma concentrations for individuals with 5% of the normal phenylalanine hydroxylase levels, the level of hydroxylase that was found for one group of atypical PKU patients (Kaufman and Max, 1971; Kaufman, 1967c). This rate of decline yields a $t_{1/2}$ value of 510 min, as compared with the previously discussed value of 600 min for classical PKU patients (i.e., for a patient with essentially no phenylalanine hydroxylase). The common experience with these two groups of patients is that they cannot be distinguished 4 hr after the load of phenylalanine had been given, whereas they can be after 24 hr (Blaskovics and Shaw, 1971). From Figure 7 it can be seen that at 2 hr, the phenylalanine levels differ by only a few milligrams per 100 ml, and even at 4 hr, there would only be a 5% difference in blood levels of phenylalanine. It is not surprising, therefore, that this difference does not provide a reliable basis for differential diagnosis. From the data in Figure 7, however, it can be seen that the *difference in the slopes* of the curves for patients with no phenylalanine hydroxylase and for those with 5% activity should be fairly readily distinguishable—these slopes differ by about 15%. In order to detect this small difference in slopes, however, it is clear that intravenous administration of the phenylalanine load would have important advantages over oral administration. With the latter route, the early points on the phenylalanine decay curve are usually lost and accurate slopes are more difficult to determine. Thus, it seems likely that the common use of phenylalanine tolerance tests in which phenylalanine is given orally accounts for some of the difficulty in discriminating, on this basis, between classical and atypical PKU. If the oral route is used, the common practice of collecting samples at fixed intervals, such as hourly, has little to recommend it other than an esthetically pleasing symmetry; more frequent sampling after the phenylalanine peak has been reached would be preferable. It would also be important that the phenylalanine be given as a solution of the amino acid rather than as a suspension.

Figure 7 also shows the phenylalanine tolerance curve expected for an individual with 50% of the normal amount of phenylalanine hydroxylase. The $t_{1/2}$ value is about 170 min, which is close to the value of 159 min observed for PKU heterozygotes (Bremer and Neumann, 1966b). Both this theoretical curve and the one for the hypothetical lack of phenylalanine transaminase will be discussed later in greater detail.

We come, finally, to hyperphenylalaninemia that might be caused by defects in an enzyme system other than the one responsible for phenylalanine hydroxylation. As mentioned previously, transamination is the only other metabolic pathway where a block might lead to hyperphenylalaninemia.

Two patients described by Auerbach et al. (1967) had symptoms that prompted these workers to postulate that the patients lacked phenylalanine transaminase. Although no additional data have been forthcoming on these two cases or, for that matter, on any new cases with putative transaminase deficiencies, the original cases are invariably cited as examples of hyperphenylalaninemia due to transaminase deficiency. It would seem that a critical evaluation of this postulate is long overdue.

Defects in phenylalanine metabolism were detected in both infants on the basis of the semiquantitative Guthrie test. Subsequently, hyperphenylalaninemia was confirmed by a chemical test for phenylalanine. The infants were distinguished from classical PKU patients by two criteria: (1) neither one excreted much phenylpyruvate (less than 10 mg/100 ml urine) or o-hydroxyphenylacetic acid; and (2) with both children, hyperphenylalaninemia was only manifest when protein intake was high. The possibility that the two infants might have been PKU heterozygotes was considered, but oral phenylalanine loading tests gave patterns that were indistinguishable from classical PKU patients, i.e., essentially no fall in phenylalanine blood levels within the first 4 hr of phenylalanine administration. On the basis of these results, a deficiency in phenylalanine transaminase was postulated.

There are reasons for being skeptical about this proposal. The strongest evidence against it comes from results obtained by the authors themselves. Auerbach and co-workers tested for the presence of transaminase by giving phenylpyruvate to the patients. They observed a prompt rise in blood levels of phenylalanine, providing strong evidence that an active phenylalanine transaminase was present in both patients. To reconcile this finding with their previous conclusion that the transaminase is missing, they offered two proposals. In the first, they stated "that phenylalanine transaminase might be altered in such a manner in these infants that the reactive site can no longer bind the amino acid but is still competent as far as the keto acid is concerned." This proposal is difficult to reconcile with the laws of thermodynamics, and is therefore unlikely. The second proposal offered was that tyrosine transaminase might be the enzyme responsible for the transamination observed with phenylpyruvate. Citing the fact that the K_m of tyrosine transaminase for phenylalanine is 50 times higher than it is for tyrosine (Jacoby and LaDu, 1964), the authors implied that the enzyme might nonetheless function well with phenylpyruvate. This proposal violates, indirectly, the same laws as does the first one. As far as the properties of the enzyme itself are concerned, if tyrosine transaminase were active starting with phenylpyruvate, it must also be active with phenylalanine, and as the considerations outlined in Table 3 show, it is most unlikely that tyrosine transaminase contributes significantly to phenylalanine transamination in either direction.

Another reason the proposed transaminase deficiency is unlikely is that such a deficiency almost certainly cannot account for the phenylalanine tolerance data reported by Auerbach and co-workers, or for the observed hyperphenylalaninemia. As mentioned, after an oral dose of phenylalanine, the blood levels of phenylalanine did not decline significantly over a 4-hr period, a pattern that is indistinguishable from that of a classical PKU patient. By contrast, the phenylalanine tolerance curve expected for a hypothetical deficiency of transaminase is shown in Figure 7. Because transamination makes such a small contribution to phenylalanine catabolism (at least up to 10–15 mg/100 ml, the peak phenylalanine concentration reached in these two children) the lack of transaminase would lead to a phenylalanine tolerance curve that would be close to that of normal individuals with a $t_{1/2}$ of 114 min, as compared with a $t_{1/2}$ of 93 min for normals.

Finally, recent data on the excretion of phenylpyruvate and o-hydroxyphenylacetate as a function of age and plasma phenylalanine concentrations have demonstrated that in children under 3 months of age, there is practically no excretion of phenylpyruvate or of o-hydroxyphenylacetate as long as the level of phenylalanine in plasma is below 1.5 mM (25 mg/100 ml) (Rey et al., 1974). Since the urinary excretion data that have been published on the two infants studied by Auerbach and co-workers were restricted to the early postnatal period, and since the peak phenylalanine blood levels in both infants were in the 25–30 mg/100 ml range, it seems likely that their urinary excretion pattern, rather than being unusual, was typical of that of most infants with hyperphenylalaninemia.

It can be concluded, therefore, that the case for a genetically determined deficiency of phenylalanine transaminase as a cause for hyperphenylalaninemia is without substantial support. Furthermore, on the basis of theoretical considerations (see Figures 6 and 7), it can be predicted that a transaminase deficiency, if it were to exist, would be unlikely, by itself, to lead to pronounced hyperphenylalaninemia. This last conclusion is supported by the observation that in rats, inhibition of aspartate-α-ketoglutarate aminotransferase, the enzyme that is probably responsible for most of phenylalanine transamination *in vivo* (see Table 3), does not lead to elevations in phenylalanine levels in either plasma or liver (Gressner, 1974).

Although it is not difficult to marshal evidence against the proposed transaminase deficiency, it is more difficult to suggest with confidence what might have been wrong with the two infants studied by Auerbach and coworkers. It seems likely that they lacked a fully functional phenylalanine hydroxylase system during the first few months of life. Attempts to make a more quantitative estimate are hampered by apparently conflicting data. On the one hand, as mentioned, the phenylalanine tolerance results are like those for a classical PKU patient with very little, if any, phenylalanine hydroxylase activity; on the other hand, the infants did not have high phenylalanine blood levels with normal protein intake, an indication of considerable hydroxylase activity. The subsequent normal development of both children indicates that whatever the degree of impairment in phenylalanine hydroxylation during the first few months of life, it was a temporary one. It may be that these infants had a transient dihydropteridine reductase deficiency, a possibility that will be discussed in greater detail in a later section.

Even if the case for hyperphenylalaninemia caused by a genetically determined phenylalanine transaminase deficiency is essentially nonexistent, the more general questions of why phenylpyruvate and *o*-hydroxyphenylacetate are not usually excreted in PKU infants during the first months of life (Armstrong and Binkley, 1956; Rey *et al.*, 1974) remains to be answered. The explanation often invoked is that of delayed maturation of phenylalanine transaminase (Rey *et al.*, 1974). If the conclusion reached earlier—that aspartate aminotransferase is responsible for the bulk of phenylalanine transamination *in vivo*—is correct, it is difficult to believe that an enzyme that is involved with such vital functions as urea synthesis, gluconeogenesis, and pyrimidine synthesis, would not be functional during the first few months of postnatal life. In this regard, it is significant that in rats, not only is phenylalanine transaminase present in the livers of the neonate, but also a peak activity is reached during the first 10 days of life that is twice as high as the adult level (Kenney and Kretchmer, 1959). In rats, therefore, there is no evidence of delayed maturation of this enzyme.

The evidence for delayed maturation of the transaminase in humans would be stronger if it were known with certainty that no transamination products derived from phenylalanine are excreted in the urine during the immediate postnatal period. Unfortunately, the compounds that are commonly measured during this early period are phenylpyruvate and o-hydroxyphenylacetate (Zelnicek and Slama, 1971; Rey et al., 1974). Before it could be concluded that phenylalanine transaminase is not functional, it would have to be shown that none of the other transaminated products (phenyllactate, phenylacetylglutamine, and, perhaps, phenylacetate) are excreted. Unless those compounds are also measured, it is possible that the rate at which phenylpyruvate is formed is slower than the rate at which it is metabolized to phenyllactate and phenylacetylglutamine (see Figure 2); under these conditions, little phenylpyruvate or o-hydroxyphenylacetate would be excreted, but this lack of excretion would not indicate lack of transaminase.

Another factor that could affect the amount of phenylpyruvate and o-hydroxyphenylacetate (OHPAA) excreted is the amount of tyrosine that is being metabolized. As shown in Figure 2, the first step in tyrosine catabolism is transamination to p-hydroxyphenylpyruvate, followed by oxidation to homogentisate. The enzyme that catalyzes the latter conversion, p-hydroxyphenylpyruvate hydroxylase, has been shown to be the same enzyme that is responsible for the conversion of phenylpyruvate to OHPAA (Fellman et al., 1972; Taniguchi and Armstrong, 1963; Taniguchi et al., 1964). As a consequence of this relationship, when the content of p-hydroxyphenylpyruvate is elevated, as it is in tyrosinemia, there is increased excretion of phenylpyruvate (Kennaway and Buist, 1971), presumably because the oxidation of phenylpyruvate to OHPAA is blocked by the higher content of its competing substrate, p-hydroxyphenylpyruvate.

That some transamination of phenylalanine can occur very early in life is shown by the data of Vollmin et al. (1971). These workers found that both of the PKU infants that were included in their study, ages 11 days and 14 days, excreted phenylalanine transamination products and that with both infants, phenyllactate constituted a significant fraction of the total of these products.

Even if measurements of all the phenylalanine transamination products were to establish a *relative* deficiency of this process during the first few months of life, there are alternative interpretations that are more attractive than the delayed maturation hypothesis. As already mentioned, this transaminase appears to be involved in too many vital metabolic processes not to be functional. Indeed, one of the reasons it may not appear to function at a high rate with phenylalanine as a substrate is that it is functioning predominantly in one of its other capacities during the neonatal period. Data consistent with this idea have been published for rabbit tissues (Stave and Armstrong, 1973), where it has been found that the ratio of aspartate to phenylalanine in muscle and kidney from 3-day-old rabbits is much higher than for these same tissues from adult rabbits. These data would indicate that in these two tissues, phenylalanine transamination would be more inhibited by the relatively higher content of the competing substrate, aspartate, in the neonate than in the adult. Another factor that would work in the same direction is the lower content of muscle tissue (per kilogram of body weight) in the infant compared with the adult. Thus muscle tissue, which, in the adult human (Fellman *et al.*, 1969) appears to contribute about one-half the total phenylalanine transaminase, is not fully developed in the neonate, and what muscle enzyme there is is probably not fully active with phenylalanine as a substrate. It is possible, therefore, that the apparently delayed maturation of the enzyme may in reality be a reflection, in part, of delayed maturation of a tissue that contains the enzyme.

The first case of hyperphenylalaninemia due to dihydropteridine reductase deficiency was recently described (Kaufman *et al.*, 1975a). The infant was diagnosed as a classical phenylketonuric on the basis of elevated blood levels of phenylalanine (53 mg/100 ml) and an abnormal phenylalanine tolerance test and was placed on a low phenylalanine diet at the age of 12 days. At 7½ months, he developed neurological symptoms (frequent seizures, abnormal EEG pattern). The onset of these symptoms, despite excellent control of his blood phenylalanine levels, indicated that this infant had a new form of PKU.

This indication was confirmed by the results of assays of the individual components of the phenylalanine hydroxylase system that were carried out on a liver biopsy sample when the baby was 14 months old. The phenylalanine hydroxylase activity was 20% of the average normal adult level. By contrast, no dihydropteridine reductase activity was detected under conditions where less than 1% of the normal activity could have been measured. The same reductase deficiency was manifest in his cultured skin fibroblasts and brain tissue. In addition to this clear-cut deficiency, the activity of dihydrofolate reductase may also be somewhat decreased in the patient's liver (Kaufman *et al.*, 1975a).

Since the role of dihydropteridine reductase is to keep biopterin in the tetrahydro form (Kaufman, 1971), it could be predicted that a deficiency of this enzyme might lead to low tissue levels of tetrahydrobiopterin. As expected, it was found that in normal human liver, the pterin cofactor exists predominantly in the tetrahydro or fully active form. By contrast, there was little or no detectable tetrahydrobiopterin in this patient's liver. This finding provided independent evidence that dihydropteridine reductase is affected in this patient's liver. Furthermore, these results supplied strong evidence that the reductase in this tissue was not functional *in vivo*.

The conclusion that the reductase is the affected component was supported by results of immunological studies. Antibodies to pure sheep liver dihydropteridine reductase showed a single precipitin line in immunodiffusion tests with normal human liver extracts. By contrast, no detectable line was found with an equivalent amount of the extract from the patient's liver (Kaufman *et al.*, 1975*a*). These results provided independent evidence that the reductase is affected in this patient's liver and also showed that in this tissue there are no cross-reacting proteins present that can form a precipitin line with this specific antiserum.

Since, as already mentioned, the reductase is an essential component of the tyrosine and tryptophan hydroxylating systems, a lack of the reductase would be expected to lead to defects in the synthesis of dopamine, norepinephrine, and serotonin. Preliminary results support the conclusion that synthesis of dopamine and serotonin may be defective. Thus, lower than normal levels of homovanillic acid (HVA), a metabolite of dopamine, and of 5-hydroxyindoleacetic acid (5HIAA), a metabolite of serotonin, were found in the patient's CSF. Added support for the conclusion that dopamine turnover is severely impaired was obtained with the probenicid technique. Lumbar CSF was analyzed before and after an 18-hr period on a high oral dose of probenicid. Egress of HVA and 5HIAA from CSF is prevented by this drug and accumulation of these metabolites is believed to reflect turnover of dopamine and serotonin, respectively, during this period of time. A marked impairment in HVA formation was evident in the patient; his initial HVA concentration was 26 μg/ml, which increased to 36 μg/ml after 18 hr. This increase is markedly lower than the five- to tenfold increase noted in control patients. It should be mentioned that the defect in dopamine metabolism in this patient, as reflected by HVA accumulation after probenicid, is much more severe than that observed with classical PKU patients (in whom, presumably, phenylalanine hydroxylase is the affected component). Thus, McKean (1972) has reported that there is a relatively mild decrease in the accumulation of HVA in the CSF of PKU patients during a 24-hr period on probenicid and that this decrease correlates with plasma phenylalanine concentrations: patients with high plasma phenylalanine (25–32 mg/100 ml) showed a 4.3 to 4.6-fold rise in HVA, whereas patients with a low plasma phenylalanine (3–5 mg/100 ml) showed a 5.6–6.4 fold rise in HVA.

Even though the *in vitro* activity measurements and the immunochemical tests indicate that the patient's lack of reductase is essentially complete, there remains some uncertainty about whether the pterin-dependent hydroxylases would be completely nonfunctional *in vivo*. The reason for this uncertainty is the observation that reducing agents, such as ascorbate and mercaptans (at concentrations of about 15 mM), can substitute effectively for dihydropteridine reductase *in vitro* (Kaufman, 1959). Of the naturally occurring compounds that might fill this role *in vivo,* glutathione and ascorbate appear to be likely candidates. The content of reduced glutathione in cat brain has been reported to be 0.32–1.56 μmol/g (Perry *et al.,* 1972) and in livers of various species to vary between 2.3 and 8.5 μmol/g (Bhattacharya *et al.,* 1955). Although these values are below the optimum levels that were shown to be effective with purified phenylalanine hydroxylase, they are in the range, especially in liver, where significant rates of hydroxylation might be expected even in the absence of the reductase. The same conclusion can be reached for ascorbate, where rat brain values of 2.8–4.7 μmol/g (Adlard *et al.,* 1973) and rat liver values of 1–2 μmol/g have been reported (Giroud *et al.,* 1938*a*). The value for ascorbate in adult human liver, however, appears to be significantly lower than in other species, being only 0.4 μmol/g (Giroud *et al.,* 1938*a*).

Another possibility for bypassing the dependence on dihydropteridine reductase in pterin-dependent hydroxylases is through the dihydrofolate-reductase-catalyzed reduction of 7,8-dihydrobiopterin to the tetrahydro level (Kaufman, 1967*a*). The low activity of this hepatic enzyme with 7,8-dihydrobiopterin as a substrate, however, indicates that this enzyme would not be able to sustain high rates of phenylalanine hydroxylation. Furthermore, dihydrofolate reductase has been reported to be absent in brain (Makulu *et al.,* 1973), so that it could not fill this role in this tissue. The finding that there was no detectable tetrahydrobiopterin in this patient's liver indicates that in this tissue, none of these alternate modes of regenerating tetrahydrobiopterin was able to keep pace with the reactions that lead to its oxidation. It has already been mentioned that the activity of dihydrofolate reductase also appeared to be somewhat decreased in the patient's liver. It is possible that this relative deficiency contributed to the lack of tetrahydrobiopterin in his liver.

The considerations just outlined suggest that the phenotype resulting from a genetic lack of dihydropteridine reductase could vary considerably. Thus, if individuals lacking the enzyme had higher-than-normal hepatic levels of reducing compounds, such as ascorbate, or of dihydrofolate reductase, their impairment in phenylalanine hydroxylation might be much less severe, while the CNS abnormalities (assuming these are a consequence of blocks in tyrosine and tryptophan hydroxylation) might be just as severe, as it is in the present patient. It is possible that these alternate pathways for tetrahydrobiopterin regeneration can account for the peculiar symptoms of this patient's sister: she appeared to have no signs of a severe impairment in phenylalanine metabolism (as measured *in vivo*), and yet she died following a progressive, degenerative neurological disease. Another possibility is that she suffered from a second metabolic defect.

If the neurological symptoms seen in this patient are a consequence of his lack of dihydropteridine reductase, there are several reasonable approaches to therapy. One would be to administer large doses of a reducing agent in an attempt to accelerate the nonenzymatic reduction of the quinonoid-dihydropterin (Kaufman, 1959) to the point where this reaction no longer limits the rate of the pterin-dependent hydroxylases.

Test-tube experiments with purified rat liver phenylalanine hydroxylase and $DMPH_4$ showed that when the concentration of reducing agents was raised to 15 mM, phenylalanine hydroxylation was no longer dependent on dihydropteridine reductase (Kaufman, 1959). Based on these results, a preliminary trial of ascorbate therapy at a level of 5 g/day for 19 days was carried out. This dose of ascorbate failed to correct either the patient's abnormal phenylalanine tolerance test or his neurological symptoms. This apparent failure indicates that it may not be possible to achieve tissue ascorbate levels that are high enough for the rapid, nonenzymatic reduction of quinonoid-dihydrobiopterin. Furthermore, although the *in vitro* results showed that reducing agents at a concentration of 15 mM were effective in supporting high rates of phenylalanine hydroxylation, it is possible that much higher levels might be needed *in vivo*. The reason for the higher requirement is that the nonenzymatic tetrahydropterin-regenerating reaction is bimolecular and depends, therefore, on the concentration of both the reductant (ascorbate) and the oxidant (quinonoid-dihydropterin). In the *in vitro* experiments mentioned above, 15 mM reductant was effective with 0.12 mM $DMPH_4$. In human liver, with a tetrahydrobiopterin content of about 0.15 mM, it is likely that much higher concentrations of reductant—perhaps as much as 10 times higher—would probably be required than the tissue content of this substance that could have been reached with an intake of ascorbate of 5 g/day.

Still another possibility for therapy would be the administration of tetrahydrobiopterin or one of its active analogues, such as 6-methyltetrahydropterin or 6,7-dimethyltetrahydropterin. In the absence of a tetrahydropterin-regenerating system, however, this therapy would require the administration of an amount of tetrahydropterin equal to, at a minimum, the total quantity of the aromatic amino acids that are hydroxylated per day. The amount of phenylalanine converted to tyrosine per day by infants can be approximated from nutritional data (i.e., from the known minimum daily requirement for phenylalanine and the fraction of it that can be spared by tyrosine) and is equal to about 300 μmol per kilogram per day. If none of the administered tetrahydropterin were excreted or degraded by other pathways [e.g., mitochondrial oxidation (Rembold and Buff, 1972)], therefore, 300 μmol per kilogram per day of tetrahydropterin (equal to about 100 mg pterin per kilogram per day) would have to be supplied merely to keep pace with the phenylalanine hydroxylation reaction.

Besides the enormous doses that would be required, another reason for being pessimistic about the use of pterin therapy in dihydropteridine reductase deficiency (where the neurological damage is presumably due to the lack of the enzyme in the brain) is the evidence that after intravenous administration in rats, tetrahydrobiopterin does not readily enter the brain (Kettler *et al.*, 1974).

Based on the considerations discussed above, it would appear that the recent suggestion by Smith *et al.* (1975*a*) that pterin administration would be of value in PKU variants who lack the reductase may not be a practical one. Furthermore, their qualified recommendation that a dose of 0.5–1.0 mg pterin per kilogram might be sufficient is probably far too low. [It should be noted that administration of a tetrahydropterin may be able to lower blood levels of phenylalanine. Indeed, based on the finding that rats that had been given a tetrahydropterin had markedly increased activity of phenylalanine hydroxylase in their livers (Milstien and Kaufman, 1975*a*), it is highly likely that tetrahydropterin administration, by increasing the activity of hepatic phenylalanine hydroxylase, can lower blood levels of phenylalanine. The recent brief report (Danks *et al.*, 1975) that administration of tetrahydrobiopterin to a patient with a variant form of PKU led to a decrease in blood phenylalanine levels is therefore in accord with these expectations. What is puzzling about this report is not the effect of the pterin, which was to be expected from these results of animal studies, but the characterization of this effect as a "treatment of a variant form of phenylketonuria." Although the nature of the enzyme defect in this patient was not established, the implicit assumption was made that she suffered from a "defect in pteridine metabolism," presumably due to a deficiency of dihydropteridine reductase. Since, as previously mentioned, it has been shown that tetrahydrobiopterin does not readily enter the brain from the blood, it is difficult to see how the intravenous administration of tetrahydrobiopterin can be regarded as a treatment for this condition. Unfortunately, these considerations make it likely that "treatment" of this variant form of PKU by intravenous administration of a tetrahydropterin may improve the abnormality in the blood while leaving the abnormality in the brain undisturbed.]

Another logical approach to therapy would be the administration of the products that are beyond the putative metabolic block, i.e., 5-hydroxytryptophan or DOPA, or both. This type of therapy is currently being evaluated.

The first example of hyperphenylalaninemia (without accompanying hypertyrosinemia) caused by a primary defect in another metabolic pathway that indirectly affects the activity of the phenylalanine hydroxylase system (type IB, p. 31) may have recently been uncovered.

In 1974, a preliminary report (Bartholomé, 1974) described a child who appeared to have classical PKU (blood phenylalanine of 46 mg/100 ml and an abnormal phenylalanine tolerance test). Although blood phenylalanine was well-controlled by a phenylalanine-restricted diet from the age of 2 weeks, at 9 months the child was mentally retarded and had severe "tetraspasm." From these developments, it was clear that the child did not have classical PKU. Since this case appeared to be similar to the patient studied by Kaufman and co-workers, in whom dihydropteridine reductase was shown to be lacking, it seemed that this child, too, would be deficient in the reductase. Surprisingly, however, it was found that levels of phenylalanine hydroxylase, dihydropteridine reductase, phenylalanine hydroxylase stimulating protein, and tetrahydropterin in a liver biopsy sample were all within the normal range (Milstien, Kaufman, and Bartholomé, unpublished data).

From these results, it is clear that even though all the components of the phenylalanine hydroxylase system are active under the somewhat idealized conditions that are used for the *in vitro* assays, one of them is essentially inactive *in vivo*.

Such a difference between *in vitro* and *in vivo* enzyme activities could reflect: (a) A normal enzyme that is unable to function in an abnormal environment, i.e., an altered intracellular environment that somehow prevents the hydroxylase activity from being expressed *in vivo*. Examples of the altered environment might be an accumulation of an inhibitory compound or a defect in uptake of one of the enzyme's substrates. According to this idea, the primary defect is one that affects an enzyme-catalyzed reaction other than phenylalanine hydroxylation and only indirectly affects phenylalanine metabolism. (b) An alteration in one of the properties of either the hydroxylase or the reductase so that it is active under *in vitro* assay conditions but not active in tissues. The alteration could, for example, lead to greater sensitivity to inhibition by normal concentrations of a metabolite.

Additional considerations lead to the conclusion that the first type of explanation—normal enzyme, abnormal environment (e.g., abnormally high concentrations of an inhibitor)—is the more likely one. This conclusion follows if one accepts the idea that the neurological symptoms in this infant are caused by defects in tyrosine and/or tryptophan hydroxylation, just as seems to be the case in the child with dihydropteridine reductase deficiency. If one accepts that idea, it seems clear that the putative inhibitor must be capable of inhibiting all three pterin-dependent hydroxylation reactions. A straightforward explanation of this generalized effect would be that the inhibitor acted on a component that is common to all three hydroxylases, such as dihydropteridine reductase. That this is probably not the explanation is indicated by the finding that the tetrahydrobiopterin levels in the patient's liver are within the normal range. It should be recalled that this finding is in contrast to that in the liver of the patient with dihydropteridine reductase deficiency, where essentially none of the biopterin was present in the tetrahydro form. Since normal hepatic tetrahydropterin levels tell us that dihydropteridine reductase is almost certainly active in this patient's liver, the putative inhibitor is probably not working on a common component of the hydroxylating system but, rather, on the hydroxylases themselves. And from what has already been said, one must further conclude that the putative compound can inhibit all three hydroxylases. At first glance, this might seem to be an unreasonable requirement. As already discussed, however (Kaufman and Fisher, 1974), these hydroxylases have enough properties in common that they can be profitably regarded as a family of enzymes. The requirement that a compound be capable of inhibiting all of them therefore does not weaken the conclusion stated above. Indeed, a variety of naturally occurring compounds, such as catechols and other metal complexing agents, do inhibit all three of the hydroxylases (see McGeer and McGeer, 1973 for review).

This line of reasoning makes unlikely the alternative hypothesis that this patient has a mutant form of phenylalanine hydroxylase that is abnormally sensitive to *normal concentrations* of an inhibitory compound. This hypothesis does make an unattractive demand: not only phenylalanine hydroxylase, but also tyrosine and tryptophan hydroxylases, must be modified in such a way that they also show enhanced sensitivity to inhibition by the putative metabolite.

Whatever the eventual biochemical explanation for the symptoms in this particular infant, it is now clear that this new syndrome—hyperphenylalaninemia with neurological deterioration that is unresponsive to dietary restriction of phenylalanine—is not a single disease entity that is always associated with a deficiency of dihydropteridine reductase. This last infant's phenotype is essentially the same as that of the one with hyperphenylalaninemia due to dihydropterin reductase deficiency, and yet it is evident that the genotypes of these two patients are not the same.

In view of the experience with the last patient, it should be evident that although the three patients described by Smith *et al.* (1975a) appeared to have similar features to those in the patient described by Kaufman *et al.* (1975a), the underlying enzyme defect in these three patients remains to be determined.

The mild hyperphenylalaninemia that is seen in some premature babies may be another example of a type IB disease, i.e., hyperphenylalaninemia due to a primary defect in a system other than the phenylalanine hydroxylase system.

This condition was described in 1941 by Levine and co-workers, who showed that premature babies receiving a high-protein diet (5 g or more of protein per kilogram of body weight per day) without ascorbate supplementation excrete tyrosine and its metabolites, *p*-hydroxyphenyllactate and *p*-hydroxyphenylpyruvate (Levine *et al.*, 1941a, 1941b). This abnormal excretion pattern, which is consistent with a defect in the enzyme that oxidizes p-hydroxyphenylpyruvate to homogentisic acid, is promptly corrected by ascorbate administration (Levine *et al.* 1941*b*) as is the hyperphenylalaninemia (Light *et al.*, 1966).

Although the nature of the defect in tyrosine metabolism in these infants is fairly clear, the defect in phenylalanine metabolism has not been established. Based on the report that phenylalanine hydroxylase is not active in livers from fetal and neonatal rats, rabbits, pigs, and a human premature infant (Kenney and Kretchmer, 1959), it had been assumed that the hyperphenylalaninemia of prematurity is due to this lack of phenylalanine hydroxylase (Menkes and Avery, 1963). Later, however, it was shown that phenylalanine hydroxylase is present in livers from a variety of different newborn animals (Brenneman and Kaufman, 1965). In the same study, livers from 1-day-old rats were shown to have a partial deficiency (about 50% of normal) of the hydroxylation cofactor and of dihydropteridine reductase. These findings indicated that the newborn rat might have a partial deficiency in phenylalanine hydroxylation due to the partial deficiency of reductase or cofactor, not to a deficiency of hydroxylase itself.

As far as human phenylalanine hydroxylase is concerned, the hydroxylase is present in premature infants as well as in the fetus (Friedman and Kaufman, 1971; Jakubovic, 1971; Perry *et al.*, 1972). Unfortunately, in none of these studies were quantitative measurements of dihydropteridine reductase or the pterin cofactor carried out. It is not known, therefore, whether the premature infant resembles the 1-day-old rat in having a modest impairment in phenylalanine hydroxylation due to a partial deficiency of cofactor and/or reductase. There is evidence, however, based on phenylalanine tolerance data, that premature infants do have an impairment in phenylalanine catabolism (Bremer and Neumann, 1966*a*) that, according to the previously discussed analysis of phenylalanine metabolism (see Figure 6), most likely means an impairment in phenylalanine hydroxylation.

What remains to be determined is whether the hyperphenylalaninemia of prematurity can be accounted for by a direct defect in the phenylalanine hydroxylase system, such as delayed maturation of dihydropteridine reductase (type IA), or whether this is an indirect effect (type IB). The finding that ascorbate administration can decrease the blood phenylalanine concentrations in these infants (Light *et al.*, 1966) strongly suggests that their hyperphenylalaninemia is not due entirely to a direct defect in the phenylalanine hydroxylase system.

The only known way in which ascorbate could directly stimulate phenylalanine hydroxylation would be by its ability to keep the pterin cofactor in the active, tetrahydro form (Kaufman, 1959). The amount of ascorbate used, however, 50–100 mg per infant (about 0.3 μmol per gram of body weight), would give tissue ascorbate levels that would be far below the 15 mM required (Kaufman, 1959) for maximum *in vitro* activity with phenylalanine hydroxylase.

Based on the reported effectiveness of ascorbate, therefore, it would seem more likely that the major part of the hyperphenylalaninemia associated with prematurity is secondary to the established defect in tyrosine metabolism in these infants. The mechanism by which phenylalanine catabolism is impaired probably involves inhibition of phenylalanine hydroxylation by tyrosine or one of its metabolites.

Phenylalanine hydroxylase can be inhibited by tyrosine in two different ways.* First, it has been shown that rat liver phenylalanine hydroxylase that has been activated either by partial proteolysis or by certain phospholipids can utilize tyrosine as "pseudosubstrate" (Fisher and Kaufman, 1973a); i.e., in the presence of tyrosine, the enzyme catalyzes a nonproductive oxidation of tetrahydrobiopterin. Under these conditions, when there is no net utilization of tyrosine, tyrosine inhibits the activity of the hydroxylase toward phenylalanine (50% inhibition at 0.1 mM phenylalanine, 0.8 mM tyrosine). As pointed out earlier, there is no indication that a significant fraction of rat liver enzyme normally exists *in vivo* in the activated state (Milstien and Kaufman, 1975a). Nothing is known, however, about the state of activation of the human liver hydroxylase.

Tyrosine can also inhibit the purified nonactivated rat liver hydroxylase. The inhibition has been observed in the presence of the synthetic cofactor analogue, DMPH$_4$. The inhibition appears to be competitive with both phenylalanine and with DMPH$_4$. At 0.2 mM phenylalanine, 0.01 mM DMPH$_4$, 1.2 mM tyrosine inhibits 35% (Chang and Kaufman, unpublished). Either of these modes of tyrosine inhibition of phenylalanine hydroxylase could account for the hyperphenylalaninemia of prematurity and its relief by ascorbate administration.

Although this cause of mild hyperphenylalaninemia is the most common one (Avery, 1967), it can be readily distinguished from all the others by the accompanying hypertyrosinemia, urinary excretion of tyrosine-derived metabolites, and by its responsiveness to ascorbate administration.

Before concluding the discussion of PKU variants, it should be mentioned that temporal modulations of any of these conditions are possible. In this regard, there have been a few reports of a mild form of hyperphenylalaninemia that appears to be transient (see Hsia, 1970 for review). The condition improves sometime during infancy or early childhood and is

---

* Theoretically, excess tyrosine could interfere with phenylalanine metabolism by interfering with the cellular uptake of phenylalanine. In the liver, the organ in which the major part of phenylalanine catabolism occurs, there is no evidence that tyrosine interferes with the uptake of phenylalanine. The lack of a competitive relationship for the two amino acids in liver is indicated by the experience with PKU patients: in the presence of massive concentrations of phenylalanine, there is no sign of any serious interference with hepatic tyrosine uptake.

distinct from the (just discussed) hyperphenylalaninemia of prematurity in that it is not associated with hypertyrosinemia. To distinguish it from the latter condition it has been called "transitory hyperphenylalaninemia of long duration" (Bremer, 1971). Since assays for the individual components of the phenylalanine hydroxylating system have not been carried out on liver samples from infants with transient hyperphenylalaninemia, the component that is apparently delayed in maturation is unknown. The occurrence of such variants emphasizes the need to reevaluate periodically (e.g., by repeating the phenylalanine tolerance test) the ability of patients with hyperphenylalaninemia, especially of the mild type, to handle a challenging load of phenylalanine.

In most cases of transient hyperphenylalaninemia, it has been found that as the blood levels of phenylalanine return to normal, the phenylalanine tolerance test also becomes normal (Hsia, 1970; Koch *et al.*, 1974). This parallelism is, of course, the expected one if the reactions by which a challenging dose of phenylalanine is catabolized are the same ones that the body utilizes to dispose of normal, dietary "doses" of phenylalanine. When these two events do not occur in parallel, as appears to be the case with a patient described by Bremer (1971), this bizarre course is unexpected. In this child, the high blood phenylalanine levels returned to normal by the time the infant was 3 months old, and yet her response to a challenging dose of phenylalanine indicated that at this time she still had a marked deficit in phenylalanine hydroxylase (decline in blood phenylalanine levels of about 1 mg/100 ml over a 150-min period). It would appear as though this infant's phenylalanine hydroxylase could handle the low levels of phenylalanine to which it was exposed when the phenylalanine was derived from normal dietary sources but not the high levels that were reached during the phenylalanine tolerance test.

There are several ways in which this peculiar pattern of phenylalanine metabolism could be explained. First, this pattern would be expected if this infant's phenylalanine hydroxylase showed greater-than-normal sensitivity to inhibition by excess phenylalanine. In order to relate this proposed explanantion to known properties of phenylalanine hydroxylase, it is necessary to review what is known about the sensitivity of the enzyme to inhibition by phenylalanine, an aspect of hydroxylase enzymology that has had a confused history.

The confusion started with studies of phenylalanine metabolism in mice that are homozygous for the gene, "dilute lethal." These mice, which show some of the characteristics of PKU—dilute pigmentation, neurological deterioration—were reported to have less than the normal amount of hepatic phenylalanine hydroxylase (Coleman, 1960; Rauch and Yost, 1963). These findings indicated that the dilute lethal mutation in mice might, indeed, provide a useful animal model for PKU.

Woolf *et al.* (1970) and Zannoni *et al.* (1966) failed to confirm these earlier findings. The latter workers and Zannoni and Moraru (1969), however, made an even more provocative claim, i.e., that the phenylalanine hydroxylase in the homozygous dilute lethal mice was abnormal in that it had a higher K_m for DMPH$_4$ and a greater sensitivity toward inhibition by phenylalanine and phenylpyruvate than the enzyme from control mice. Unfortunately, these claims were not confirmed by subsequent work (Woolf *et al.*, 1970; Treiman and Tourian, 1973). It seems likely that these unsubstantiated claims for the existence in dilute lethal mice of a genetically determined alteration in phenylalanine hydroxylase were based on results obtained with an assay for phenylalanine hydroxylase that was simplified to the point where it was no longer valid. As pointed out earlier, the use of such assay conditions can generate results that defy interpretation.

Although there is as yet no convincing evidence for the occurrence of a mutant form of phenylalanine hydroxylase with enhanced sensitivity to phenylalanine inhibition, the enzyme can be modified in ways that do lead to this change.

When rat liver phenylalanine hydroxylase is exposed to low concentrations of certain phospholipids, such as lysolecithin, or to partial proteolysis, the enzyme is markedly activated when assayed with tetrahydrobiopterin and, concomitantly, the activated enzyme exhibits much greater sensitivity to inhibition by excess phenylalanine (Fisher and Kaufman, 1973b). Human liver phenylalanine hydroxylase can also be activated by phospholipids (Friedman and Kaufman, 1973) and probably by limited proteolysis. Although the phospholipid activation is not as great as that seen with the rat liver enzyme, the activated human liver enzyme is also inhibited by excess phenylalanine, the enzyme from both sources being inhibited about 50% at 0.5 mM phenylalanine.

Based on these properties of phenylalanine hydroxylase, the pattern of phenylalanine metabolism seen in the infant studied by Bremer (1971) could be explained if one assumed that the infant's hydroxylase had been modified, perhaps through proteolysis, so that it became sensitive to inhibition by the high phenylalanine levels to which it was exposed during the acute phenylalanine tolerance test. In addition, to counterbalance the greater hydroxylase activity (due to the concomitant activation), one would be forced to also assume that the total amount of hydroxylase had been decreased. If the modification of phenylalanine hydroxylase that is being considered were due to proteolysis, this could easily lead, if it proceeded too far, to a decrease in total phenylalanine hydroxylase.

An alternative explanation for how an individual might have normal blood levels of phenylalanine and yet show signs of a serious impairment in the ability to metabolize high levels of phenylalanine would involve a change in the normal balance between the levels of hydroxylase and dihydropteridine reductase.

Thus, although the specific activity of the reductase in rat liver extracts is much higher than that of the hydroxylase, there are strong indications that *in vivo,* there is no functional excess of the reductase. These indications are based on results obtained with methotrexate, an inhibitor of dihydropteridine reductase, with a K_i value of 38 μM (Craine *et al.,* 1972). When methotrexate was given to rats in amounts that, according to this K_i value, should inhibit the reductase by only 10–20% (assuming uniform tissue distribution of the drug), 50% inhibition of the *in vivo* activity of phenylalanine hydroxylase was observed (Milstien and Kaufman, 1975*b*). If the activity of the reductase were far in excess of that of the hydroxylase, one would not expect the hydroxylase to be inhibited at all by this degree of inhibition of the reductase. Other lines of evidence also indicate that in the rat, the activity of the reductase is not present in great excess compared to the activity of the hydroxylase, expecially when the hydroxylating system is challenged by a large loading dose of phenylalanine (Fuller and Baker, 1974; Brand and Harper, 1974).

The same conclusion appears to be valid for humans. Thus, methotrexate did not appear to lead to increases in normal blood levels of phenylalanine, an indication that normal dietary amounts of phenylalanine were still capable of being metabolized at a normal rate, whereas the drug led to a marked impairment of the ability of the patients to handle challenging doses of phenylalanine (Goodfriend and Kaufman, 1961).*

From the considerations discussed above, it seems likely that any condition that leads to a partial deficiency of the reductase could lead to the pattern of phenylalanine metabolism that is similar to the one that was observed in the infant studied by Bremer (1971), i.e., phenylalanine metabolism that appeared to be normal when the metabolism was assessed by

---

* One reason the reductase may appear to be in excess of the hydroxylase with respect to basal phenylalanine metabolism, but not in excess with respect to the metabolism of a large loading dose of phenylalanine, could be related to the peculiar kinetics of phenylalanine hydroxylation in the presence of tetrahydrobiopterin. It has been reported that at low phenylalanine concentrations (below 0.1 mM), the rate of phenylalanine hydroxylation increases faster than linearly as phenylalanine concentrations are increased; i.e., the relationship between the initial rate of hydroxylation and phenylalanine concentration is sigmoidal (Fisher and Kaufman, 1973*b*). On the other hand, the relationship between the rate of quinonoid dihydropteridine reduction and concentration of quinonoid dihydropterin is hyperbolic (Craine *et al.,* 1972). As a result of this relationship, as phenylalanine concentration is increased (up to about 0.1 mM) the rate of hydroxylation will increase faster than the rate of quinonoid dihydropterin reduction and the balance between these two enzymatic activities could be altered; i.e., a given level of reductase may be able to keep pace with the rate of phenylalanine hydroxylation at low phenylalanine concentrations, whereas it may not be able to keep up with the higher rate of phenylalanine hydroxylation at higher phenylalanine concentrations.

resting phenylalanine blood levels, but appeared to be abnormal when the metabolism was assessed by a challenging dose of phenylalanine. It is possible that this pattern of phenylalanine metabolism is characteristic of heterozygotes for dihydropteridine reductase deficiency.

Finally, it should be noted that the pattern of phenylalanine metabolism that is under consideration, and that is characteristic of the infant studied by Bremer (1971), has features in common with those of the two infants described by Auerbach *et al.* (1967). As already discussed, these patients also had a high tolerance to dietary protein, and yet results of acute phenylalanine challenge tests indicated that they had a marked impairment in phenylalanine metabolism. Also discussed previously are the reasons that the conclusion reached by Auerbach and co-workers, that these two infants lacked phenylalanine transaminase, is untenable. It seems likely that these two infants, like the infant described by Bremer, might have had a partial deficiency of dihydropteridine reductase.

7. HETEROZYGOTES FOR HYPERPHENYLALANINEMIA DUE TO PHENYLALANINE HYDROXYLASE DEFICIENCY

Assays for phenylalanine hydroxylase in liver samples from classical PKU heterozygotes have not been reported. Data are available, however, for atypical PKU heterozygotes, i.e., for parents of children with persistent mild hyperphenylalaninemia due to partial deficiency of phenylalanine hydroxylase (Kaufman *et al.*, 1975*b*).

The results for three sets of parents, together with those from their affected children and from a group of control patients, are shown in Table 5. As can be seen from the table, the patients have an average of 5% of the normal hepatic phenylalanine hydroxylase activity. For the parents, the values show a considerable scatter, raising the possibility that they do not constitute a homogeneous group. In particular, the relatively high hydroxylase activity of Father Ke (31% of the average control value) suggests that his genotype may be different from that of the other parents (in which case, the genotype of J. Ke would also be different from that of the other patients). It should be noted that it is unlikely that Father Ke is a normal, since both his fasting plasma phenylalanine levels and his phenylalanine levels 4 hr after an oral phenylalanine load were elevated when compared with normal controls (Kang *et al.*, 1970*a*). All the other parents also showed signs of abnormal phenylalanine metabolism (either elevated plasma phenylalanine-to-tyrosine ratios or abnormally high phenylalanine levels after oral phenylalanine loading) (Kang *et al.*, 1970*a*). Excluding

TABLE 5.   Human Liver Phenylalanine
Hydroxylase Activity (All Values Expressed
as Micromoles Tyrosine per Gram Protein
per Hour)

Hyperphenylalaninemics	
J. Ke	4.7
S. P.	3.8
P. G.	2.7
Parents of hyperphenylalaninemics	
Father Ke	23.0
Mother Ke	7.2
Father P	4.4
Mother P	7.7
Father G	5.8
Mother G	11.4
Normal controls	
1. Extrahepatic block	96
2. Choledocholithiasis	59
3. Cholelithiasis	71
4. Biliary atresia	49
5. Biliary atresia	95
6. Head injury	83

Father Ke from the group, the parents have an average of 7.3% of normal hepatic hydroxylase activity; including him, the average value is 10% of normal. This low value for the parents is of considerable interest, for although there have been other examples of heterozygotes having less than the expected 50% of normal activity—such as in cystathionine synthetase deficiency, where the heterozygotes have about 34% of the normal level of the affected enzyme (Finkelstein et al., 1964; Laster et al., 1965; Gaull and Sturman, 1971), and adenine phosphoribosyl transferase deficiency, where the heterozygotes have between 21 and 37% of the normal activity (Kelley et al., 1968)—in most other cases where enzyme activities have been assayed in individuals heterozygous for enzyme deficiencies, proportionality with gene dosage has been observed (Harris, 1970). Therefore, if the hyperphenylalaninemia patients synthesize a hydroxylase variant with 5% of the normal activity, one might expect the parents (presumably heterozygotes) to show about 50% of the normal activity. Two interpretations may explain the deviation from proportionality of activity with gene dosage in the heterozygotes.

1. Regulatory gene mutations have been the basis of models proposed in other cases of enzyme deficiencies where heterozygote activity levels differed from expected values. Against this explanation for hyperphenylalaninemia heterozygotes is the evidence that phenylalanine hydroxylase in

livers from hyperphenylalaninemic patients is distinguishable from the normal enzyme in the following properties (Friedman, 1972b): (a) it has a lower apparent K_m value for phenylalanine in the presence of 6,7-dimethyl-tetrahydropterin; (b) its activity is stimulated less by lysolecithin in the presence of the natural cofactor, tetrahydrobiopterin; and (c) it appears to be somewhat more labile to heat inactivation. These results suggest that the low hydroxylase activity in hyperphenylalaninemic patients is due to a mutation in the gene coding for the structure of the hydroxylase, rather than in a gene that regulates the rate of synthesis of the normal enzyme.

2. An alternative explanation would involve negative interallelic complementation, a phenomenon for which there is some precedent in microbial systems (Fincham, 1966), which involves protein–protein interaction between subunits in a multimeric enzyme. In order to evaluate this possibility, one must consider what is known about the structure of phenylalanine hydroxylase. As discussed earlier, there is evidence that rat liver phenylalanine hydroxylase is a multimeric protein composed of two electrophoretically distinguishable subunits (mol. wt. 51,000–55,000) (Kaufman and Fisher, 1970). In addition to the monomers, the enzyme exists as dimers (mol. wt. 110,000) and tetramers (mol. wt. 210,000). It is not known with certainty, however, which form is catalytically active. Nonetheless, the fact that the enzyme shows cooperative kinetic behavior (in the presence of tetrahydrobiopterin) (Fisher and Kaufman, 1972b, 1973b) strongly suggests that a polymeric form of the enzyme is active. It is also known that limited proteolysis of the enzyme leads to a highly active form that can exist only as monomers and dimers (Fisher and Kaufman, 1973b). Since there is kinetic evidence that a polymeric form of the enzyme is active, this last finding indicates that the dimer is catalytically active.

With these considerations in mind, and on the assumption that the structure of the human liver enzyme is essentially the same as the rat liver enzyme, the interallelic complementation model would lead to the following picture for the phenotype of patients and their parents. According to this model, the normal hydroxylase is a dimer composed of two different subunits, α and β, each subunit under the genetic control of an A locus and a B locus. The product of the mutant hyperphenylalaninemia gene would be a modified hydroxylase subunit, α' or β'. The hydroxylase in homozygous patients would then have the structure $\alpha'\beta$ or $\alpha\beta'$ (genotype $A'A'BB$ or $AAB'B'$); these mutant dimers would have 5% of the normal hydroxylase activity. If the parents are simple heterozygotes (genotype $A'ABB$ or $AAB'B$), their tissues would have equal amounts of two types of enzyme molecules: $\alpha\beta$ and $\alpha'\beta$ for the first genotype, and $\alpha\beta$ and $\alpha\beta'$ for the second, and their hydroxylase activity would be expected to be about 50% of normal. To explain the finding that parents of patients with mild hyperphenylalaninemia have an average of 10% of the normal activity, one must

assume nonrandom combination of the subunits (leading to selective formation of $\alpha'\beta$ or $\alpha\beta'$ mutant dimers) and/or more rapid synthesis (or slower
degradation) of the mutant subunit compared with that of the normal
subunits; this situation could lead to a preponderance of mutant dimers
over the normal, fully active dimers.

These arguments are based in part on the finding that phenylalanine
hydroxylase is composed of nonidentical subunits. It is possible that the
enzyme is a multimer of identical subunits, the observed physical differences being due to partial proteolysis of some of them during the isolation
procedure. In this case, the analogous model of negative *intraallelic* complementation would lead to the same conclusions for the hydroxylase
activity in heterozygotes.

If the active form of the hydroxylase were the tetramer $(\alpha_2\beta_2)$, the
same line of reasoning would lead to the prediction that the parents would
have a minimum of about 29% of the normal hydroxylase activity (i.e., for
the case where the α subunit is the modified one, their tissues would
contain three types of enzyme molecules, $\alpha_2\beta_2$, $\alpha\alpha'\beta_2$, and $\alpha'_2\beta_2$, constitut-
ing, respectively, 25, 50, and 25% of the hydroxylase population; if the
presence of even a single mutant subunit, α', in the tetramer led to an
enzyme with only 5% of the normal catalytic activity, three-fourths of the
molecules would have 5% and one-fourth would have 100% of the hydroxylase activity, giving an average of 29% of the normal activity for this
population of enzyme molecules). As mentioned, the hepatic phenylalanine
hydroxylase levels in heterozygotes for classical PKU (i.e., parents of
patients with classical PKU) have not been reported. It is not known with
certainty, therefore, whether their phenylalanine hydroxylase activity will
show the same deviation from proportionality with gene dosage as does the
phenylalanine hydroxylase activity of mild hyperphenylalaninemia heterozygotes (i.e., parents of patients with atypical PKU).

At first glance, one might think that because classical PKU homozygotes have a far greater deficit in phenylalanine hydroxylase activity (0.27%
of normal) (Friedman *et al.,* 1973) than do patients with mild hyperphenylalaninemia (5% of normal, see Table 5), classical PKU heterozygotes
would then have less phenylalanine hydroxylase activity than do heterozygotes for mild hyperphenylalaninemia. What should be evident from the
models that have been discussed, however, is that there need be no strict
correlation between hydroxylase activity in homozygotes and in heterozygotes. Thus, classical PKU heterozygotes could well have higher phenylalanine hydroxylase activity than do mild hyperphenylalaninemic
heterozygotes.

Indeed, there are indications that they do. Bremer and Neumann
(1966*b*) have found that after a challenging dose of phenylalanine, the rate
of fall of blood phenylalanine levels in classical PKU heterozygotes is about

50% that of group of controls ($t_{1/2}$ for controls, 89 min; $t_{1/2}$ for heterozygotes, 159 min). On the basis of the theoretical analysis of phenylalanine metabolism that was outlined earlier, it can be estimated that Bremer and Neumann's results with heterozygotes are close to the expected ones if the heterozygotes have 50% of the normal level of phenylalanine hydroxylase (see Figure 7, where an individual with 50% of the normal phenylalanine hydroxylase level was estimated to have a $t_{1/2}$ of 170 min, as compared with the value of 159 min that was actually found by Bremer and Neumann). [Woolf et al. (1967) previously reported that controls metabolized a challenging dose of phenylalanine almost exactly twice as fast as did heterozygotes for classical PKU; i.e., $t_{1/2}$ values calculated from their data are 67 min for controls and 131 min for PKU heterozygotes. From their results, these authors concluded that controls have almost exactly twice as much phenylalanine hydroxylase in each liver cell as do PKU heterozygotes. Since Woolf and co-workers assumed that hydroxylation was the only route by which a dose of phenylalanine is metabolized, an assumption that is unwarranted because it neglects transamination (see Figure 6), this last conclusion is probably in error. If the contribution to phenylalanine metabolism of transamination is included, it can be estimated that the results of Woolf and co-workers are consistent with PKU heterozygotes having 35–40% of the normal level of phenylalanine hydroxylase.]

The results of these two tolerance studies on classical PKU heterozygotes indicate that they probably have between 35 and 50% of the normal phenylalanine hydroxylase levels, a value that is much higher than the 10% of normal found by direct assay for parents of patients with persistent mild hyperphenylalaninemia. Phenylalanine tolerance test results obtained with the latter group tend to support the conclusion that these individuals may have a more severe deficit in phenylalanine hydroxylase than do parents of classical PKU patients. Thus, Rampini (1973) reported that after an oral dose of phenylalanine, the peak concentration of phenylalanine in the plasma of parents of classical PKU patients may be somewhat higher than in the parents of patients with mild hyperphenylalaninemia. The rate of decline of plasma phenylalanine levels, however, is faster in the parents of the classical PKU patients. As already suggested, the rate of decline in peak phenylalanine plasma concentrations after a challenging dose is a better indication of in vivo phenylalanine hydroxylase activity than is the peak phenylalanine level that is reached. Rampini's results, therefore, are consistent with the possibility that classical PKU heterozygotes have more phenylalanine hydroxylase activity than do atypical PKU heterozygotes. The final test of this conclusion, however, must wait on results of in vitro assays of phenylalanine hydroxylase activity in these two groups of parents.

8. PATHOGENESIS

The steps leading from the deficiency of phenylalanine hydroxylase to the development of the characteristic pathology of PKU—mental retardation and other behavioral abnormalities—remain obscure. Indeed, the very first link in what might be expected to be an expanded causal chain has not yet been identified with certainty, that first link presumably being the demonstration of a relationship between the degree of the enzyme defect (as reflected by elevation of blood levels of phenylalanine or urinary excretion of phenylalanine-derived metabolites) and the degree of mental disability.

Before the pathogenesis of PKU can be discussed, it should be evident that any such discussion will be shaped and limited by certain assumptions. Implicit in the following remarks is the bias that the mental defect in PKU is caused by alterations in structure. This first assumption is based on the evidence that the mental defect in PKU, once it becomes manifest, appears to be irreversible. Furthermore, it is assumed that these structural changes are caused by disturbed physiological processes, which, in turn, are caused by alterations in metabolism, which ultimately can be traced to a genetically determined deficiency of phenylalanine hydroxylase. Obviously, other biases, e.g., that elevated concentrations of phenylalanine or one of its metabolites "directly depress the activity of the higher mental center" (Woolf and Vulliamy, 1951), could lead to a different kind of discussion.

To be even more explicit, in the following discussion it will be assumed that phenylalanine or one of its metabolites is a toxic or damaging compound in the sense that it interferes—by mechanisms that are yet to be defined—with the development, and ultimately with the functioning, of the CNS.

As discussed earlier, in addition to the irreversible changes, classical PKU is also characterized by symptoms (e.g., EEG abnormalities, irritability) that appear to be reversible by phenylalanine restriction long after the developing brain has been permanently damaged. The mechanism by which excess phenylalanine leads to these reversible changes may well be different from that involved in the permanent damage. In the following discussion, the focus will be almost entirely on the pathogenesis of the irreversible changes.

Perhaps the word "toxic" should be avoided because it is a term, like "toxicity," with connotations of quantitative or acute potency as a poison. Within this context, a toxic substance is often equated with something that is poisonous at low concentrations. Perhaps it was from this pharmacological perspective that scientists have for years been searching, unsuccessfully, for toxic effects of phenylalanine and its major metabolites. Other than a report that phenylacetate, when ingested by man at the enormous dose of 0.27 g/kg in 2 hr, led to dizziness followed either by drowsiness or nervousness (Sherwin and Kennard, 1919), these metabolites seem to be relatively innocuous. [On the other hand, minor metabolites, such as phenylethylamine, are known to be vasoactive and may disturb normal brain function (Oates *et al.*, 1963, and references therein).] Indeed, it was this lack of success, together with the early failures to find a correlation between the intelligence quotient of PKU patients and the concentration of phenylalanine in their blood (or its metabolites in their urine) that led to the conclusion that "these major metabolites are not involved in a direct toxic action" (Jervis, 1963). Even more recently, this view was still being expressed. Thus, Udenfriend, referring to these metabolites, stated that "although these are interesting metabolites and can be detected in the urine, they are to my knowledge without effect on the central nervous system. As a pharmacologist I don't consider them to be active compounds, although they could conceivably produce effects in some unknown manner" (Udenfriend, 1967).

As will be discussed later, some evidence for a correlation between the severity of the mental defect and the abnormality in phenylalanine metabolism has been reported recently. Even without such evidence, however, it is difficult to see why the lack of such a correlation should have led to the conclusion that exonerated phenylalanine or its major metabolites as toxic compounds. Rather, the lack of correlation may merely indicate that the toxic effects can be modulated by other substances (e.g., if phenylalanine can inhibit an enzyme competitively, the normal substrate of the enzyme will modulate this toxic effect of phenylalanine). In any case, results of the last 10 years have taken away much of the cogency of the question about whether phenylalanine or its metabolites are "toxic." As will be seen, it is now apparent that phenylalanine and phenylpyruvate can inhibit a very large group of enzymes.

Data relevant to a consideration of the pathogenesis of the mental defect in PKU come from two separate sources, i.e., from studies with patients as well as with animals. Within that latter category are those studies aimed at producing an animal model for the disease. To evaluate the results of these animal studies, it is necessary to examine the fidelity of these so-called models of PKU to the human disease.

The study of animal models for PKU was stimulated in 1958 by the report that phenylalanine-fed rats showed some impairment in learning ability (Auerbach *et al.*, 1958). Although it was originally claimed that phenylalanine feeding led to a severe decrease in the activity of hepatic phenylalanine hydroxylase, later work showed the decrease to be much more modest (Woods and McCormick, 1964).

This line of investigation took a critical turn in 1967, when Lipton *et al.* (Lipton *et al.* 1967), concerned because the phenylalanine-feeding model for PKU differed from the clinical condition in at least one obvious respect, i.e., phenylalanine feeding led to elevated blood levels not only of phenylalanine, but also of tyrosine (whereas tyrosine levels are somewhat depressed in PKU), developed a model for PKU based on the administration to rats of a combination of phenylalanine and PCPA (*p*-chlorophenylalanine). The latter compound had previously been shown to be a potent inhibitor of both phenylalanine hydroxylase and tryptophan hydroxylases (Koe and Weissman, 1966).

There is no doubt that the use of PCPA did partly correct this most evident biochemical deviation of the animal model from clinical PKU; what is less evident is how many more subtle deviations were introduced into the model with the use of PCPA and whether the tradeoff was a favorable one. These questions remain largely unexamined. Instead, advocates of this particular model have emphasized the similarities between it and the disease in humans (Andersen and Guroff, 1972; Andersen *et al.*, 1974; Berry *et al.*, 1975). A more balanced view might be useful.

At the outset, it should be noted that any model for PKU that uses rats as the experimental animal will differ from the disease in humans in at least two respects: one developmental and the other biochemical. First, there are profound differences in the developmental maturity of the brains of newborn rats as compared to those of newborn humans. In rats, the main phase of cerebral neuronal multiplication occurs before birth, while most glial mitosis occurs postnatally (Dobbing, 1975). Since glial proliferation precedes myelination and circuitry development (dendritic growth and branching and synapse formation), the newborn rat brain is relatively immature. It should be noted that the brain "growth spurt," believed to be the time when the growing brain is most vulnerable to undernutrition (Dobbing, 1975) and perhaps to other metabolic disturbances, corresponds to this period of rapid glial proliferation and myelination. In the rat, the brain growth spurt occurs postnatally (Dobbing, 1975).

By contrast, neuronal multiplication in humans is the main component of brain growth in the second intrauterine trimester and the growth spurt, which, as mentioned, includes myelination, covers the period from the beginning of the third intrauterine trimester until near the end of the second postnatal year. Because of these differences, it has been estimated that the equivalent human brain age of the rat brain at birth cannot be much later than 18 weeks of human gestation (Dobbing, 1975).

As far as PKU models are concerned, the greater immaturity of the rat brain at birth suggests that it might be more sensitive to nonspecific biochemical damage than the neonatal human brain.

In view of this greater immaturity, and hence vulnerability, of the rat brain at birth, one might have expected that the phenylalanine-PCPA-treated neonatal rat would show profound deficits in learning. That the deficits do not seem to be dramatic is perhaps the most significant finding to come out of these animal studies.

Biochemically, rats and humans also differ in the way phenylacetate is conjugated. In humans, as mentioned (see Figure 2), this metabolite of phenylalanine is excreted as the glutamine derivative. By contrast, rats excrete phenylacetate conjugated with glycine (Meister, 1965). Although this may be a trivial metabolic difference, Perry *et al.* (1970) have raised the possibility that the lower-than-normal plasma glutamine levels seen in PKU patients may be causally related to the development of the mental defect. Furthermore, they have postulated that the deficiency of glutamine might be due to its excretion in the urine as the phenylacetyl derivative. If this postulate should prove to be correct, the pathogenesis of the brain abnormalities in the phenylalanine-PCPA-treated rats would obviously be different from that of the clinical condition in humans.

One of the most cogent questions that can be raised about the model is one of specificity. The advocates of the model seem to have accepted the following scenario: PCPA inhibits phenylalanine hydroxylase, which, together with the administration of phenylalanine, leads to elevated blood and tissue levels of phenylalanine, which leads to learning deficits and other behavioral changes. At the same time, the advocates have ignored the following scenario: PCPA inhibits phenylalanine hydroxylase *as well as other enzyme-catalyzed reactions,* and it is the inhibition of one of these other enzymes that adversely affects the brain. According to the first scenario, the model may be a good one for PKU; according to the second, it may be a good model for some other disease.

As far as the specificity question is concerned, it is known that PCPA inhibits at least one other enzyme in addition to phenylalanine hydroxylase. In 1966, Koe and Weissman reported that the drug depletes brain serotonin and that it inhibits hepatic hydroxylation of both phenylalanine and tryptophan (Koe and Weissman, 1966). Subsequently, Jéquier *et al.* (1967) reported that it also inhibits brain tryptophan hydroxylase.

Whereas there is evidence that the activity of tryptophan hydroxylase is somewhat inhibited in classical PKU patients, the extent of inhibition—about 35% (Jéquier, 1968)—is much less than the approximate 90% inhibition of the brain-stem enzyme in PCPA-treated rats (Jéquier *et al.,* 1967). The depletion of brain serotonin in these rats is also much more severe (Koe and Weissman, 1966; Jéquier *et al.,* 1967) than is the decrease reported in the necropsied brains of untreated classical PKU patients (McKean, 1972).

As far as serotonin and tryptophan hydroxylases are concerned, therefore, this PKU animal model produces much more marked changes than those seen in the clinical condition.

That PCPA treatment has deleterious effects beyond those that can be traced to inhibition of either phenylalanine hydroxylase or tryptophan hydroxylase is strongly suggested by the finding that the daily injection of the drug into neonatal rats leads to the development of cataracts in a large number of the treated animals (Watt and Martin, 1969), a lesion that is not a characteristic symptom of classical PKU in humans.

Certainly one of the most significant deviations of this animal model from classical PKU in humans is in the extent of the defect in phenylalanine hydroxylase activity. The maximum inhibition of hepatic phenylalanine hydroxylase in rats treated with PCPA has been reported to be 80–85% (Koe and Weissman, 1966). By contrast, classical PKU patients have essentially no hepatic phenylalanine hydroxylase activity. Surprisingly, advocates of the model seem to have discounted the significance of this difference, "rounding off" the residual hydroxylase activity so that it now equals zero. Once the rounding-off has been accomplished, the phenylalanine hydroxylase activity in the PCPA-phenylalanine treated rats becomes equivalent to "an absence of phenylalanine hydroxylase activity," permitting the conclusion that in respect to the defect in phenylalanine hydroxylase activity, the model is the "same as clinical characteristics" (Andersen and Guroff, 1972). In reality, however, the 15–20% residual phenylalanine hydroxylase activity of PCPA-treated rats indicates that, in this respect, they bear a closer resemblance to patients with mild hyperphenylalaninemia (atypical PKU) than they do to patients with the classical form of the disease. It may be recalled that atypical PKU patients have about 5% normal phenylalanine hydroxylase (Kang et al., 1970a; Kaufman and Max, 1971), and significantly, this level of hydroxylase appears to be adequate to permit normal brain development. This quantitative comparison raises disturbing questions about the aptness of this animal model to the classical human disease. If, on the basis of residual phenylalanine hydroxylase activity, the animal model resembles atypical PKU, a condition that even if untreated does not lead to mental retardation, how can this animal model be used to study the pathogenesis of mental retardation in classical PKU?

Indeed, it seems highly likely that this residual phenylalanine hydroxylase activity is related to the other very significant deviation of the animal model from classical PKU, namely, that the learning deficit in the animal model is so moderate when compared with that seen in the clinical condition. In this regard, it should be recalled that untreated classical PKU produces mainly idiots (I.Q. < 50), two-thirds of the patients being unable to talk (Knox, 1972). By contrast, most of the enduring changes that have been reported in these animals are in behavior rather than in learning (Andersen and Guroff, 1972; Andersen et al., 1974). For example, in one recent study, neonatal rats treated with PCPA showed no evidence of any learning deficit but, rather, showed "enduring behavioral deficits best described as a defective arousal function" (Hole, 1972b; see also, Hole, 1972a). It is not obvious that this defect in arousal function has been factored out of any of the studies designed to show impaired learning ability in PCPA-phenylalanine treated rats.

The significance of this relatively high amount of residual phenylalanine hydroxylase activity in the PCPA-phenylalanine-treated rats is probably not in the hydroxylase activity *per se,* but in its metabolic consequences, the most important of which may be the degree of amino acid imbalance, e.g., phenylalanine-to-tyrosine ratios, that characterizes the PCPA-phenylalanine treated rat. Although it is true that this ratio is higher in the PCPA-phenylalanine-treated rat than it is in rats treated with phenylalanine alone, the reported ratios of 14:1 (Lipton *et al.,* 1967) and 6:1 (Berry *et al.,* 1975) for PCPA-phenylalanine-treated rats are far lower than the 50:1 ratio found for a large group of patients with classical PKU (Perry *et al.,* 1970). Also, there is no evidence (Berry *et al.,* 1975) that these animals show the pattern of decrease in the plasma levels of some of the other amino acids that is characteristic of classical PKU (Perry *et al.,* 1970; Linneweh and Ehrlich, 1962). The significance of these differences between the animal model and the human disease is not known.

It is apparent from this brief survey that the PCPA–phenylalanine model for PKU differs from the clinical condition in many respects. And even where there are similarities, the fact that PCPA is known to affect reactions other than phenylalanine hydroxylation makes it hazardous to conclude that any change seen in the PCPA-phenylalanine-treated rat is necessarily a consequence of inhibition of phenylalanine hydroxylase by the drug.

Even if doubts can be raised about the usefulness of the PCPA-phenylalanine-treated rat as a high-fidelity model for PKU, studies on the effects of phenylalanine and its metabolites on a variety of enzymes continue to be carried out with the hope that these studies may provide a clue to the pathogenesis of PKU. The problem with this approach, as will become evident, is not that deleterious effects were hard to find, but rather that phenylalanine and its metabolites were found to inhibit too many enzyme-catalyzed reactions.

In dealing with this mass of data, especially those from the animal studies, one needs a sieve to help sort the dross from the ore. For this purpose, one of the few aids available is the great specificity of tissue damage, mentioned earlier, that is characteristic of the disease. A proposed mechanism for the pathogenesis of the mental defect in PKU that does not account for this specificity must be regarded as less attractive than one that does.

What is the basis for this specificity? There are several aspects of brain morphology, physiology, and metabolism that are unique and that may be related to this question. Because the central nervous system of mammals appears to require an exquisitely stable internal environment to function normally, special control systems have evolved that serve to regulate the transport of a variety of substances into the brain. This system of transport mechanisms, called collectively "the blood–brain barrier," includes a unique one that operates at the level of the endothelial cells of brain capillaries. Together, this system of *multiple* barriers controls the traffic of molecules between the blood and the brain more stringently than that between the blood and other organs of the body.

Although the exact mechanism for this stringent control is not known, the entry of most molecules into the brain appears to be regulated through the operation of a system of saturable carriers. And since natural carriers are usually not specific for single substances but, rather, for families of structurally related substances, these systems characteristically show competition among members of the family. Competition among various amino acids, e.g., between tyrosine and aromatic amino acids, has been demonstrated in whole animals (Chirigos *et al.*, 1960) as well as in brain slices (Neame, 1961). That the brain is particularly sensitive to this kind of competition—perhaps because of its system of *multiple* barriers, which, when overwhelmed, can lead to amplification of deleterious effects—is shown by the fact that inhibition of L-tyrosine uptake by L-tryptophan is much less in muscle than in brain (Guroff and Udenfriend, 1962). That such competition actually occurs in PKU patients is indicated by the findings of Oldendorf *et al.* (1971), who showed reduced uptake of [^{75}Se]selenomethionine into the brains of such patients, presumably due to their elevated blood levels of phenylalanine.

In addition to the unusually stringent systems that regulate the exchange of substances between the blood and the brain, there are unique aspects of brain development that may be the basis for the specificity of tissue damage in PKU. As already discussed in another context, development of the brain is characterized by a growth spurt, which in humans occurs between the period shortly before birth and the second year of postnatal life. No other major tissue or organ system has a comparable postnatal period of maximum growth. Since the brain appears to be unusually sensitive to metabolic insult during this period of rapid growth, this developmental characteristic could be the basis for the specificity of tissue damage in untreated PKU.

Finally, there are unique aspects of brain metabolism, especially those related to brain function—e.g., neurotransmitter turnover and myelin synthesis—that could account for the specific and characteristic pathophysiology of PKU.

As mentioned, it has proved to be surprisingly difficult to identify the first link between the hydroxylase defect and the brain damage, i.e., to establish a relationship between levels of phenylalanine in blood, or of phenylalanine-derived metabolites in the urine, and the degree of mental disability in PKU. Although the ability to discriminate with some success atypical PKU patients (most of whom will not be retarded) from classical PKU patients (most of whom will be retarded) on the basis of plasma phenylalanine levels [< 15 mg/100 ml, atypical PKU; > 25 mg/100ml, classical PKU (Hsia, 1970)] implies the existence of such a relationship, attempts to establish it within a group of classical PKU patients have been unsuccessful (Borek *et al.*, 1950).

More recently, Andrews *et al.* (1973) reported that there is a relationship in classical PKU patients between phenylalanine concentration in granulocytes and degree of "neuropsychiatric" damage. Significantly, in confirmation of the earlier negative findings, the correlation broke down when plasma phenylalanine concentrations were used instead of granulocyte concentrations. In addition to this modification, Andrews and co-workers did not rely on I.Q. scores as a measure of brain damage, claiming that it lacked discriminating power, especially with severely retarded patients. Rather, they used an "ability score," a composite measure that includes not only I.Q. score but also such behavioral parameters as mutism and epilepsy. In a separate study on the same group of patients, the excretion of phenylpyruvate and the sum of the excretion of 4-hydroxyphenyllactate and 4-hydroxyphenylpyruvate were found to correlate with the "ability score" (Chalmers and Watts, 1974).

These results support the conclusion that it is not the deficiency of phenylalanine hydroxylase *per se* that leads to the mental defect in PKU but, rather, one of the metabolic consequences of this deficiency. Furthermore, of the consequences that could be listed, it is the accumulation of phenylalanine that is probably the critical event. These results support the assumption, stated earlier, that phenylalanine or one of its metabolites is a toxic or damaging substance that interferes with normal brain development and function.

If the work of Andrews and co-workers has finally established the first link in the chain of events that connects enzyme deficiency to mental defect, the next step in the analysis of pathogenesis would be to determine whether phenylalanine or one of its metabolic products, such as phenylpyruvate, is closer to the offending compound.

One approach to deciding between phenylalanine and phenylpyruvate would be to take advantage of the fact that the two compounds are interconvertible *in vivo* by a transamination reaction. It is known, for example, that urinary excretion of phenylpyruvate, and presumably its content in tissues, can be decreased by feeding glutamine or glutamate to PKU patients (Meister *et al.*, 1956; Bowman and King, 1961). [In another study, the effect of glutamate feeding on phenylpyruvate excretion was quite variable (Woolf and Vulliamy, 1951).] The decrease in phenylpyruvate excretion is believed to be due to Reaction 3—being driven from right to left by high concentrations of glutamate.

$$\text{phenylalanine} + \alpha\text{-ketoglutarate} \rightleftarrows \text{phenylpyruvate} + \text{glutamate} \quad (3)$$

If phenylpyruvate were the toxic compound, or were closer to it than phenylalanine, glutamate or glutamine administration would be expected to prevent, at least partially, the development of mental retardation in PKU. On the other hand, if phenylalanine were the toxic compound, or were closer to it than phenylpyruvate, feeding glutamine or glutamate (by increasing blood concentrations and tissue contents of phenylalanine) might actually exacerbate the symptoms of the disease.

As just mentioned, there have been several studies in which glutamate and other amino donors have been given to PKU patients (Woolf and Vulliamy, 1951; Meister *et al.,* 1956; Bowman and King, 1961). Only in the last of these studies was the treatment reported to have led to slight improvement in the patient's mental ability. On the basis of their results, Woolf and Vulliamy (1951) concluded that "glutamate feeding was unlikely to produce beneficial effects in PKU patients." In the light of today's knowledge of PKU, there is little basis for this pessimistic conclusion, since the two patients used in this study, ages 20 and 23 months, were almost certainly beyond the age when the most severe damage to the brain seen in classical PKU can still be prevented. The widely used low-phenylalanine diet, for example, would also be largely ineffective if started at this age (Knox, 1972). As far as the other two studies are concerned, the patients used were all between 7 and 21 years old, an age range when it was highly unlikely that the glutamate feeding could have improved their mental abilities. A decisive experiment that might indicate whether phenylalanine or phenylpyruvate were closer to the toxic compound has yet to be done. The conclusion (Bessman, 1964) that "phenylpyruvic acid alone is not the toxic agent in phenylketonuria" (reached so confidently in 1964) must therefore be regarded today as premature.

One must hasten to add, however, that the opposite conclusion—that "clinically, phenylpyruvate is more toxic to the brain than phenylalanine on an equimolar basis" (Patel *et. al.,* 1973)—is also without convincing support. The clinical evidence cited by the authors is based mainly on their earlier observations on two infants with presumed phenylalanine aminotransferase deficiency. These patients, who developed normally, did not excrete large quantities of phenylpyruvate even when, in response to a high-protein diet, their plasma phenylalanine levels were as high as 30 mg/100 ml. Since high phenylalanine in the absence of high phenylpyruvate did not appear to lead to mental retardation in these two infants, these results have been used to support the conclusion that phenylpyruvate is the more toxic compound.

As already discussed at length, there are reasons for seriously questioning the original conclusion that these infants did in fact lack phenylalanine aminotransferase. The present use of the biochemical results obtained with the patients to support the conclusion that the keto acid is closer to the toxic compound than is phenylalanine can, however, be questioned on separate grounds. It is not known whether these infants would have developed normally if their blood phenylalanine levels had been maintained in the range of 30 mg/100 ml by deliberately feeding them a high-protein diet. Almost certainly, this was not done. Instead, they developed normally while their phenylalanine blood levels were presumably maintained close to the normal range simply by not giving them high-protein diets. These studies, therefore, provide no basis for concluding that high phenylalanine blood levels (in the range of 30 mg/100 ml) are innocuous as long as they are not accompanied by elevated phenylpyruvate levels. These results, therefore, cannot sustain the conclusion that phenylpyruvate is more toxic to brain development than is phenylalanine.

Another argument that has been used in support of the conclusion that the keto acid is more toxic than the amino acid is that blood phenylalanine levels in the range of 8–12 mg/100 ml are generally believed to be innocuous to the developing brain. This is the range seen, for example, in atypical PKU patients, whose development is normal without dietary treatment. Citing results from Armstrong and Low (1957), Patel *et al.* (1973) have recently stated that there is little or no transamination to phenylpyruvate at these concentrations of phenylalanine. The implication here, as before, is that phenylalanine concentrations up to this level can be tolerated by the developing brain because at this concentration of phenylalanine there is little or no phenylpyruvate formed. Hence, it is not high phenylalanine but, rather, the accompanying high phenylpyruvate that is toxic.

As with the previous line of reasoning, this one is also unsupported by the available data. The cited results of Armstrong and Low did not show that there was no phenylpyruvate *formed* by transamination at these phenylalanine blood concentrations but, rather, that none was *excreted*. (This excretion threshold, incidentally, was reported by Armstrong and Low to be in the 15–20 mg/100 ml range of serum phenylalanine concentrations, and not in the 8–12 mg/100 ml range stated by Patel and co-workers.) Indeed, the results of Wadman *et al.* (1971) clearly show that large amounts of phenylpyruvate and products derived from it are still excreted by classical PKU patients when their phenylalanine serum concentrations are in the 14–19 mg/100 ml range. Also relevant to this point are the results of Partington and Vickery (1974), who showed that the concentration of phenylpyruvate in the plasma of atypical PKU patients (mean phenylalanine plasma concentration of 13.0 mg/100 ml) was not zero, as would be expected if the statement of Patel and co-workers were correct, but was 0.17 mg/100 ml, a value that is 20–25% of that found in a group of classical PKU patients (plasma phenylalanine concentration of 31–33 mg/100 ml). Thus, the claim that there is little or no transamination of phenylalanine to phenylpyruvate at the blood phenylalanine concentrations seen in atypical PKU patients is not in accord with the data. The conclusion that phenylpyruvate is more toxic to the brain than is phenylalanine is therefore based on two different misconceptions and is just like the opposite conclusion—premature. Regrettably, this view, that phenylpyruvate is more toxic than phenylalanine, has recently been given wider dissemination in *Nutrition Reviews* (September 1975, Vol. 33, p. 276) where it was stated that "variants of the disease occur where only plasma phenylalanine is elevated but phenylpyruvate remains normal and brain development is not retarded."

In the absence of a convincing experimental answer to the question of whether phenylalanine or one of its metabolic products is closer to the offending compound in PKU, one can only speculate about the answer. We will return to this question later.

Most hypotheses for the pathogenesis of the mental defect in PKU assume that the deleterious effects on the brain are caused by metabolic disturbances in the periphery. Before considering these multi-tissue mechanisms, there is one simpler model that can be considered, which assumes that the deleterious effects on the brain are a direct consequence of defects in brain metabolism.

Based on the finding that fetal rat brains have phenylalanine hydroxylase activity that increases to a maximum before birth and then decreases to the low adult level shortly before birth (Wapnir *et al.*, 1971), Bessman has proposed that the gestational role of phenylalanine hydroxylase is to supply tyrosine to the fetal brain (Bessman, 1972). If this were true, it would support a relatively simple model for the pathogenesis of the mental defect in PKU: the brain in the PKU fetus does not develop normally because a genetically determined lack of phenylalanine hydroxylase in the brain leads to a deficiency of tyrosine, a deficiency that would limit normal protein synthesis in the brain.

Other than its simplicity, this hypothesis has little to recommend it. In the first place, implicit in this proposal is the idea that the brain of the newborn infant with PKU has already been damaged *in utero*. If this were the case, there would be little reason a low phenylalanine diet initiated *after* birth should be effective in preventing mental retardation. The evidence that the phenylalanine-restricted diet is effective, when started early enough, is therefore not consistent with the idea that the brain of the PKU infant has been damaged irreverisbly *in utero*.

Besides the fact that it is not in accord with the effectiveness of postnatal dietary therapy of classical PKU, the Bessman proposal was shown to have been based on biochemical results that could not be confirmed by subsequent work. In 1974, Abita *et al.* (1974) showed that rat brain (either fetal or adult) does not have phenylalanine hydroxylase activity other than the minute amounts due to tyrosine and tryptophan hydroxylases. Essentially all the high phenylalanine hydroxylase activity found in rat brain by Wapnir *et al.* (1971) was shown to be due to nonenzymatic hydroxylation of phenylalanine that occurred under their assay conditions.

The proposal of Bessman that brain phenylalanine hydroxylase normally functions to supply tyrosine to the fetal brain, therefore, proved to be without support.

As mentioned, most theories of how hyperphenylalaninemia leads to brain damage in PKU accept the idea that pathogenesis involves metabolic interactions between the periphery and the CNS. They also assume that excess phenylalanine interferes with normal metabolism. Not surprisingly, the first examples of the inhibition by excess phenylalanine of normal biochemical reactions in PKU patients came from the area of the metabolism of the other aromatic amino acids (tyrosine and tryptophan). The decreased pigmentation that is characteristic of phenylketonuria was probably the earliest indication that this type of metabolic interference could take place in this disease. A probable mechanism for this defect was indicated when it was shown that phenylalanine could inhibit mushroom tyrosinase (Dancis and Balis, 1955) as well as mammalian tyrosinase (Miyamoto and Fitzpatrick, 1957). The idea that competition between phenylalanine and tyrosine actually takes place in PKU patients was supported by the finding that the new hair of these patients can be darkened when either tyrosine intake is increased (Snyderman et al., 1955) or phenylalanine intake is decreased (Armstrong and Tyler, 1955).

Although this inhibition by phenylalanine of melanogenesis probably has no pathogenic significance in this disease, these early results were nonetheless noteworthy because they helped establish a principle that is critical to our understanding of the pathogenesis of the mental defect in PKU: excess phenylalanine can interfere with normal metabolism.

There were other early indications that excess phenylalanine could inhibit both normal tyrosine and tryptophan metabolism. Considerable quantities of p-hydroxyphenyllactate and p-hydroxyphenylacetate (Boscott and Bickel, 1953), almost certainly derived from the normal intermediate in tyrosine oxidation, p-hydroxyphenylpyruvate (see Figure 2), were found in the urine of PKU patients. Shortly afterward, the corresponding tryptophan metabolites, indolelactate and indoleacetate, were also found in greatly increased amounts (Armstrong and Robinson, 1954). It should be noted that the urinary excretion in PKU patients of the keto acids from which these lactate and acetate derivatives are probably formed, i.e., p-hydroxyphenylpyruvate and indolepyruvate, has been reported (Rees, 1955; Schreier and Flaig, 1956), although these findings have not yet been confirmed (Armstrong, 1963).

More recently, indican has been identified as another tryptophan-derived metabolite that is excreted in abnormal amounts in these patients (Bessman and Tada, 1960).

For many years, the excretion of these tyrosine and tryptophan metabolites was one of the most puzzling biochemical aspects of the disease. Indeed, these observations were so puzzling that merely to explain the increased excretion of the phenolic compounds, a second metabolic block in the breakdown of tyrosine metabolites was proposed (Boscott and Bickel, 1953).

More recent results have obviated such an unlikely and unparsimonious explanation. It is now known that the metabolism of the aromatic amino acids is characterized by a surprising degree of cross-reactivity. Almost every one of the individual enzymes involved in the metabolism of these amino acids (or their derivatives) shows some activity with at least one of the other aromatic amino acids (or its corresponding derivative). Thus, the ability of a single enzyme, p-hydroxyphenylpyruvate hydroxylase, to utilize both p-hydroxyphenylpyruvate and phenylpyruvate as substrates (Fellman *et al.*, 1972) provides a plausible explanation for the excretion of p-hydroxyphenylpyruvate and the phenolic compounds derived from it by PKU patients; i.e., since phenylpyruvate and p-hydroxyphenylpyruvate are competing substrates for the same enzyme, the high concentrations of the former compound inhibit the utilization of the hydroxyphenyl compound, some of which is then excreted.

The exact mechanism that leads to the increased formation and excretion of the deaminated tryptophan metabolites is not known. It has been assumed (Armstrong, 1963) that the indolepyruvate is formed by transamination between phenylpyruvate and tryptophan as shown in Reaction 4. Indolelactate and indoleacetate would be formed from the pyruvate derivative by reduction and oxidative decarboxylation, respectively.

phenylpyruvate + tryptophan $\rightleftarrows$ phenylalanine + indolepyruvate (4)

Although these effects of phenylalanine on tyrosine and tryptophan metabolism clearly demonstrate that excess phenylalanine or its metabolic derivatives can, and do, interfere with the way in which the organism handles other vital compounds, with the exception of the increased indicanuria, no pathogenic significance has ever been attributed to these disturbances. As far as the increased excretion of indican is concerned, Bessman and Tada (1960) have calculated that it could account for a loss of 50% of ingested tryptophan and have proposed that this loss could be pathogenic. As has been pointed out (Jervis, 1963), however, there is no evidence that PKU patients suffer from a tryptophan deficiency. The calculation above, therefore, is probably unrealistic. It has also been proposed that increased amounts of indole derivatives in the blood and tissues of PKU patients could lead to impaired functioning of the CNS (Bessman and Tada, 1960). Jervis (1963), however, could find no evidence in support of this proposal.

With respect to the increased excretion of p-hydroxyphenylpyruvate and its derivative, it has been noted that this represents lost tyrosine (Knox and Hsia, 1957) for the PKU patients. This idea, however, is based on a misconception, since it implies that there is in phenylketonurics an increased conversion of tyrosine to its keto derivative. As mentioned previously, however, it is now clear that the enhanced excretion of p-hydroxyphenyl derivatives is a result not of excess conversion of tyrosine to these deaminated compounds but, rather, of inhibition by phenylpyruvate of the subsequent oxidation of the p-hydroxyphenylpyruvate. The increased excretion of the latter compound and its derivatives in PKU patients, therefore, does not lead to an increased loss of tyrosine but only to an increased loss of an intermediate that would normally be further oxidized to CO_2 and water.

The interference by phenylalanine in more specialized aspects of tyrosine and tryptophan metabolism may be of greater pathogenic significance in PKU. Beyond their incorporation into proteins, a fate they share with all the other amino acids that are found in proteins, both tyrosine and tryptophan have more unique metabolic functions of special importance to the CNS: they are both precursors of putative neurotransmitters, dopamine and norepinephrine being formed from tyrosine, and 5-hydroxytryptamine (serotonin) being formed from tryptophan.

There is convincing evidence that the conversion of tryptophan to serotonin is impaired in phenylketonuric patients. The first indication of this disturbance was the finding that the daily excretion of 5-hydroxyindoleacetate, one of the major metabolic products of serotonin, is decreased in phenylketonurics (Armstrong and Robinson, 1954). In addition, the blood concentration of serotonin itself is decreased in the patients (Pare *et al.*, 1957). Based on these results, Pare and co-workers proposed that tryptophan hydroxylation was impaired in classical PKU patients (1957). These results were confirmed and extended by Perry *et al.* (1964), who also concluded that tryptophan hydroxylation was defective in these patients. It has also been reported that the concentration of serotonin in certain areas of the brains (brain stem, occipital cortex, caudate nucleus) of PKU patients is lower than that in the corresponding brain areas of control patients (McKean, (1972).

The conversion of tryptophan to serotonin involves two successive transformations: hydroxylation, catalyzed by the pterin-dependent tryptophan hydroxylase (Reaction 5), followed by decarboxylation (Reaction 6):

$$\text{tryptophan} \rightarrow \text{5-hydroxytryptophan} \qquad (5)$$
$$\text{5-hydroxytryptophan} \rightarrow \text{5-hydroxytryptamine} + CO_2 \qquad (6)$$

There is evidence that both these steps may be impaired in PKU.

Pare *et al.* (1958) found that after the intravenous administration of 5-hydroxytryptophan, phenylketonurics excreted less 5-hydroxyindoleacetate and serotonin than did a group of normal controls. These results were confirmed by Jervis (1963) who, in addition, showed that PKU patients had no defect in their capacity to metabolize serotonin to 5-hydroxyindoleacetate. Based on these findings, Pare *et al.* (1958) abandoned their original idea that tryptophan hydroxylation (Reaction 5) was the site of the impaired serotonin metabolism and proposed instead that Reaction 6, the decarboxylation of 5-hydroxytryptophan to serotonin, is impaired in these patients.

That phenylalanine and its metabolites can inhibit the decarboxylation of 5-hydroxytryptophan was shown with rat and guinea pig kidney homogenates (Deguchi and Barchas, 1972; Huang and Hsia, 1963). With these crude enzyme preparations, phenylalanine was found to be the least active inhibitor, phenylacetate was the most active, and phenylpyruvate and phenyllactate were in between. It should be noted, however, that even with the most potent compound, phenylacetate, 50% inhibition required a concentration of 17 mM (Huang and Hsia, 1963). This relatively low potency raises questions about whether inhibition of the decarboxylase provides an adequate explanation for the defect in serotonin metabolism in PKU.

Another indication that inhibition of the decarboxylase cannot completely account for the defect in serotonin metabolism in PKU is the evidence that the rate-limiting reaction in the conversion of tryptophan to serotonin is the hydroxylation step (Reaction 5), rather than the decarboxylation step (Jéquier *et al.,* 1967). If this view is correct, it is unlikely that the modest levels of inhibition of the decarboxylase that would be expected from the already cited *in vitro* results of Huang and Hsia (1963) would inhibit the conversion of tryptophan to serotonin. In such a situation, inhibition of the rate-limiting step in the pathway, in this case, the hydroxylation step, seems like a more probable way to achieve inhibition of the overall conversion. Relevant to this point, phenylalanine has been shown to be a competitive inhibitor of crude brain tryptophan hydroxylase and this inhibition has been proposed to explain the decrease in serotonin synthesis that is seen in PKU (Lovenberg *et al.,* 1968). Subsequently, it was shown that phenylalanine is not an inert inhibitor of brain tryptophan hydroxylase but, rather, that it inhibits the enzyme by virtue of its being a competing substrate; i.e., brain tryptophan hydroxylase can catalyze the conversion of phenylalanine to tyrosine (Tong and Kaufman, 1975).

In addition to these proposals that are based on the ability of phenylalanine and its metabolites to inhibit the two enzymes involved in serotonin biosynthesis, inhibition by phenylalanine of transport into the brain of tryptophan or 5-hydroxytryptophan could also contribute to impaired serotonin metabolism. Phenylalanine (as well as some other amino acids) has been shown to inhibit the uptake of the serotonin precursor, 5-hydroxytryptophan into rat brains (McKean *et al.,* 1962). This inhibitory effect of phenylalanine has been proposed as an explanation for the defect in serotonin metabolism that is seen in classical PKU patients. Although it seems clear that phenylalanine can inhibit the cerebral uptake of 5-hydroxytryptophan in rats, it is much less clear that this mechanism has any relationship to the defect in central serotonin metabolism that has been observed in PKU patients. The implication of this mechanism is that serotonin in the central nervous system is derived from 5-hydroxytryptophan that is synthesized in the periphery. Since brain contains all the enzymes needed to synthesize its own serotonin, i.e., an active tryptophan hydroxylase system and aromatic amino acid decarboxylase, it seems unlikely that this organ would rely on a peripheral supply of 5-hydroxytryptophan. And, if the brain makes its own serotonin precursor, inhibition of transport of this compound into the brain would be of no consequence as far as serotonin levels in the brain are concerned.

Even if the brain did normally supplement its endogenous supply of 5-hydroxytryptophan with some of this material derived from the periphery, it seems likely that there would be less of this peripheral supply in PKU patients than in normals. This conclusion is based on the fact that one of the likely sources of peripheral 5-hydroxytryptophan, hepatic phenylalanine hydroxylase [which can catalyze the conversion of tryptophan to 5-hydroxytryptophan (Renson *et al.*, 1962)], is precisely the enzyme that is missing in classical PKU. It should be noted that the other potential source of peripheral serotonin, i.e., pineal tryptophan hydroxylase, is unlikely to play a major role in supplying the brain with serotonin. It is known that brain stores of serotonin are depleted by treatment of rats with PCPA in spite of the fact that their pineal tryptophan hydroxylase remains active (Deguchi and Barchas, 1972). Hence, of the two known peripheral enzymes capable of supplying the brain with 5-hydroxytryptophan, pineal tryptophan hydroxylase and hepatic phenylalanine hydroxylase, it is clear (in the rat at least) that the pineal enzyme does not play this role to a significant extent. Since the other enzyme, phenylalanine hydroxylase, is essentially inactive in PKU, the proposal that the decrease in central serotonin seen in PKU patients is due to an interference by phenylalanine of cerebral uptake of 5-hydroxytryptophan is implausible.

While inhibition by phenylalanine of transport into the brain of 5-hydroxytryptophan seems unlikely to be of pathogenic significance, the same is not true of inhibition by phenylalanine of tryptophan uptake.

There is ample evidence to support the conclusion that high phenylalanine levels can decrease the tryptophan content in brain tissue (McKean *et al.*, 1968; Graham-Smith and Parfitt, 1970; Barbosa *et al.*, 1971). That this type of interference actually does take place in PKU is indicated by the finding that samples of brain cortex from these patients have lower levels of tryptophan than do samples from control patients (McKean, 1972). Finally, since brain tryptophan hydroxylase is probably not normally saturated with respect to tryptophan (Fernstrom and Wurtman, 1971; Friedman *et al.*, 1972*a*), any decrease in the tryptophan content in brain will likely lead to a decreased rate of hydroxylation of tryptophan and, hence, a decreased rate of serotonin synthesis.

Measurements of the rate of conversion of tryptophan to 5-hydroxyindoleacetate in classical PKU patients support the conclusion that one of the steps in the pathway of serotonin synthesis (i.e., either Reaction 5 or 6) is inhibited in these patients (Jéquier, 1968). The degree of inhibition, however (about 30%), was regarded as too mild to contribute to the pathogenesis of the disease.

Using a different method to assess the rate of serotonin synthesis, McKean (1972) reported results that indicate a much more severe impairment in this metabolic pathway in classical PKU patients. In this study, probenicid was given to the patients to reduce the efflux from the CSF of amine catabolites. Under these conditions, the amount of 5-hydroxyindoleacetate that accumulates in the CSF in a given period of time is thought to be a measure of the rate of central serotonin synthesis. Not only was a decreased accumulation of this serotonin-derived metabolite found in the CSF of the PKU patients, but also this deficit was largely corrected when the patient's plasma phenylalanine concentration was reduced by dietary restriction of phenylalanine. These results complement the earlier ones of Pare *et al.* (1958) in that they show that the inhibition of serotonin synthesis that is seen in classical PKU patients can be relieved by a decrease in phenylalanine concentrations.

Despite the clear evidence pointing to an impaired synthesis of serotonin in classical PKU patients, the case for this impairment being related to the etiology of the irreversible mental defect is weak. The vigor of the case probably reached its zenith in 1965, when Woolley and van der Hoeven (1965) reported that mice treated from birth with DL-phenylalanine and L-tyrosine to induce phenylketonuria, a regimen that will also lead to a decrease in the serotonin content in the brain (Boggs *et al.*, 1963; Culley *et al.*, 1962; Yuwiler and Louttit, 1961), performed less well than the controls in a maze test. Significantly, this apparent learning deficit could be largely prevented if the phenylalanine–tyrosine diet was supplemented with 5-hydroxytryptophan. Since the administration of this serotonin precursor probably corrected the serotonin deficiency, the authors concluded that the serotonin deficiency was the cause of the learning deficit.

Related to this hypothesis, Hole (1972*a*) has recently shown that the decrease in brain weight that is caused by administration to neonatal rats of either phenylalanine or PCPA can be largely prevented by the simultaneous administration of 5-hydroxytryptophan, which also corrects the serotonin deficiency. Unfortunately, there was no attempt to determine whether this effect of 5-hydroxytryptophan was specific; i.e., no other compound was tested. If future studies should prove that this beneficial effect is a specific one, these results would only suggest a role for serotonin in normal brain growth; they would not establish a link between serotonin deficiency and the characteristic pathology of classical PKU: irreversible mental retardation.

Indeed, it has been demonstrated with young rats that deficiencies of brain serotonin much more severe than those seen in classical PKU patients do not lead to irreversible learning defects. Watt and Martin (1969) showed that rats treated with PCPA during the neonatal period performed poorly in a water-maze test designed to measure the acquisition of learning. This defect, however, was not permanent, and 26 days after the last treatment with PCPA, there was no difference in the maze performance between the treated and the untreated controls.

Hole confirmed and extended these results (1972c). Again, in this study, no permanent learning deficit could be detected in young rats whose brain serotonin levels had been reduced to 20% of control values by treatment with PCPA. Hole did report, however, that the drug-treated animals showed behavioral changes, i.e., decreased arousal levels, that persisted for at least 4 weeks after the last PCPA injection.

Thus, these results with PCPA-treated rats offer no support for the idea that serotonin depletion is the cause for mental retardation in classical PKU. On the other hand, they do suggest that depletion of this neurotransmitter may be causally related to the particular deficit in attention that has been detected in children with classical PKU (Anderson et al., 1969).

One of the few results obtained with PKU patients that is relevant to what might be called the serotonin-depletion hypothesis was reported by Perry et al. (1970). They compared the urinary excretion of serotonin and 5-hydroxyindoleacetate in two adult brothers, both with untreated, classical PKU: one was retarded, and the other had a higher-than-normal I.Q. The amounts of these tryptophan metabolites excreted were approximately the same in each brother and were only 10–20% of those excreted by normal adults under equivalent conditions. Since the brother with the superior I.Q. appeared to suffer from the same defect in tryptophan hydroxylation as did the severely retarded brother, Perry and co-workers concluded that this disturbance in tryptophan metabolism could not explain the mental defect found in the retarded brother.

As far as the relationship between serotonin depletion and other behavioral abnormalities is concerned, it should be noted that there was no indication (in the case report) that the unretarded brother displayed the attention deficit, referred to earlier, that is characteristic of many children with classical PKU. Thus, these results do not support the idea that serotonin depletion is related to either the mental retardation or the behavioral abnormalities that are seen in classical PKU. These conclusions, of course, teeter on the assumption that the status of tryptophan hydroxylation in these adults is the same as it was when they were infants.

High levels of phenylalanine or one of its metabolites also interfere with the conversion of tyrosine to the neurotransmitters, dopamine and norepinephrine. Just as with the impairment in serotonin synthesis in the brain, this metabolic disturbance could be of pathogenic significance in PKU.

The first indication that the synthesis of the catecholamine neurotransmitters might be depressed in this disease came from Weil-Malherbe (1955), who reported that the plasma levels of both epinephrine and norepinephrine were lower in PKU patients than in other mentally retarded subjects.

These results were confirmed by Nadler and Hsia (1961), who also showed that classical PKU patients excreted less dopamine, norepinephrine, and epinephrine in their urine than did either controls or non-PKU retarded patients. It was also shown that these decreases could be reversed when the patients were treated with a low-phenylalanine diet. Based on these findings, Nadler and Hsia concluded that inhibition of dopa decarboxylase could account for the defect in catecholamine synthesis.

Although these findings suggested that in PKU patients there was a disturbance in the synthesis of these biogenic amines, they provided no basis for identifying the affected enzyme.

The conversion of tyrosine to norepinephrine proceeds according to Reactions 7, 8, and 9.

$$\text{tyrosine} \rightarrow \text{dopa} \tag{7}$$
$$\text{dopa} \rightarrow \text{dopamine} + CO_2 \tag{8}$$
$$\text{dopamine} \rightarrow \text{norepinephrine} \tag{9}$$

Up to the formation of dopamine, this pathway is strictly analogous to that followed by tryptophan in its conversion to serotonin; i.e., first, the amino acid is hydroxylated by a pterin-dependent enzyme, tryosine hydroxylase (Reaction 7), followed by decarboxylation (Reaction 8). In the present case, dopamine is then converted to norepinephrine by a side-chain hydroxylation catalyzed by the ascorbate-dependent enzyme, dopamine β-hydroxylase (Reaction 9).

Just as with the serotonin biosynthetic pathway, the rate-limiting step here is also believed to be the initial hydroxylation step (Reaction 7), rather than the decarboxylation reaction (Udenfriend, 1966). And, as has just been discussed with respect to serotonin biosynthesis, with this pathway too, it is unlikely that inhibition by phenylalanine of the step that is not rate limiting—unless the inhibition were severe—can adequately explain the defective catecholamine synthesis that is seen in PKU. Rather, as in the case with serotonin synthesis, it seems more likely that the affected step in dopamine synthesis is the one catalyzed by tyrosine hydroxylase.

Convincing evidence that the conversion of tyrosine to dopa is decreased was obtained in experiments where deuterated L-tyrosine was given to PKU patients (Curtius *et al.*, 1972*a*). It was found that the excretion of deuterated metabolites derived from both dopamine and norepinephrine was decreased in the patients. Significantly, the excretion of these compounds was markedly increased when L-dopa was given, which is an indication that the conversion of tyrosine to dopa is the reaction that is inhibited.

While these results show that catecholamine synthesis is impaired in PKU patients, they do not indicate whether the disturbance occurs in the periphery or within the CNS. Evidence that catecholamine synthesis in the CNS is impaired was presented by McKean (1972), who showed that the levels of dopamine and norepinephrine in the brains of PKU patients, obtained at autopsy, is lower than that in the brains of control patients. Using the same probenicid technique with which he documented a defect in central serotonin synthesis, and as part of that same study, McKean also showed that the accumulation in the CSF of the dopamine metabolite, homovanillic acid, is decreased in PKU patients. Just as the depressed accumulation of the serotonin metabolite returned toward normal when blood phenylalanine concentrations were decreased, so the accumulation of homovanillic acid also very significantly increased by lowering the blood phenylalanine concentration. These results support the conclusion that high levels of phenylalanine interfere with the biosynthesis in the CNS of dopamine and probably also of norepinephrine.

The mechanism by which high levels of phenylalanine inhibit dopa (and, hence, dopamine and norepinephrine) formation is not known with certainty. Phenylalanine has been reported to be a competitive inhibitor of (as well as a weak substrate for) tyrosine hydroxylase from adrenal medulla and from brain (Ikeda *et al.*, 1967), and this inhibition has been suggested as an explanation for the decreased catecholamine biosynthesis in PKU. Although there was no reason to question the adequacy of this explanation at the time it was offered, the finding that phenylalanine is actually an excellent substrate for tyrosine hydroxylase when the activity is measured in the presence of the naturally occurring cofactor, tetrahydrobiopterin (Shiman *et al.*, 1971), does raise a question about whether inhibition by phenylalanine of tyrosine hydroxylase can completely account for the impairment in catecholamine synthesis that is characteristic of the disease.

Even less is known about the pathogenic significance of this defect in catecholamine metabolism than about the just discussed abnormality in serotonin synthesis. There is evidence, however, that suggests that classical PKU patients do not show the pathology to be expected from severe defects in central serotonin and dopamine synthesis. The evidence comes from a study of neurotransmitter metabolism in a single infant with hyperphenylalaninemia due to dihydropteridine reductase deficiency, where preliminary results indicate a marked impairment in the synthesis of these two neurotransmitters (Butler *et al.*, 1975). As already discussed, these metabolic disturbances are probably a direct consequence of the lack of dihydropteridine reductase in the CNS. If one also accepts the assumption that the defects in central neurotransmitter synthesis are causally related to the neurological deterioration in this patient (see Kaufman *et al.*, 1975a), then the defects in dopamine and serotonin synthesis that have been documented in classical PKU patients are clearly not severe enough to lead to the neurological deterioration that is seen in this variant form of PKU. These considerations place a limit (admittedly, a tentative one) on the severity of the defects in neurotransmitter synthesis in classical PKU. Whether defects below this limit are severe enough to contribute substantially to the pathophysiology of classical PKU remains to be determined.

It has also been proposed that the biosynthesis of another putative neurotransmitter, γ-aminobutyric acid (GABA), is impaired in PKU and that this metabolic disturbance is involved in the etiology of the mental defect in this disease. The proposal is based on the finding that metabolites of phenylalanine inhibit brain glutamate decarboxylase (Hanson, 1959; Tashian, 1961). Compared with the levels at which these metabolites occur in brain tissue, however, it does not seem likely that significant inhibition of this enzyme would take place *in vivo*. Tashian, for example, found that *o*-hydroxyphenylacetate was the most potent inhibitor of the decarboxylase *in vitro*, whereas this compound has not even been detected in the brains of rats made phenylketonuric by administration of phenylalanine and PCPA (Edwards and Blau, 1972). Phenylacetate, another of the more potent inhibitors of the decarboxylase (Tashian, 1961), has been found in the brains of these phenylalanine-PCPA treated animals, but its level is three orders of magnitude lower than the amount needed to inhibit the enzyme (Edwards and Blau, 1972).

Another mechanism for the genesis of the mental defect in PKU that also involves a disturbance in the metabolism of a glutamate derivative was put forth by Perry *et al.* (1970). As part of a study (which has already been discussed in another context) of the biochemistry of two adult brothers with untreated classical PKU, one who is retarded and one with superior intelligence, they found that the plasma glutamine concentration of the retarded brother was significantly lower than that of the nonretarded one. The plasma levels of glutamate in another untreated PKU subject with normal intelligence was also found to be higher than that of a group of 12 untreated classical PKU patients. On the basis of these findings, Perry *et al.* (1970) proposed that chronic deficiency of glutamine during the period of rapid brain growth plays a part in producing the mental defect of PKU. The mechanism of the glutamine depletion in the disease is not known with certainty, but large amounts are lost in the urine by the daily excretion of phenylacetylglutamine.

The proposal of Perry and his colleagues suffers from several weaknesses. First, as the authors themselves have wondered, can this modest reduction of glutamine level—to 65% of control values—be of any pathogenic significance? And, is the still more modest difference between the glutamine levels in the retarded brother and those in the intelligent one—a difference of about 25%—sufficient to account for the vast differences in their I.Q. scores? Second, in view of the wide range of biosynthetic reactions in which the amide nitrogen atom of glutamine is used, it is difficult to see why the tissue damage in PKU should be restricted to the brain. Even more destructive than these considerations to the glutamine-depletion hypothesis are the recent findings of McKean and Peterson (1970). First, they were able to confirm the findings of others that blood glutamine levels are lower in classical PKU patients than in controls. They then measured the serum glutamine concentrations in three siblings, all with classical PKU, and all with serum phenylalanine concentrations of about 30 mg/100 ml. But in contrast to the results of Perry *et al.* (1970), McKean and Peterson found that the sibling with the highest I.Q. (I.Q. = 80) had lower glutamine concentrations than did the two severely retarded children (I.Q.s = 35 and less than 20). In addition, they found that the glutamine concentration in brain tissue from PKU patients (obtained at autopsy) was actually somewhat higher than that from control patients. Although the finding of increased glutamine concentration in the brain tissue from PKU patients is paradoxical when compared with the decreased concentration in their blood, these results of McKean and Peterson suggest that there may be no factual basis for the glutamine-depletion hypothesis.

In addition to the neurotransmitters and their special role in the brain, there are, as already discussed, other unique aspects of brain structure and function that might account for the fact that this organ is rather specifically affected in PKU. As far as structure is concerned, neuropathology, until recently provided no clue to the etiology of the mental defect in PKU. Other than a moderate reduction in weight, the brain of the PKU patient looked discouragingly normal—discouraging, because the job of trying to understand why a normal-looking brain fails to function seems more formidable than that of analyzing why a grossly abnormal-looking brain does not function properly.

The first report that there might be a defect in myelination in the brains of PKU patients (Alvord *et al.,* 1950), therefore, signalled a significant development in the field. These workers reported that the brains of three of five PKU patients that they examined showed signs of "a marked lack of myelinization of the nervous system," with gliosis present in all. Based on these observations, Alvord and co-workers suggested that a possible pathogenic mechanism in PKU was a retardation of myelination caused by a defect in myelin anabolism. These findings were significant because here at last was a structural or anatomical defect that might well be causally related, as Alvord and co-workers had suggested, to the mental defect. And, if this causal relationship could be established, then the problem of pathogenesis would be reduced from one where we are trying to explain how high phenylalanine or one of its metabolites leads to mental retardation to one where the same metabolic disturbance leads to a defect in myelin metabolism. The latter problem seems less formidable than the former.

The early histological evidence pointing to a defect in myelination was confirmed by subsequent neuropathological data (Poser and Van Bogaert, 1959; Malamud, 1966; Crome, 1971) and extended by results of chemical analysis. In general, the affected brains show not only a decreased content of myelin *per se,* but also decreased amounts of such typical myelin components as cerebroside, cholesterol, and proteolipids (Gerstl *et al.,* 1967; Menkes, 1966; Menkes, 1968; Crome *et al.,* 1962; Prensky *et al.,* 1968; Shah *et al.,* 1972*b*). Despite the substantial reduction in proteolipid protein, that which is present has been reported to have a normal amino acid composition (Prensky *et al.,* 1968). Foote *et al.* (1965) did not find a lower cerebroside content or evidence of demyelination in the two phenyl-ketonuric brain samples that they examined. They did find, however, a decrease in the amount of major monoenoic acids relative to the corresponding saturated acids in the cerebroside fraction, as well as in other lipid fractions. Based on their findings, Foote *et al.* (1965) discussed the possibility that there might be an impairment in the biosynthesis of oleic acid in the disease.

The finding of decreased amounts of unsaturated fatty acids in brain tissue from PKU patients has been confirmed by Gerstl *et al.* (1967), who also found a disproportionate loss of polyunsaturated fatty acids, as well as of the long-chain saturated acids. It should be noted that the relative loss of both unsaturated and long-chain fatty acids has been observed in the brains of neonatal rats that have been made hyperphenylalaninemic by treatment with phenylalanine (Johnson and Shah, 1973).

In addition to this last study, hyperphenylalaninemic animals have been used to investigate other aspects of the manner by which high phenylalanine levels might interfere with myelin metabolism as well as with other aspects of brain development. Not only is the amount of myelin that can be isolated from the brains of hyperphenylalaninemic rats (neonatal rats treated with phenylalanine at 3 mg/g body weight per day from day 5 to day 20) reduced by 20–25% as compared with controls (Shah *et al.*, 1972*a*), but also the rate of *in vivo* incorporation of [U^{14}C]glucose into brain lipids is also reduced (Shah *et al.*, 1970). In addition, it has been found that high levels of phenylalanine reduced the *in vivo* incorporation of mevalonic acid into cerebral sterols (Shah *et al.*, 1969). Because of this last observation, Shah and his co-workers proposed that the defect in myelination is due to a reduction in the rate of cholesterol synthesis by high levels of phenylalanine or one of its metabolites (Shah *et al.*, 1972*a*). Subsequently, these workers modified this view and suggested that a defect in the elongation of fatty acids also contributes to the hypomyelination that is seen in the brains of phenylalanine-treated rats (Johnson and Shah, 1973). Whichever specific reaction is ultimately pinpointed as the primary one that is affected in the brains of hyperphenylalaninemic rats, these animal studies support the early conclusion of the neuropathologists, which indicated that the defect in cerebral myelination in PKU is probably due to a defect in the synthesis of myelin rather than to an excessive breakdown or normally formed myelin; i.e., neuropathologically, PKU has been classified as a dysmyelinating disease rather than as a demyelinating disease (Poser and Van Bogaert, 1959).

Since myelin is composed of both lipid and protein moieties, it would be an important advance if it could be established whether hyperphenylalaninemia affects primarily the synthesis of the protein or the lipid portion of myelin. Although the answer to this question is not known for clinical PKU, there is evidence that in hyperphenylalaninemic rats the primary effect may be on cerebral protein synthesis. It has been found in young rats made acutely hyperphenylalaninemic by a single injection of phenylalanine that there is a severe inhibition of incorporation of [^{35}S]methionine and [U-^{14}C]leucine into both myelin and nonmyelin cerebral protein fractions. In the same animals, the incorporation of acetate into cerebral lipids was unaffected (Agrawal *et al.*, 1970). These workers also found that the high levels of phenylalanine interfered with the transport of other amino acids, such as methionine and leucine, into the acid-soluble pool in the brain. Because of this finding, they concluded that the defect in cerebral protein synthesis was due to inhibition by high phenylalanine of transport of other amino acids into the brain and that the defect in lipid synthesis was secondary to this defect in protein synthesis. The authors made no attempt to explain how a defect in amino acid transport could account for the defects in fatty acid elongation and desaturation that have been reported for cerebral lipids both from patients with PKU and from hyperphenylalaninemic animals.

In addition to these studies on the effects of hyperphenylalaninemia on protein and lipid synthesis in the developing brain, neuroanatomical studies have detected pathological changes in the cerebellum of rats that have been subjected to chronic neonatal hyperphenylalaninemia. These changes—cytoplasmic lesions in Purkinje cells and nuclear lesions in granule cells—led Adelman *et al.* (1973) to suggest that neurons of the cerebellum undergoing mitosis and differentiation are especially vulnerable to the deleterious effects of sustained hyperphenylalaninemia.

The exact mechanism by which high concentrations of phenylalanine inhibit myelin synthesis—or, to be less specific, inhibit brain growth—is unknown. As noted earlier, although the connection has not been established, it seems entirely plausible that these structural defects are causally related to the genesis of the mental defect in PKU. It seems likely, therefore, that if we understood how disturbed phenylalanine metabolism leads to these structural defects in the brain, we would also be closer to understanding the cause of the mental defect.

Studies of the effects of phenylalanine or its metabolites on various enzyme activities would have been expected to yield cogent clues to the mechanism of the pathogenesis. The problem with this approach, as pointed out earlier, is not that it proved to be difficult to find enzymes that are inhibited by phenylalanine or its derivatives but, rather, that it proved to be too easy: so many enzymes were shown to be sensitive to these compounds that this whole approach appears to have been overwhelmed—and undermined—by its own success. Given the long list of sensitive enzymes, it has not been possible to decide which of these effects may be relevant to the pathogenesis of the mental defect in PKU.

The problem is illustrated by the data in Table 6, adopted from recent publications (Patel and Arinze, 1975; Land and Clark, 1973), showing a list of brain enzymes that have been reported to be inhibited by either phenylalanine or phenylpyruvate. The inhibition of any of the first group of enzymes could, by interfering with energy production, i.e., the synthesis of ATP that is coupled to carbohydrate oxidation, indirectly inhibit the synthesis in the brain of any macromolecule. Inhibition of any of the last group of enzymes could account for a defect in the synthesis of cerebral lipids, such as cholesterol and fatty acids, since these enzymes are involved in the supply of either the carbon skeleton or the hydrogen atoms of these lipids.

In addition to inhibition of specific enzymes, phenylpyruvate has been reported by Land and Clark (1974) to inhibit the cerebral metabolism *in vitro* of both pyruvate and β-hydroxybutyrate by inhibition of the transport of these metabolites across the brain mitochondrial membrane. Since the pathology of PKU resembles, to some extent, that seen in maple syrup urine disease (MSUD), it is worth noting that it was also shown in this study that α-ketoisocaproate, the keto compound that occupies the same pivotal position (both metabolically and conceptually) in that disease as does phenylpyruvate in PKU, also inhibits the oxidation of both pyruvate and β-hydroxybutyrate. As pointed out by Land and Clark, inhibition of the ·

TABLE 6.   Enzymes in Brain Inhibited by Phenylalanine and Phenylpyruvate

Enzyme	Source of brain	Inhibitor	Ki/mM	Reference
Hexokinase	Rat	Phenylpyruvate	2.3–13.0	Weber et al. (1970)
	Human	Phenylpyruvate	2.0–6.8	Weber et al. (1970)
6-Phosphogluconate dehydrogenase	Rat	Phenylpyruvate	3–6	Weber et al. (1970)
Pyruvate kinase	Rat	Phenylalanine	7.5–10.0	Weber et al. (1970)
	Human	Phenylalanine	8.5–11.0	Weber et al. (1970)
Pyruvate dehydrogenase complex	Rat	Phenylpyruvate	6	Bowden and McArthur (1972)
Pyruvate carboxylase	Rat	Phenylpyruvate	—	Patel (1972); Patel et al. (1973)
	Human	Phenylpyruvate	—	Patel et al. (1973)
Citrate synthase	Rat	Phenylpyruvate	0.7	Land and Clark (1973)
α-Ketoglutarate dehydrogenase complex	Rat	Phenylpyruvate	—	Patel and Arinze (1975)
	Human	Phenylpyruvate	—	Patel and Arinze (1975)
NADP-malate dehydrogenase	Rat	Phenylpyruvate	—	Patel and Arinze (1975)
Acetyl-CoA carboxylase	Rat	Phenylpyruvate	10	Land and Clark (1973)
Fatty acid synthetase	Rat	Phenylpyruvate	0.25	Land and Clark (1973)

metabolism of the latter compound may be of special significance to the development of the brain because of the ability of the immature brain to utilize ketone bodies as a respiratory fuel.

Despite the finding that some of the individual enzymes of glycolysis, such as hexokinase and pyruvate kinase, are inhibited by phenylalanine or by phenylpyruvate (Table 6), measurements of the content of glycolytic intermediates in quick-frozen brain samples from hyperphenylalaninemic rats have not supported the idea that cerebral energy production in these animals is inhibited. Thus, Miller et al. (1973) found that although the pattern of change in the content of cerebral glycolytic intermediates after administration of phenylalanine to rats was consistent with a partial inhibition of brain pyruvate kinase, the levels of ATP, ADP, and creatine phosphate were unchanged. Similar findings for both brain and liver were reported with rats made hyperphenylalaninemic by administration of the inhibitors of phenylalanine hydroxylation, PCPA, and esculin (Gimenez et al., 1974).

Unless an inhibition of ATP production by hyperphenylalaninemia were just balanced by a decrease in the utilization of ATP, these results do not indicate that brains of hyperphenylalaninemic animals suffer from a lack of ATP.

One of the compounds that was determined in these quick-frozen brain samples was pyruvate. Miller *et al.* (1973) found pyruvate content to be unchanged, whereas Gimenez *et al.* (1974) reported a modest 20% decrease in the brains of hyperphenylalaninemic rats. These results are relevant to another proposed mechanism for the way in which elevated phenylalanine levels might lead to brain damage. In 1957, Korey (1957) proposed that substrates such as pyruvate and α-ketoglutarate might be siphoned out of the tricarboxylic acid cycle *via* transamination with phenylalanine. If the pyruvate were then converted to alanine, these reactions could lead to "a chronic deficiency syndrome of the Krebs cycle constituent, pyruvic acid."

The relatively minor change in cerebral pyruvate content found in hyperphenylalaninemic rats is not consistent with this idea. Although brain α-ketoglutarate levels were not measured in these studies, it is known that the amount of glutamate, the amino acid precursor of α-ketoglutarate, is not consistently lower than normal in the blood of PKU patients (Linneweh and Ehrlich, 1962; Efron *et al.*, 1969; Perry *et al.*, 1970) or in the brains of hyperphenylalaninemic infant rats (Lowden and LaRamee, 1969). It is not likely that the steady-state level of cerebral α-ketoglutarate could remain decreased for long in the presence of normal levels of glutamate.

Also against the "pyruvate depletion" theory is the fact that the rate at which the brain oxidizes glucose (via pyruvate) is enormous (about 60 μmol per 100 g brain per minute for infants; see Sokoloff, 1960) compared with the total amount of phenylalanine that is ingested per day. Thus, it can be calculated that even if one made the unrealistic assumption that all the phenylalanine that is ingested per day by an infant with classical PKU were to be transaminated in its brain, the amount of pyruvate that would be siphoned off by this process would only equal the amount of pyruvate that is generated from glucose in a 10–20 min period. In other words, the amount of glucose that is oxidized per day in the brain is 70–140 times greater than the total amount of phenylalanine that is ingested per day, the latter quantity being the maximum amount of phenylalanine that is available per day for transamination. Since most phenylalanine transamination probably occurs not in the brain but, rather, in the periphery, it is most unlikely that significant pyruvate depletion could occur in the brain by virtue of phenylalanine transamination.

Although it does not seem likely that the operation of the Krebs cycle could be seriously disrupted in the brain by phenylalanine-mediated depletion of pyruvate, disruption could occur by other mechanisms, some of which will be discussed later. Here, we will consider the possibility that the cycle can be disrupted by depletion, or siphoning off, of an essential coenzyme, such as coenzyme A (CoA). In 1958 (Kaufman, 1958c), the possibility was discussed that the oxidative decarboxylation of phenylpyruvate to phenylacetyl CoA (See Figure 2), if it occurred in brain, might be more rapid than reactions (e.g., condensation of phenylacetyl-CoA with glutamine and hydrolytic deacylation) that lead to regeneration of the CoA. Under these conditions, a significant fraction of the CoA in brain tissue might be tied up as the useless phenylacetyl derivative and other CoA-dependent reactions, such as the synthesis of acetylcholine, would be inhibited. Since lipid synthesis is also CoA-dependent, this hypothesis would provide an explanation for the findings of decreased concentration of cholesterol and other lipids in brains of PKU patients and hyperphenylala-ninemic rats. In addition, since glucose oxidation is also dependent on CoA, this proposal would be consistent with a decrease in the rate of cerebral oxygen consumption in PKU patients. Such a decrease has been indicated by the finding of lower oxygen arteriovenous differences in the cerebral circulation in these patients (Himwich and Fazekas, 1940). As pointed out (Knox, 1972), however, blood-flow measurements were not carried out in this study, leaving open the possibility that, because of their reduced brain weights, oxygen consumption per gram of tissue may be normal in patients with classical PKU.

According to the CoA-depletion hypothesis, the levels of cerebral pyruvate and α-ketoglutarate, the two intermediates that are oxidatively metabolized primarily via CoA-dependent reactions, might be expected to increase in the brains of hyperphenylalaninemic rats, an expectation that does not appear to be in accord with the observations cited. It is possible, however, that inhibition of the rate of pyruvate generation from glucose (by inhibition of pyruvate kinase) could compensate for decreased pyruvate oxidation and thereby leave the steady-state concentration of cerebral pyruvate essentially unchanged.

Depletion of vitamin-B_6-derived coenzymes in the brain has also been considered as a possible cause for brain damage in PKU. Loo and her co-workers have reported that hyperphenylalaninemic rats excrete a metabolite that is derived from pyridoxine and phenylalanine (Loo and Ritman, 1967). The finding that the carboxyl carbon atom of phenylalanine was not incorporated into this new compound suggested that phenylethylamine, rather than phenylalanine itself, was the direct precursor of this compound, a suggestion that was confirmed when the structure of the new metabolite was determined; degradation studies showed that the compound is a derivative of phenylethylamine: 2-methyl-3-hydroxy-5-hydroxymethyl-4-pyridylmethylene-β-phenylethylamine (pyridoxylidene-β-phenylethyl-amine) (Loo, 1967). The compound has also been found in the brains of hyperphenylalaninemic rats as well as in urine from patients with untreated classical PKU (Loo, 1967).

Although these findings raised the attractive possibility that hyper-phenylalaninemia might lead to the depletion in the brain of vitamin-B_6-derived coenzymes, and that this depletion might be a pathogenic factor in

PKU, the finding of normal contents of both pyridoxal phosphate and pyridoxamine phosphate in the brains of immature hyperphenylalaninemic rats (Loo and Mack, 1972) did not support this idea.

It should be mentioned that clinical experience with PKU patients is also inconsistent with the idea that vitamin B_6 deficiency contributes to the CNS damage seen in the disease. First, there is no indication that these patients are deficient in the vitamin (Tischler and McGeer, 1958). Second, the administration of large doses (up to 150mg/day) of vitamin B_6 to patients with untreated PKU was without effect on their clinical condition, their behavior, or their intellectual capabilities (McGeer and Tischler, 1959). (In view of the ages of the patients during the period of B_6 supplementation— 8–41 years old—it is highly likely that CNS damage had already occurred, and therefore, it was not surprising that the treatment had no effect on intelligence. Of greater interest was the finding that the behavior of the patients was also unaffected by the B_6 treatment.)

Even though the formation of pyridoxylidine-β-phenylethylamine does not lead to depletion of pyridoxal in the brains of hyperphenylalaninemic rats, formation of this compound may still be a pathogenic factor in PKU. This possibility has been suggested because of the finding that mice injected with the amine derivative show signs of neurotoxicity (Loo, 1967). Unfortunately, it is not clear that pyridoxylidine β-phenylethylamine is any more potent in this respect than is phenylethylamine itself, or, for that matter, that the toxicity is not due to release of free phenylethylamine. As far as the neurotoxicity of either the amine or its pyridoxylidine derivative is concerned, there is no evidence that either compound leads to any irreversible CNS effects that resemble those seen in PKU patients.

In addition to its direct effects on brain metabolism, phenylpyruvate has been found to inhibit enzymes in the periphery. In view of the critical dependence of the brain on glucose synthesized in the liver and kidney, the inhibition by phenylpyruvate of gluconeogenesis in these tissues (Krebs and deGasquet, 1964; Arinze and Patel, 1973) is especially noteworthy since it could have serious consequences for brain development. Against this possibility is the finding that at the levels of phenylpyruvate that are found in the plasma of patients with untreated classical PKU, 0.05 mM (Partington and Vickery, 1974), there is very little inhibition in kidney slices of gluconeogenesis from lactate (Krebs and deGasquet, 1964). It is possible, however, that tissue concentrations of phenylpyruvate could reach inhibitory concentrations. An additional argument against the idea that gluconeogenesis is impaired in PKU is the lack of evidence that these patients generally suffer from hypoglycemia (Dodge et al., (1959).

It should be noted that although there are no signs in PKU of impaired gluconeogenesis or any defect in peripheral glucose metabolism, the oxidation of glucose, β-hydroxybutyrate, or any other fuel could still be impaired within the CNS. In this connection, there is evidence that a high-glucose diet may protect the developing rat brain from some of the biochemical (Glazer and Weber, 1971; McKean *et al.*, 1962) and behavioral (McKean *et al.*, 1962) defects caused by hyperphenylalaninemia. The mechanism of this glucose protection is not known. It may be working directly by alleviating a defect in cerebral energy metabolism that it caused by hyperphenylalaninemia. Alternatively, the beneficial effects of glucose may be related to the decreased rate of cerebral serotonin synthesis that, as discussed previously, is caused by hyperphenylalaninemia. Thus, dietary carbohydrate has been reported to increase the rate of serotonin synthesis by increasing the cerebral concentration of the serotonin precursor, tryptophan (Fernstrom and Wurtman, 1971).

From the foregoing discussion, it is evident that phenylalanine or phenylpyruvate inhibit a wide variety of enzymes in the CNS. It is also clear that phenylpyruvate is a far more general inhibitor than is phenylalanine (see Table 6). Of special significance in trying to relate results of these animal-model studies to the neuropathological findings in PKU patients, such as a defect in myelinization, is the finding that phenylpyruvate, but not phenylalanine, inhibits enzymes that are involved in fatty acid synthesis in brain (Land and Clark, 1973). It may also be significant that of all the enzymes known to be inhibited by phenylpyruvate, fatty acid synthetase appears to be the most sensitive (Land and Clark, 1973) (see Table 6). Because of this sensitivity, Land and Clark (1973) have proposed that cerebral fatty acid synthetase is the primary site of inhibition and that this inhibition is responsible for the defect in myelinization that is seen in PKU.

This pathogenic mechanism, as well as any other that is based on inhibition of an enzyme by phenylpyruvate, obviously rests on the assumption that phenylpyruvate rather than phenylalanine is the compound that damages the developing nervous system in PKU. As already discussed, there are partisans on both sides of this question, but despite their claims the question remains unanswered.

A major hurdle that must be overcome before a neurotoxic role for phenylpyruvate can be accepted is the paucity of evidence for the presence of this compound in the CNS. Thus, although it has been asserted (Land and Clark, 1973) that phenylpyruvate was found by Edwards and Blau (1972) in the brains of hyperphenylalaninemic rats, this assertion is incorrect. Moreover, this metabolite was not detected by Partington and Vickery (1974) in the CSF of patients with untreated classical PKU. It should be noted that these last authors have concluded that their findings do not support the idea that phenylpyruvate is the "key metabolic factor in the pathogenesis of PKU."

If phenylpyruvate proves not to be the "key metabolic factor," phenylalanine itself probably is. In contrast to the keto acid, however, it is clear that the amino acid is not a potent inhibitor of enzyme reactions (see Table 6). If high concentrations of phenylalanine are toxic to the developing brains, it is most likely that this deleterious effect is due to inhibition by phenylalanine of transport into the brain of other amino acids.

The first suggestion that the excess phenylalanine in the tissues of phenylketonurics might interfere with the ability of their cells to "capture" other amino acids was made by Christensen (1953). In 1961, Neame reported that such competition occurs with brain tissue. He found that phenylalanine could interfere with the uptake by rat brain slices of other amino acids such as tyrosine, histidine, and arginine (Neame, 1961). Significantly, the concentration of phenylalanine used, 2 mM, was well within the range found in the blood of PKU patients. Neame suggested that this competition might be a pathogenic factor in PKU. He also pointed out that brain tissue, which transports amino acids like tyrosine very actively, would be more severely affected by this competition than would a tissue like liver.

One of the earliest findings which indicated that excess phenylalanine did interfere with transport of other amino acids in PKU patients was reported by Linneweh and Ehrlich (1962). They found that the plasma concentration of all the essential amino acids was lower than normal in these patients. Subsequently, it was shown that this amino acid imbalance in the blood was probably due to the inhibition by excess phenylalanine of the absorption from the intestine of amino acids like leucine and arginine (Linneweh *et al.*, 1963). This inhibition was reversed when the level of phenylalanine was lowered.

That inhibition by one amino acid of the uptake into the brain of another one can occur *in vivo* was shown by Chirigos *et al.* (1960). Although it has been stated (Knox, 1972) that these workers showed that phenylalanine inhibits transport of tyrosine and other amino acids in the brain, what was actually shown was that amino acids such as tryptophan, leucine, isoleucine, valine, histidine, and cysteine inhibit the uptake of tyrosine into the brain. On the basis of differences between results obtained *in vivo* and those obtained with brain slices (Guroff *et al.*, 1961), differences that showed there was a greater discrimination of the brain *in vivo*, e.g., that D-tyrosine is concentrated by brain slices but not by the brain *in vivo*, it was suggested that in the whole organism there was an additional barrier, "the blood–brain barrier," operating to restrict entry of tyrosine into the brain (Guroff *et al.*, 1961). Inhibition by high phenylalanine of uptake into the brain of another amino acid, leucine, was shown and reported in 1963 (Linneweh *et al.*, 1963).

Later, high blood levels of phenylalanine were shown to cause depletions of threonine, valine, methionine, leucine, isoleucine, histidine, tryptophan, and tyrosine in immature rat brains. With the exception of tyrosine, the same pattern of depletion by phenylalanine was seen in adult brains. The branched-chain amino acids were most affected, being reduced by 38–64% (McKean et al., 1968).

Using a technique in which the amount of amino acid taken up from the carotid artery is measured in a 15-s period, Oldendorf found a pattern of inhibition by phenylalanine in rats that was very similar to that reported by McKean et al. (1968): phenylalanine interfered with the cerebral uptake of the large neutral amino acids (tyrosine, leucine, isoleucine, methionine, tryptophan, histidine, valine, and threonine); the uptake of the basic amino acids, arginine, ornithine, and lysine was unaffected (Oldendorf, 1973b).

A somewhat different picture of the effects of excess phenylalanine emerges from the work of Lowden and LaRamee (1969). They claimed that hyperphenylalaninemia in young rats caused a marked decrease in the cerebral content of the nonessential amino acids. Because of the divergent effects of acute and chronic hyperphenylalaninemia, however, a meaningful interpretation of their results is difficult. The problem is well illustrated with alanine, the amino acid that they found to be most affected. Thus, chronic hyperphenylalaninemia in 10-day rats resulted in a 50% *increase* in cerebral alanine content, whereas acute administration of phenylalanine led to a *decrease* in alanine content.

Fortunately, the need to wrestle with the significance of these divergent effects is not compelling because hyperphenylalaninemia has not been found to lead invariably to decreased cerebral levels of the nonessential amino acids. Acute hyperphenylalaninemia in 7-day-old rats had no significant effect on the cerebral levels of aspartate, glutamate, or alanine (Roberts and Morelos, 1976); the levels of the branched-chain amino acids, however, were depressed. Very similar results were reported for the effects of chronic hyperphenylalaninemia on the amino acids in the brain of an infant monkey, the most pronounced effect being a decrease in the level of the branched-chain amino acids (O'Brien and Ibbot, 1966). The levels of the nonessential amino acids were also measured but they showed no consistent pattern; e.g., in response to excess phenylalanine, alanine was not significantly changed, glutamate was decreased, and aspartate was increased.

Only limited comparisons can be made between these results obtained with animals and those obtained with human brain tissue. Lowered levels of tryptophan and tyrosine have been found in cortex tissue from PKU patients (McKean, 1972). In another study in which the levels of threonine, serine, glutamine, tyrosine, and histidine were measured, only tyrosine was consistently lower in the PKU samples (McKean and Peterson, 1970).

These results with autopsied human brain tissue might suggest a much more limited interference of cerebral amino acid uptake in PKU than would have been expected from the animal studies. For example, as mentioned, phenylalanine has been found to interfere with the cerebral uptake of histidine *in vitro* (i.e., with brain slices) and *in vivo,* and yet this amino acid was not consistently affected in this limited study of brain tissue from PKU patients. The possibility that postmortem changes in amino acid levels might have distorted the pattern seen in the autopsied human brain tissue, however, cannot be evaluated.

Recently, Oldendorf *et al.* (1971) presented evidence that strongly suggests that the *in vivo* uptake of at least one of the large neutral amino acids, methionine, is decreased in PKU patients. Using the gamma-emitting analogue of methionine, [^{75}Se]selenomethionine, they showed that its uptake into the brains of these patients was reduced relative to the uptake in a group of retarded, nonphenylketonuric patients. It seems likely that the reduced uptake was caused by the high blood levels of phenylalanine in the PKU patients. This conclusion might have been strengthened if the study had been extended so that the PKU patients served as their own controls; i.e., if the uptake of selenomethionine by the PKU patients had been shown to be normalized when their blood phenylalanine levels had been normalized (by dietary restriction of their phenylalanine intakes). As it is, it is possible that it was some abnormality other than the high-phenylalanine levels that interfered with the cerebral uptake of the methionine analogue in the PKU patients.

Even with this mild caveat, it is clear that these *in vivo* results, when taken together with results of amino acid analysis on brain tissue from PKU patients, and the results from a large number of animal studies all point to the same conclusion: high levels of phenylalanine interfere with the cerebral uptake of other amino acids. Since this interference is selective, it must lead to an abnormal pattern of amino acid concentrations in the brain. And, since it is known that excess amounts of every amino acid (except for alanine) can inhibit growth of weanling rats (Sauberlich, 1961), it was to be expected—unless the developing brain were more tolerant of amino acid imbalances than is the rest of the organism—that this altered pattern would very likely adversely affect the development of the brain.

One of these adverse effects appears to be on protein synthesis. There is ample evidence that this metabolic process in brain is highly sensitive to changes in the concentrations of free amino acids (Munro, 1970; Roberts and Morelos, 1965). A specific mechanism by which high levels of one amino acid might interfere with protein synthesis was demonstrated by Appel (1966), who showed that there is competition between certain amino acids for attachment to tRNA molecules.

That excess phenylalanine can inhibit protein synthesis in brain was demonstrated with young rabbits: hyperphenylalaninemia was shown to decrease the incorporation of intraperitoneally injected lysine into brain ribosomal proteins (Swaiman *et al.,* 1968).

Agrawal *et al.* (1970) have presented evidence indicating that the inhibition of protein synthesis in brain by high levels of phenylalanine in the blood is secondary to the inhibition by the hyperphenylalaninemia of transport of amino acids into the brain. Thus, they found that the decreased transport of methionine and leucine into the acid-soluble pool of brain could account for the inhibition of the rates of incorporation of these amino acids into protein. Another indication that the defect in protein synthesis is secondary to a defect in transport is the finding that hyperphenylalaninemia did not inhibit the incorporation into protein of glycine, an amino acid the transport of which into brain was unaffected by the high levels of phenylalanine (Agrawal *et al.,* 1970).

This sequence of events was strongly supported by the recent results of Antonas and Coulson (1975), who compared the incorporation into brain protein of intraperitoneally and intraventricularly administered leucine in rats made hyperphenylalaninemic by administration of PCPA. They found that high levels of phenylalanine interfered with incorporation of leucine into brain protein when the leucine was given intraperitoneally, but not when it was injected directly into the brain. When leucine was given by the intraperitoneal route, they found, as have many others, that the hyperphenylalaninemia inhibited its transport into the brain. It should be noted that with these chronically hyperphenylalaninemic rats, incorporation of lysine into cerebral protein was not inhibited by hyperphenylalaninemia. This result is in contrast to the results obtained by Swaiman *et al.* (1968) with rabbits in which blood phenylalanine levels were acutely elevated by injection of phenylalanine.

Although these results of Agrawal and co-workers and of Antonas and Coulson indicate that the interference by high phenylalanine of transport of amino acids into the brain can adequately account for inhibition of cerebral protein synthesis, excess phenylalanine may also interfere more directly. In this regard, the finding that acute hyperphenylalaninemia caused by a single intraperitoneal injection of phenylalanine into young rats (but not in older ones) can depress the rate of protein synthesis in cell-free preparations of cerebral cortex (Siegel *et al.,* 1971) suggests a deleterious effect of hyperphenylalaninemia that cannot be ascribed simply to an imbalance in the supply of amino acids that are available for protein systhesis. Based on the finding that the polyribosomes from the brains of these phenylalanine-treated animals were largely disaggregated, these authors postulated that this defect in polyribosome formation, which is perhaps caused by a phenylalanine-mediated decrease in cerebral tryptophan, could account for the decreased protein synthesis (Siegel *et al.,* 1971).

The disaggregating effect of excess phenylalanine on brain polyribosomes has been confirmed in mice (MacInnes and Schlesinger, 1971). These workers concluded, however, that polyribosome disaggregation did not lead to inhibition of protein synthesis but, rather, that inhibition of protein synthesis led to polyribosome disaggregation.

Recently, Roberts and Morelos (1976) proposed that phenylalanine-induced disaggregation of cerebral polyribosomes was secondary to an effect of excess phenylalanine on the activity of ribonuclease. They observed that ribonuclease activity was elevated in the brains of young rats that had been treated *in vivo* with phenylalanine and suggested that phenylalanine or one of its metabolites labilized cerebral lysosomal membranes. As a consequence of this labilization, ribonuclease would be released during homogenization of the tissue, resulting in polyribosome disaggregation. Based on their results, Roberts and Morelos concluded that disruption of cerebral polyribosomes does not occur *in vivo*. This conclusion is in accord with the observations of Hartman and Becker (1973), who found no evidence for polyribosomal disaggregation in electron micrographs of brain sections prepared from rats that had been given a single injection of phenylalanine. Unfortunately, it was not actually shown that this treatment did in fact lead to hyperphenylalaninemia in the animals that were used for the electron-micrographic study.

In trying to decide whether phenylalanine or phenylpyruvate is the "key metabolic factor in the pathogenesis of PKU" or, at least, which of these compounds is closer to this factor, it is clear that certain aspects of the chemical pathology can be more readily traced to the keto compound, whereas others can be more easily traced to excess concentrations of the amino acid. Thus, the available evidence strongly supports the conclusion that excess phenylalanine, by interfering with the transport into the brain of a group of other amino acids, can inhibit cerebral protein synthesis. It seems likely that the decreased levels of DNA (Chase and O'Brien, 1970) and RNA (MacInnes and Schlesinger, 1971; Castells *et al.*, 1971) in the brains of hyperphenylalaninemic animals is secondary to the inhibition of protein synthesis.

On the other hand, the defects in myelinization and, in particular, in myelin lipid synthesis can be explained readily by the known inhibitory effects of phenylpyruvate (see Table 6). As previously discussed, however, the weakness of this pathogenic mechanism is the lack of evidence for the presence in the brain of inhibitory concentrations of phenylpyruvate. While it is possible that there is no single neurotoxic metabolite in PKU, and that it will turn out that the excess phenylalanine causes some of the irreversible symptoms, that phenylpyruvate causes others, and perhaps that phenylethylamine or serotonin deficiency causes still others, the lure of a more parsimonius explanation remains.

If a choice of a single damaging substance had to be made, a strong case can be made for that substance being phenylalanine. By interfering with protein synthesis, excess phenylalanine could, in theory, inhibit the synthesis of any enzyme, and, therefore, given enough time, could interfere with any metabolic process. It is possible that the defects in cerebral lipid metabolism seen in infant hyperphenylalaninemic rats are due to this effect. Against this interpretation, however, is the finding that incorporation of radioactivity from [U^{14}C]glucose into brain lipids was depressed by acute hyperphenylalaninemia (Shah *et al.*, 1970). Unless the enzymes involved in cerebral lipid synthesis had extremely high turnover rates, this inhibition occurred much too quickly for it to be due to inhibition of the synthesis of an enzyme. The available evidence is not consistent with high turnover rates for these enzymes. The $t_{1/2}$ of fatty acid synthetase in the brains of 4-day-old rats has been reported to be about 2 days (Volpe *et al.*, 1973). With this relatively low turnover, it is unlikely that the level of the synthetase could change rapidly enough to account for this acute effect of excess phenylalanine. The other brain enzyme that is involved in fatty acid synthesis, acetyl-CoA carboxylase, probably also does not turn over rapidly, since its activity is not depressed by either acute starvation or chronic undernutrition (Gross and Warshaw, 1975).

It should be noted that there are data that do not support the idea that the hyperphenylalaninemia-induced defect in cerebral lipid synthesis is an immediate consequence of the excess phenylalanine. For example, Agrawal *et al.* (1970) found that acute hyperphenylalaninemia in infant rats did not depress the incorporation of acetate into cerebral lipids, whereas it did depress the incorporation of leucine into cerebral protein. This differential effect of acute hyperphenylalaninemia on cerebral protein and lipid synthesis led Agrawal *et al.* (1970) to conclude that the decrease in lipid synthesis caused by excess phenylalanine is secondary to an impairment in myelin protein synthesis. They confirmed the many previous findings that hyperphenylalaninemia inhibits the transport of other amino acids into the brain, and therefore accepted the popular idea that inhibition of cerebral protein synthesis by hyperphenylalaninemia is due to the consequent alterations in the pools of free amino acids in the brain. These workers did not, however, discuss mechanisms by which inhibition of protein synthesis might lead to secondary inhibition of lipid synthesis. Implicit in such a proposal is the notion that synthesis of myelin proteins and myelin lipids is somehow coupled.

Knowledge of how a complex structure like myelin is built is far too scanty to allow detailed speculation on the mechanism of such a coupling process. This coupling may occur at the cellular level and may simply reflect the fact that both protein and lipid components of myelin are products of glial cell metabolism. The effect of excess phenylalanine, which can lead to decreased concentration within the brain of other amino acids, which in turn can inhibit cerebral protein synthesis, may be akin to the effects on myelinization of neonatal malnutrition, where it has been shown that accretion of myelin protein and lipid are both inhibited. In that situation, there is evidence that precursors of glial cells fail to undergo the normal differentiation and migration that appear to be necessary for myelinization (Bass *et al.*, 1970).

More specific mechanisms for this coupling are suggested by the recent studies of myelin metabolism in peripheral nerves (Gould and Dawson, 1976). These results serve to emphasize not only that migration of cells may be involved in the process of myelinization, but also that lipid molecules themselves must be translocated from their sites of synthesis within the cytoplasm of glial cells to specific sites on the outer portions of the myelin sheaths. There are many steps in this journey, including the transport process itself, as well as that of the final deposition on the myelin, where specific lipid–protein interaction may be involved and where defects in protein synthesis could lead to deficits in the amount of lipid that is ultimately associated with myelin. Furthermore, defects in this transport and deposition process could, by leading to increased concentrations of free lipid, i.e., lipid that is not complexed or associated with protein, inhibit the synthesis of that lipid.

That hyperphenylalaninemia also affects lipids by mechanisms that cannot be explained simply by lipid–protein interaction is strongly indicated by the findings that hyperphenylalaninemia leads not only to decreases in the amounts of cerebral lipids, but also (as discussed earlier) to alterations in their structure; i.e., the lipids appear to be of shorter chain length and of a lower degree of unsaturation than normal cerebral lipids. A possible mechanism for these effects will be discussed later.

In addition to inhibiting protein synthesis, there are other ways in which excess phenylalanine, by interfering with the uptake of amino acids into the brain, can upset cerebral metabolism. The inhibition by hyperphenylalaninemia of cerebral neurotransmitter synthesis, at least in part due to depressed uptake of the precursor amino acids, tryptophan and tyrosine, has already been discussed.

A clue to still another pathogenic mechanism may be provided by the observation that of all the amino acids the transport of which into the brain is inhibited by excess phenylalanine, that of the branched-chain amino acids may be the most affected. This effect, as already discussed, is almost certainly part of the mechanism by which excess phenylalanine inhibits cerebral protein synthesis.

But beyond their participation in protein synthesis, the branched-chain amino acids almost certainly have other roles in the metabolism of the brain. There are numerous indications that point to this possibility. There is evidence, for example, that the essential amino acids are taken up by the brain more rapidly than are the nonessential ones and that within the former group, the branched-chain amino acid may be taken up most rapidly

(Oldendorf, 1971; Oldendorf, 1973a; Baños *et al.*, 1973; Baños *et al.*, 1975; Battistin *et al.*, 1971; Felig *et al.*, 1973). This rapid uptake by brain, when taken together with the fact that valine, leucine, and isoleucine are released from muscle, but not extracted by liver, makes it likely that the brain is an important site of utilization of this group of amino acids (see Felig, 1975, and references therein).

That they are used in the brain for reactions in addition to those involved in protein synthesis is strongly suggested by the finding that brain tissue from rats can metabolize [1-^{14}C]leucine four times faster than either muscle or liver (Odessey and Goldberg, 1972). Furthermore, of all tissues studied, brain had the highest ratio of the amount of leucine converted to CO_2 as compared to the amount incorporated into protein (Odessey and Goldberg, 1972). There is evidence that this ratio varies with the age of the animals and with their nutritional status (Chaplin and Diamond, 1974). That leucine metabolism in brain is not limited to reactions that lead to liberation of its carboxyl group as CO_2 was shown by the rapid oxidation of leucine to dicarboxylic amino acids (Roberts and Morelos, 1965).

Of relevance to the defect in myelin lipid synthesis in PKU is the possible role of leucine as a significant precursor of cerebral lipids. It is known that leucine is catabolized in animal tissues to acetyl CoA and acetoacetate via β-hydroxy-β-methylglutaryl CoA, a precursor of cholesterol. That these reactions occur in brain is indicated by the report that brain slices can incorporate carbon from [U-^{14}C]leucine into both myelin and nonmyelin lipids (Smith, 1974). These findings support the notion that the branched-chain amino acids, especially leucine, may be important precursors of certain cerebral lipids.

This idea may receive added support from the finding that in maple syrup urine disease (MSUD), in which there is a primary defect in the catabolism of this group of amino acids, there is, just as in PKU, a decreased content of lipids in the brain, which, in part, reflects a defect in myelinization (see Prensky *et al.*, 1968, and references therein). Just as we have been discussing with PKU, it is likely that with MSUD, too, this defect in lipid metabolism is secondary to a defect in protein synthesis (caused in this case by a leucine-induced imbalance of amino acids in the brain). It is also possible, however, again along lines similar to those that we have been considering with PKU, that part of the defect in cerebral lipid synthesis in this disease is a more direct consequence of the primary metabolic block. Here, the primary defect in leucine metabolism happens to be an early step in the same pathway that ultimately leads to lipid synthesis.

In this context, it must be acknowledged that the experience with patients with MSUD who had been treated with diets that restrict their consumption of the branched-chain amino acids sets a limit on the quantitative importance of these amino acids as precursors of cerebral lipids. The finding that the lipid composition in the cerebral white matter of these treated patients was probably normal (Linneweh and Socher, 1965; Menkes and Solcher, 1967; Prensky *et al.*, 1968) indicates that the branched-chain amino acids cannot be the sole source of cerebral lipids. Clearly, if this were the case, restricting the intake of amino acids such as leucine could not correct the deficiency in leucine-derived lipids. But beyond being the sole source, the results with MSUD patients do not rule out the possibility that these amino acids normally can contribute importantly to the synthesis of certain cerebral lipids.

As far as the specific lipid, cholesterol, is concerned, however, the finding that it is present in normal amounts in brain tissue from untreated MSUD pateints (Menkes and Solcher, 1967; Prensky and Moser, 1966), as pointed out by Smith (1974), is inconsistent with the idea that leucine is a major precursor for its synthesis. As already discussed, the primary defect in leucine metabolism in this disease would be expected to prevent the conversion of the amino acid to the cholesterol precursor, β-hydroxy-β-methyl-glutaryl CoA, and thereby prevent the conversion of the amino acid to cholesterol.

In addition to serving as precursors of cerebral proteins, and perhaps of lipids, amino acids, especially the nonessential ones, may be an important storage form of readily oxidizable fuel in the brain (McIlwain, 1966). Indeed, because they found that the acute administration of phenylalanine to 10-day-old rats decreased the cerebral concentration of some nonessential amino acids, Lowden and LaRamee (1969) have suggested that hyperphenylalaninemia produces brain damage in infant animals by decreasing energy production, presumably from the nonessential amino acids.

This suggestion receives little support from the available data, including, as discussed previously, those of Lowden and LaRamee themselves. Also, as pointed out earlier, the results of other studies with hyperphenylalaninemic infant rats (Roberts and Morelos, 1976) and monkeys (O'Brien and Ibbot, 1966) do not confirm the finding that excess phenylalanine decreases the cerebral concentration of the nonessential amino acids.

Although these results provide no experimental support for the idea that excess phenylalanine causes a sustained decrease in the cerebral concentration of the nonessential amino acids, the metabolism of this group of amino acids may nonetheless be disturbed in classical PKU. One reason this possibility is still open is that the effect of excess phenylalanine on the nonessential amino acids has been reported so far only for animals made hyperphenylalaninemic by administration of phenylalanine alone. The effect of excess phenylalanine on cerebral amino acid metabolism in such animals might be quite different from its effect in patients with classical PKU. Thus, in the animal experiments, the metabolic consequences of high levels of phenylalanine will almost certainly be modified by the presence of an active phenylalanine hydroxylating system. In this situation, it is likely that the elevated levels of phenylalanine would increase the amount of tyrosine that is available to the brain (Lo *et al.*, 1970; O'Brien and Ibbot, 1966). And since tyrosine can be metabolized in this tissue to glutamate and aspartate (Van den Berg, 1970, 1971), the increase in tyrosine would probably lead to increased amounts of these nonessential amino acids. In animals with an active phenylalanine hydroxylating system, therefore, even if excess phenylalanine did cause a depletion of nonessential amino acids in the brain, this effect might be masked by the increased synthesis of these amino acids caused by the increased amounts of tyrosine. The last effect, of course, could not take place in patients with classical PKU because they do not have an active phenylalanine hydroxylating system. In short, the effect of hyperphenylalaninemia on the cerebral concentration of the nonessential amino acids probably depends on whether or not phenylalanine hydroxylase is active: when the enzyme is essentially inactive, as in classical PKU, hyperphenylalaninemia might decrease the concentration of the nonessential amino acids, whereas when the enzyme is active, this effect might not be seen. Measurements of the concentration of amino acids such as aspartate and glutamate in the brains of animals treated with both phenylalanine and PCPA could clarify this point.

With these considerations in mind, it is worth examining another possible role for the essential amino acids in the brain, i.e., that they serve as important precursors for the nonessential ones. And, despite the findings that chronic hyperphenylalaninemia in animals does not lead to sustained decreases in the levels of the nonessential amino acids, the possibility exists that in classical PKU, the excess phenylalanine, by inhibiting the transport into the brain of essential amino acids such as tyrosine, valine, and isoleucine leads to a deficiency of the nonessential ones. Deficiency of these amino acids could contribute to a slower rate of protein synthesis in the growing brain. According to this idea, hyperphenylalaninemia could lead to a deficiency of both the essential and the nonessential amino acids in the brain, the former by direct interference with their transport into the brain, and the latter as a secondary consequence of this decreased transport of their essential precursors. (It should be noted that even if these relationships exist, static measurements of the levels of the nonessential amino acids in the brains of PKU patients may not show a lower concentration: the decreased synthesis may be balanced by decreased utilization for protein synthesis, leaving their steady-state levels relatively unchanged.)

Implicit in this proposal is the idea that the immature brain does not rely exclusively on a supply of the nonessential amino acids coming into it from the periphery, but depends importantly on its own ability to synthesize these compounds from the essential amino acids. This idea is consistent with the findings, mentioned previously, that the transport into the brain of the essential amino acids is, in general, faster than that of the nonessential ones.

Since amino acids such as glutamate and aspartate can certainly be synthesized from glucose, the idea that they can also be synthesized from some of the essential amino acids raises a question about the relationship between these two kinds of precursors. This question, in turn, brings us to a more general role that some of the essential amino acids may play, and that is to prevent the Krebs citric acid cycle from running down by replenishing the supply of intermediates that have been tapped from the cycle and diverted to the synthesis of compounds such as glutamate and aspartate. This anaplerotic function [Kornberg has introduced the term *anaplerotic sequence* to describe metabolic pathways that lead to the net formation of cycle intermediates (1966)] would probably be particularly important for those amino acids, such as isoleucine, valine, methionine, and tyrosine (an essential amino acid in the amounts needed by the growing brain for protein synthesis), that are metabolized to the Krebs cycle intermediates: succinyl CoA and fumarate. As already discussed, these amino acids are all members of the group of large neutral amino acids whose uptake into the brain is depressed by excess phenylalanine. By reducing their concentration in brain, therefore, excess phenylalanine would interfere indirectly with their ability to replenish citric acid cycle intermediates.

Whether amino acids such as valine are converted to the nonessential amino acids within the same compartment where the bulk of glucose is being oxidized to CO_2 and water, or whether these two metabolic processes occur in separate compartments, is not known. There is evidence, however, indicating that separate compartments may be involved (Baxter, 1976; Berle and Clarke, 1969).

It should be noted that the possibility has been raised previously, but without any consideration of specific mechanisms, that the brain, especially when immature, can use certain amino acids for biosynthetic processes. It has also been suggested, again without any elaboration, that interference with the metabolism of these alternative substrates may lead to impairment of the normal development of the brain (Van den Berg, 1970, 1971).

Another link between amino acid metabolism and energy-yielding reactions in the brain is related to a unique aspect of brain metabolism, namely, the sequence of reactions by which the neurotransmitter, γ-amino-butyric acid (GABA), is synthesized. This route, from α-ketoglutarate via glutamate, GABA, and succinic semialdehyde to succinate, constitutes a shunt around the usual step in the Krebs citric acid cycle wherein α-ketoglutarate is oxidized to succinate via succinyl CoA.

Just as we have discussed for the utilization of cycle intermediates for the synthesis of certain nonessential amino acids, the operation of the GABA shunt may affect energy metabolism in certain areas of the brain by depletion of Krebs cycle intermediates. There is evidence that GABA is lost from the brain; e.g., stimulation of the cerebellum cortex in cats results in an increased release of GABA into the fourth ventricle (Obata and Takeda, 1969). Loss of GABA means that α-ketoglutarate and succinate are being drained from the tricarboxylic acid cycle, and unless these losses are made good, the cycle may run down.

How rapidly the cycle will run down will be determined not only by the rate of loss of GABA from the brain, but also by the quantitative importance of the GABA shunt. On this latter point, it has been estimated that 8–10% of the carbon atoms going through the Krebs cycle are shunted via glutamate and GABA to succinate (Baxter, 1976, and references therein). It should be emphasized that this small fraction tells us only how rapidly the cycle will run down in the absence of replenishment of intermediates, not *whether* it will run down.

The interference by excess phenylalanine with the anaplerotic function of other amino acids in brain may underlie several different cerebral metabolic abnormalities in classical PKU. Thus, the decreased oxygen arteriovenous differences in cerebral circulation seen in PKU patients (Himwich and Fazekas, 1940), which may be due to a lower oxygen consumption (but see the earlier discussion), could be a reflection of impaired functioning of the citric acid cycle. In addition, the finding that the brains from these patients contain less of the unsaturated fatty acids relative to the corresponding saturated ones (Foote *et al.*, 1965), as well as the relative decrease in the long-chain fatty acids (Gerstl *et al.*, 1967), could also be due to the same metabolic disturbance. Since both the desaturation (Bloomfield and Bloch, 1960) and the elongation of fatty acids (Aeberhard and Menkes, 1968) require reduced pyridine nucleotide, probably mainly TPNH, and since the TPNH used for biosynthetic reactions comes in part from the oxidation of citric acid cycle intermediates such as isocitrate and malate, the relative deficit of cerebral unsaturated and long-chain fatty acids in PKU may be due to decreased availability of TPNH, a deficit that is caused indirectly by excess phenylalanine.

Another metabolic consequence of the GABA shunt is a decrease in the amount of succinyl CoA and, hence, GTP that is formed for each turn of the Krebs cycle. For this reason, even though GTP is not utilized directly for GABA synthesis, the neurotransmitter is nonetheless being made at the expense of GTP. The finding that drugs such as isonicotinic acid hydrazide decrease the cerebral concentration of GABA and increase the concentration of cyclic GMP (Mao *et al.*, 1974) is consistent with that kind of relationship.

It is possible that a significant fraction of the GTP in the brain is normally supplied by amino acids (valine, isoleucine, methionine, and tyrosine) that are metabolized to succinyl CoA or fumarate. Excess phenylalanine, by interfering with the transport into the brain of these amino acids, therefore, could decrease the amount of GTP. As a consequence of this decrease, metabolic processes that utilize GTP, e.g., protein synthesis and cyclic GMP formation, could be inhibited.

How significant a fraction of the GTP in the brain might be supplied by this group of amino acids is not known with certainty. From available data (Felig *et al.*, 1973; Sokoloff, 1960), however, it can be estimated that for adult humans, the quantity of isoleucine, valine, tyrosine, and methionine taken up by brain is only a small percentage of the amount of glucose. For whole human brain, therefore, the contribution of this family of amino acids to GTP formation, as well as to the total energy supply, would appear to be small. It should be noted, however, that these amino acids may make a larger contribution to the energy metabolism of the growing brain (Van den Berg, 1970, 1971). There is also strong evidence, as mentioned previously, that in brain there may be more than a single citric acid cycle—with separate pools of cycle intermediates and with preferential utilization of certain metabolites (for reviews, see Baxter, 1976; Berle and Clarke, 1969). With such metabolic heterogeneity, it is possible that these amino acids, by virtue of their conversion to citric acid cycle intermediates, contribute very significantly to the supply of GTP, and therefore, to energy metabolism, within one of these compartments.

Even though the secretion of several hormones (insulin, glucagon, and growth hormone) is known to be stimulated by amino acids (see Felig, 1975), the possible metabolic consequences and pathogenic significance of disturbed hormonal status in PKU have received little attention. With growth hormone (Knopf *et al.*, 1965) and insulin (Floyd *et al.*, 1966), phenylalanine has been tested specifically and been shown to be an effective stimulator of hormone release. As far as growth hormone is concerned, there is evidence that there is something unusual in its regulation in PKU patients: their basal plasma concentrations of the hormone are higher than that of controls and, in six of seven patients, the plasma level is unresponsive to a loading dose of phenylalanine. By contrast, in both normals and PKU heterozygotes, the phenylalanine dose elicited a three- to fourfold increase in the plasma growth hormone concentration (Castells *et al.*, 1969). The authors point out that the lack of response may be related to the low-phenylalanine diet that the patients were receiving. Whatever the explanation ultimately proves to be, these results support the idea that hyperphenylalaninemia may lead to persistent abnormalities in the status of some hormones. The pathogenic significance of such abnormalities has yet to be explored. It should be noted that both growth hormone and insulin are believed to be involved in the transport of amino acids across cell membranes. This role may not only be related to pathogenesis, but also may complicate the analysis of phenylalanine-tolerance data.

In summary, a plausible case can be made for phenylalanine itself, rather than one of its metabolites, such as phenylpyruvate, being the substance that, when present in excess, interferes with normal brain development. The main irreversible consequence of this impaired development is mental retardation. By inhibiting the transport into the brain of a group of amino acids, excess phenylalanine can interfere with cerebral protein and lipid synthesis. Perhaps related to effects on these specific anabolic processes, but also having the potential for reaching far beyond them, are possible deleterious effects of hyperphenylalaninemia on the functioning of the citric acid cycle. This last relationship may be of physiological significance only within certain compartments in the brain.

Even if the mechanism of pathogenesis were to be elucidated at the biochemical level, i.e., by unequivocally identifying the responsible compound or compounds, there would still remain the problem of how altered metabolism leads to altered brain structure and function. How does impaired protein or lipid synthesis, for example, lead to mental retardation? An answer to this question will probably require a clearer understanding of the pathological significance of the structural change that seems to be characteristic of the brains of patients with classical PKU, i.e., impaired myelinization. For even though, as discussed earlier, PKU has been classified as a dysmyelinating disease (Poser and Van Bogaert, 1959), it is not known whether this defect is primary or whether it is secondary to some still more fundamental pathological process. That the symptoms of classical PKU are not those that are characteristic of other diseases of myelin suggests that the myelin defect in PKU is incidental to some other structural alteration.

The nature of this underlying defect is not known, but an impairment in the normal development of neuronal processes, including defective synapse formation, is a reasonable possibility. This possibility is supported by results of studies with hyperphenylalaninemic neonatal rats (Nordyke and Roach, 1974), which indicate that excess phenylalanine interferes with the formation of metabolic compartments in the brain, compartments that are believed to be associated with the formation of neuronal processes (Cocks *et al.,* 1970) and with neuronal–glial interactions (Patel and Balazs, 1970).

9. PROSPECTS FOR ALTERNATE THERAPY

As discussed earlier, treatment of classical PKU by restriction of phenylalanine intake, when started early enough, appears to be effective in preventing the mental defect. This treatment, however, is far from being ideal: it is tricky, requiring careful monitoring of phenylalanine blood levels, relatively expensive, and a burden to both patient and family. There are, therefore, ample reasons for considering alternate forms of therapy.

There is a logical connection between therapy and pathogenesis. Implicit in most of the mechanisms of pathogenesis that have been discussed are different types of therapy. Just to make one of these relationships explicit, if the CoA-depletion hypothesis (which is not one of the most attractive ones) were ever to gain support, CoA (or pantothenic acid) supplementation would deserve consideration as part of therapy.

One kind of treatment that is unrelated to any specific idea of pathogenesis—except the one that assumes that the symptoms of the disease are due to hyperphenylalaninemia caused by the lack of phenylalanine hydroxylase—would be enzyme replacement therapy. To be effective, the added hydroxylase would not necessarily have to gain access to the liver. It could, presumably, still lower phenylalanine levels in the blood if it were encased in a sac that is exposed to the patient's blood supply. The sac would, of course, have to be freely permeable to phenylalanine, tyrosine, and oxygen. One of the major problems with this approach is that of supplying the encased phenylalanine hydroxylase with a constant supply of an active tetrahydropterin. Frequent exogeneous administration of the tetrahydropterin, perhaps together with a reductant such as ascorbate, might overcome this problem.

Independent of therapy based on enzyme replacement, the possible value of pterin supplementation itself has been considered previously (Milstien and Kaufman, 1975*a*). These workers found that administration of 6-methyltetrahydropterin (plus ascorbate) to rats increased three- to sixfold the rate of phenylalanine hydroxylation. It is not known whether the small amount of phenylalanine hydroxylase activity that has been detected in a classical PKU patient (Friedman *et al.,* 1973) can also be stimulated *in vivo* by administration of an active tetrahydropterin, but there is no apparent reason why it cannot. The point has been made previously (Milstien and Kaufman, 1975*a*) that even a sixfold increase in phenylalanine hydroxylase activity in these patients could be of therapeutic value.

Studies with the purified enzyme from rat liver have indicated other ways in which the activity of the hydroxylase might be enhanced. As mentioned, partial proteolysis and exposure to phospholipids such as lysolecithin can markedly activate the enzyme. It is not known, however, whether either of these modes of activation are operative in the whole organism.

As discussed in the previous section, there is support for the idea that pathogenesis of the mental defect in classical PKU is due to interference by excess phenylalanine with the transport into the brain of other amino acids. It was also pointed out, however, that the possibility is still open that phenylalanine is the direct precursor of a substance, such as phenylpyruvate, that interferes with normal brain development. According to these ideas, it should be possible to prevent the deleterious effects of hyperphenylalaninemia on the immature brain by administration of the amino acids that share the same transport system with phenylalanine, i.e., the large neutral amino acids such as tyrosine, tryptophan, valine, leucine, isoleucine, and methionine. Clearly, the amounts of each of these amino acids that would have to be supplied would have to be determined empirically, but certain limits can be suggested. If phenylalanine is the direct precursor of the toxic compound, or if high amounts of phenylalanine itself in the brain can interfere with brain development, then the goal would be to prevent excess phenylalanine from entering the brain. To accomplish this, the competing amino acids would have to be administered so that their aggregate concentration in the blood was several times greater than that of phenylalanine. Thus, if the blood level of phenylalanine in a typical, untreated classical PKU patient is 2 mM, it would be desirable to reach (and maintain) an aggregate blood concentration of these amino acids of 4–6 mM, or a concentration of each of the six or so individual amino acids of 0.7–1 mM. If, on the other hand, excess phenylalanine is toxic because it interferes with the transport into the brain of the other amino acids, then the latter compounds might have to be supplied in amounts that make each of their blood concentrations comparable to that of phenylalanine.*

As discussed in the previous section, if phenylpyruvate contributes to the pathogenesis of the mental defect in PKU, then glutamate administration would be expected to be of therapeutic value.

Several ways in which a phenylalanine-induced deficiency of some amino acids in the brain could interfere with cerebral energy metabolism have already been considered. If such a pathogenic mechanism should ever gain experimental support, glucose administration, perhaps combined with amino acid supplementation, would deserve consideration in the treatment of classical PKU. The beneficial effects of glucose in hyperphenylalaninemic animals, as well as the possibility that this effect may be an indirect one, mediated by an increased cerebral content of tryptophan and hence serotonin, have been discussed.

The prospects for therapy of hyperphenylalaninemia due to dihydropteridine reductase deficiency have already been discussed.

---

* The proposal that amino acid supplementation be considered as an alternative to dietary restriction of phenylalanine for therapy of classical PKU is also relevant to treatment of maternal PKU. This condition arises when mothers with classical PKU give birth to children who are mentally retarded and otherwise damaged, even if they are nonphenylketonuric (i.e., the children are heterozygotes) (Mabry *et al.*, 1963). Administration of a low phenylalanine diet to the PKU mothers during pregnancy has been reported to prevent damage to their offspring, but this treatment has not been invariably successful (Koch *et al.*, 1974).

Since there is evidence that there is competition between different amino acids for uptake by slices of human placenta (Miller and Berndt, 1974), it may be possible to prevent excessive amounts of phenylalanine from reaching the fetus by supplementation with other amino acids. Before this possibility can be evaluated, it will be necessary to identify the amino acids that compete with phenylalanine for transport by the human placenta.

10. REFERENCES

Abita, J.-P., Dorche, C., and Kaufman, S., 1974, Further studies on the nature of phenylalanine hydroxylation in brain, *Pediatr. Res.* **8**:714–717.

Adelman, L. S., Mann, J. D., Caley, D. W., and Bass, N. H., 1973, Neuronal lesions in the cerebellum following the administration of excess phenylalanine to neonatal rats, *J. Neuropathol. Exp. Neurol.* **XXXII**:380–393.

Adlard, B. P. F., deSouza, S. W., and Moon, S., 1973, The effect of age, growth retardation and asphyxia on ascorbic acid concentrations in developing brain, *J. Neurochem.* **21**:872–881.

Aeberhard, E., and Menkes, J. H., 1968, Biosynthesis of long chain fatty acids by subcellular particles of mature brain, *J. Biol. Chem.* **243**:3834–3840.

Agrawal, H. C., Bone, A. H., and Davison, A. N., 1970, Effect of phenylalanine on protein synthesis in the developing rat brain, *Biochem. J.* **117**:325–331.

Allen, R. J., and Wilson, J. L., 1964, Urinary phenylpyruvic acid in phenylketonuria, *J. Amer. Med. Assoc.* **188**:720–724.

Allen, R. J., Heffelfinger, J. C., Masotti, R. E., and Tsau, M. U., 1964, Phenylalanine hydroxylase activity in newborn infants, *Pediatrics* **33**:512–525.

Alvord, E. C., Stevenson, L. D., Vogel, F. S., and Engle, R. L., 1950, Neuropathological findings in phenylpyruvic oligophrenia (phenyl-ketonuria), *J. Neuropathol. Exp. Neurol.* **9**:298–310.

Andersen, A. E., and Guroff, G., 1972, Enduring behavioral changes in rats with experimental phenylketonuria, *Proc. Natl. Acad. Sci. U.S.A.* **69**:863–867.

Andersen, A. E., Rowe, V., and Guroff, G., 1974, The enduring behavioral changes in rats with experimental phenylketonuria, *Proc. Natl. Acad. Sci. U.S.A.* **71**:21–25.

Anderson, V. E., Siegel, F. S., Fisch, R. O., and Wirt, R. D., 1969, Responses of phenylketonuric children on a continuous performance test, *J. Abnorm. Psychol.* **74**: 358–362.

Andrews, T. M., McKeran, R. O., Watts, R. W. E., McPherson, K., and Lax, R., 1973, Relationship between the granulocyte phenylalanine content and the degree of disability in phenylketonuria, *Q. J. Med. New Ser.* **XLII**:805–817.

Antonas, K. N., and Coulson, W. F., 1975, Brain uptake and protein incorporation of amino acids studied in rats subjected to prolonged hyperphenylalaninemia, *J. Neurochem.* **25**:309–314.

Appel, S. H., 1966, Inhibition of brain protein synthesis: an approach to the biochemical basis of neurological dysfunction in the amino-acidurias, *Trans. N.Y. Acad. Sci.* **29**:63–70.

Arinze, I. J., and Patel, M. S., 1973, Inhibition by phenylpyruvate of gluconeogenesis in the isolated perfused rat liver, *Biochemistry* **12**:4473–4479.

Armstrong, M. D., 1963, Biochemistry, in: *Phenylketonuria* (F. L. Lyman, ed.), pp. 62–95, Charles C. Thomas, Springfield, Illinois.

Armstrong, M. D., and Binkley, E. L., Jr., 1956, Studies on Phenylketonuria. V. Observations on a newborn infant with phenylketonuria, *Proc. Soc. Exp. Biol. Med.* **93**:418–420.

Armstrong, M. D., and Low, N. L., 1957, Phenylketonuria, VIII. Relation between age, serum phenylalanine level, and phenylpyruvic acid exretion, *Proc. Soc. Exp. Biol. Med.* **94**:142–146.

Armstrong, M. D., and Robinson, K. S., 1954, On the excretion of indole derivatives in phenylketonuria, *Arch. Biochem.* **52**:287–288.

Armstrong, M. D., and Tyler, F. H., 1955, Studies on phenylketonuria. I. Restricted phenylalanine intake in phenylketonuria, *J. Clin. Invest.* **34**:565–580.

Armstrong, M. D., Shaw, K. N. F., and Robinson, K. S., 1955, Studies on phenylketonuria. II. The excretion of *o*-hydroxyphenylacetic acid in phenylketonuria, *J. Biol. Chem.* **213**:797–804.

Auerbach, V. H., and Waisman, H., 1959, Tryptophan peroxidase-oxidase, histidase, and transaminase activity in the liver of the developing rat, *J. Biol. Chem.* **234**:304–306.

Auerbach, V. H., Waisman, H. A., and Wyckoff, L. B., 1958, Phenylketonuria in the rat associated with decreased temporal discrimination learning. *Nature (London)* **182**:871–872.

Auerbach, V. H., DiGeorge, A. M., and Carpenter, G. G., 1967, Phenylalaninemia, a study of diversity of disorders which produce elevation of blood levels of phenylalanine, in: *Amino Acid Metabolism and Genetic Variation* (W. L. Nyhan, ed.), pp. 11–68, McGraw-Hill, New York.

Avery, M. E., 1967, Transient tyrosinemia of the newborn: dietary and chemical aspects, *Pediatrics* **39**:378–384.

Ayling, J. E., Helfand, G. D., and Pirson, W. D., 1975, Phenylalanine hydroxylase from human kidney, *Enzyme* **20**:6–19.

Bãnos, G., Daniel, P. M., Moorhouse, S. R., and Pratt, O. E., 1973, The influx of amino acids into the brain of the rat *in vivo:* the essential compared with some nonessential amino acids, *Proc. R. Soc. Lond. Ser. B.* **183**:59–70.

Bãnos, G., Daniel, P. M., Moorhouse, S. R., and Pratt, O. E., 1975, The requirements of the brain for some amino acids, *J. Physiol.* **246**:539–548.

Barbosa, E., Herreros, B., and Ojeda, J. L., 1971, Amino acid accumulation by brain slices: interactions among tryptophan, phenylalanine and histidine, *Experientia* **27**:1281–1282.

Barranger, J. A., Geiger, P. J., Huzino, A., and Bessman, S. P., 1972, Isozymes of phenylalanine hydroxylase, *Science* **175**:903–905.

Bartholomé, K., 1974, A new molecular defect in phenylketonuria, *Lancet* **2**:1580.

Bass, N. H., Netsky, M. G., and Young, E., 1970, Effect of neonatal malnutrition on developing cerebrum. II. Microchemical and histological study of myelin formation in the rat, *Arch. Neurol.* **23**:303–313.

Battistin, L., Grynbaum, A., and Lajtha, A., 1971, The uptake of various amino acids by the mouse brain *in vivo, Brain Res.* **29**:85–99.

Baxter, C. F., 1976, Some recent advances in studies of GABA metabolism and compartmentation, in: *GABA in Nervous System Function* (E. Roberts, T. N. Chase, and D. B. Tower, eds.), pp. 61–87, Raven Press, New York.

Berl, S., and Clarke, D. D., 1969, Compartmentation of amino acid metabolism, in: *Handbook of Neurochemistry, Vol. II* (A. Lajtha, ed.), pp. 447–472, Plenum Press, New York.

Berman, J. L., Cunningham, G. C., Day, R. W., Ford, R., and Hsia, D. Y.-Y., 1969, Causes for high phenylalanine with normal tyrosine in newborn screening programs, *Amer. J. Dis. Child.* **117**:54–65.

Berry, H., Sutherland, B. S., and Guest, G. M., 1957, Phenylalanine tolerance tests on relatives of phenylketonuric children, *Amer. J. Hum. Genet.* **9**:310–316.

Berry, H. K., Butcher, R. E., Kazmaier, K. J., and Poncet, I. B., 1975, Biochemical effects of induced phenylketonuria in rats, *Biol. Neonate* **26**:88–101.

Bessman, S. P., 1964, Some biochemical lessons to be learned from phenylketonuria, *J. Pediatr.* **64**:828–838.

Bessman, S. P., 1966, Legislation and advances in medical knowledge–acceleration or inhibition?, *J. Pediatr.* **69**:334–338.

Bessman, S. P., 1972, Genetic failure of fetal amino acids "justification": a common basis for many forms of metabolic, nutritional, and "nonspecific" mental retardation, *J. Pediatr.* **81**:834–842.

Bessman, S. P., and Tada, K., 1960, Indicanuria in phenylketonuria, *Metabolism* **9**:377–385.

Betheil, J. J., Feigelson, M., and Feigelson, P., 1965, The differential effects of glucocorticoid on tissue and plasma amino acid levels, *Biochim. Biophys. Acta* **104**:92–97.

Bhattacharya, S. K., Robson, J. S., and Stewarti, C. P., 1955, Determination of glutathione in blood and tissues, *Biochem. J.* **60**:696–702.

Bickel, H., Gerrard, J., and Hickmans, E. M., 1954, The influence of phenylalanine intake on the chemistry and behavior of a phenylketonuric child, *Acta Paediatr.* **43**:64–77.

Bickel, H., Boscott, R. J., and Gerrard, J., 1955, Observations on the biochemical error in phenylketonuria and its dietary control, in: *Biochemistry of the Developing Nervous System* (H. Waelsch, ed.), p. 417, Academic Press, New York.

Birch, H. G., and Tizard, J., 1967, The dietary treatment of phenylketonuria: not proven?, *Dev. Med. Child Neurol.* **9**:9–12.

Blaskovics, M. E., and Shaw, K. N. F., 1971, Hyperphenylalaninemia: methods for differential diagnosis, in: *Phenylketonuria and Some Other Inborn Errors of Amino Acid Metabolism* (H. Bickel, F. P. Hudson, and L. I. Woolf, eds.), pp. 98–102, Georg Thieme Verlag, Stuttgart.

Blau, K., 1970, Aromatic acid excretion in phenylketonuria. Analysis of the unconjugated aromatic acids derived from phenylalanine, *Clin. Chim. Acta* **27**:5–18.

Blau, K., Summer, G. K., Newsome, H. C., and Edwards, C. H., 1973, Phenylalanine loading and aromatic acid excretion in normal subjects and heterozygotes for PKU, *Clin. Chim. Acta* **45**:197–205.

Bloomfield, D. K., and Bloch, K., 1960, The formation of Δ^9-unsaturated fatty acids, *J. Biol. Chem.* **235**:337–345.

Boggs, D. E., Rosenberg, R., and Waisman, H. A., 1963, Effects of phenylalanine, phenylacetic acid, tyrosine and valine on brain and liver serotonin in rats, *Proc. Soc. Exp. Biol. Med.* **114**:356–358.

Borek, E., Brecher, A., Jervis, G. A., and Waelsch, H., 1950, Oligophrenia phenylpyruvica. II. Constancy of the metabolic error, *Proc. Soc. Exp. Biol. Med.* **75**:86–89.

Boscott, R. J., and Bickel, H., 1953, Detection of some new abnormal metabolites in the urine of phenylketonuria, *Scand. J. Clin. Lab. Invest.* **5**:380–382.

Bowden, J. A., and McArthur, C. L., III, 1972, Possible biochemical model for phenylketonuria, *Nature (London)* **235**:230.

Bowman, B. H., and King, F. J., 1961, Effects of glutamine and asparagine supplements in the dietary regimen of three phenylketonuric patients, *Nature (London)* **190**:417–418.

Boyd, J. W., 1966, The extraction and purification of the two isozymes of L-aspartate: 2-oxoglutarate aminotransferase, *Biochim. Biophys. Acta* **113**:302–311.

Brand, L. M., and Harper, A. E., 1974, Effect of glucagon on phenylalanine metabolism and phenylalanine-degrading enzymes in the rat, *Biochem. J.* **142**:231–245.

Bremer, H. J., 1971, Transitory hyperphenylalaninemia, in: *Phenylketonuria and Some Other Inborn Errors of Amino Acid Metabolism* (H. Bickel, F. P. Hudson, and L. I. Woolf, eds.), pp. 93–97, Georg Thieme Verlag, Stuttgart.

Bremer, H. J., and Neumann, W., 1966*a*, Phenylalanin-Toleranz bei Frühgeborenen, reifen Neugeborenen, Säuglingen und Erwachsenen, *Klin. Wochenschr.* **44**:1076–1081.

Bremer, H. J., and Neumann, W., 1966*b*, Tolerance of phenylalanine after intravenous administration in phenylketonurics, heterozygous carriers, and normal adults, *Nature (London)* **209**:1148–1149.

Brenneman, A. R., and Kaufman, S., 1965, Characteristics of the phenylalanine-hydroxylating system in newborn rats, *J. Biol. Chem.* **240**:3617–3622.

Butler, I. J., Krumholz, A., Holtzman, N. A., Koslow, S. H., Kaufman, S., and Milstien, S., 1975, Dihydropteridine reductase deficiency variant of phenylketonuria: a disorder of neurotransmitters, *Arch. Neurol.* **32**:350.

Cahalane, S. F., 1968, Phenylketonuria: mass screening of newborns in Ireland, *Arch. Dis. Child.* **43**:141–144.

Carpenter, G. G., Auerbach, V. H., and DiGeroge, A. M., 1968, Phenylalaninemia, *Pediatr. Clin. North Amer.* **15**:313–323.

Castells, S., Grunt, J. A., and Brandt, I. K., 1969, Changes in plasma growth hormone after a phenylalanine tolerance test in normal and phenylketonuric children, *J. Pediatr.* **75**:820–825.

Castells, S., Zischka, R., and Addo, N., 1971, Alteration in composition of deoxyribonucleic acid, ribonucleic acid, proteins, and amino acids in brain of rats fed high and low phenylalanine diets, *Pediatr. Res.* **5**:329–334.

Chalmers, R. A., and Watts, R. W. E., 1974, Quantitative studies on the urinary excretion of unconjugated aromatic acids in phenylketonuria, *Clin. Chim. Acta* **55**:281–294.

Chaplin, E. R., and Diamond, I., 1974, The importance of leucine oxidation in brain, *Trans. Amer. Soc. Neurochem.* **5**:167.

Chase, H. P., and O'Brien, D., 1970, Effects of excess phenylalanine and of other amino acids on brain development in the infant rat, *Pediatr. Res.* **4**:96–102.

Chirigos, M. A., Greengard, P., and Udenfriend, S., 1960, Uptake of tyrosine by rat brain *in vivo, J. Biol. Chem.* **235**:2075–2079.

Christensen, H. N., 1953, Metabolism of amino acids and proteins, *Annu. Rev. Biochem.* **22**:233–260.

Christenson, J. G., Dairman, W., and Udenfriend, S., 1970, Preparation and properties of a homogenous aromatic L-amino acid decarboxylase from hog kidney, *Arch. Biochem. Biophys.* **141**:356–367.

Civen, M., Trimmer, B. M., and Brown, C. B., 1967, The induction of hepatic tyrosine α-ketoglutarate and phenylalanine pyruvate transaminase by glucagon, *Life Sci.* **6**:1331–1338.

Cocks, J. A., Balazs, R., Johnson, A. L., and Eayrs, J. T., 1970, Effect of thyroid hormone on the biochemical maturation of rat brain: conversion of glucose-carbon into amino acids, *J. Neurochem.* **17**:1275–1285.

Coleman, D. L., 1960, Phenylalanine hydroxylase activity in dilute and nondilute strains of mice, *Arch. Biochem. Biophys.* **91**:300–306.

Cowie, V. A., 1971, Neurological and psychiatric aspects of phenylketonuria, in: *Phenylketonuria and Some Other Inborn Errors of Amino Acid Metabolism* (H. Bickel, F. P. Hudson, and L. I. Woolf, eds.), pp. 29–39, Georg Thieme Verlag, Stuttgart.

Craine, J. E., Hall, E. S., and Kaufman, S., 1972, The isolation and characterization of dihydropteridine reductase from sheep liver, *J. Biol. Chem.* **247**:6082–6091.

Crome, L., 1971, The morbid anatomy of phenylketonuria, in: *Phenylketonuria and Some Other Inborn Errors of Amino Acid Metabolism* (H. Bickel, F. P. Hudson, and L. I. Woolf, eds.), pp. 126–131, Georg Thieme Verlag, Stuttgart.

Crome, L., and Pare, C. M. B., 1960, Phenylketonuria: a review and a report of the pathological findings in four cases, *J. Ment. Sci.* **106**:862–883.

Crome, L., Tymms, U., and Woolf, L. I., 1962, A chemical investigation of the defects of myelination in phenylketonuria, *J. Neurol. Neurosurg. Psychiatry* **25**:143–148.

Culley, W. J., Saunders, R. N., Mertz, E. J., and Jolly, D. H., 1962, Effect of phenylalanine and its metabolites on the brain serotonin level of the rat, *Proc. Soc. Exp. Biol. Med.* **111**:444–446.

Curtius, H.-C., Baerlocher, K., and Vollmin, J. A., 1972*a*, Pathogenesis of phenylketonuria: inhibition of dopa and catecholamine synthesis in patients with phenylketonuria, *Clin. Chim. Acta* **42**:235–239.

Curtius, H.-C., Vollmin, J. A., and Baerlocher, K., 1972*b*, The use of deuterated phenylalanine for the elucidation of the phenylalanine-tyrosine metabolism, *Clin. Chim. Acta* **37**:277–285.

Dakin, H. D., 1911a, The catabolism of phenylalanine, tyrosine and their derivatives, *J. Biol. Chem.* 9:139–150.

Dakin, H. D., 1911b, The chemical nature of alcaptonuria, *J. Biol. Chem.* 9:151–160.

Dancis, J., and Balis, M. E., 1955, A possible mechanism for disturbance in tyrosine metabolism in phenylpyruvic oligophrenia, *Pediatrics* 15:63–66.

Danks, D. M., Cotton, R. G. H., and Schlesinger, P., 1975, Tetrahydrobiopterin treatment of variant form of phenylketonuria, *Lancet* 2:1043.

David, J. C., Dairman, W., and Udenfriend, S., 1974, On the importance of decarboxylation in the metabolism of phenylalanine, tyrosine, and tryptophan, *Arch. Biochem. Biophys.* 160:561–568.

Davison, A. N., and Sandler, M. S., 1958, Inhibition of 5-hydroxytryptophan decarboxylase by phenylalanine metabolites, *Nature (London)* 181:186–187.

Deguchi, T., and Barchas, J., 1972, The effect of *p*-chlorophenylalanine and tryptophan hydroxylase in rat pineal, *Nature (London) New Biol.* 235:92–93.

Dobbing, J., 1975, Prenatal nutrition and neurological development, in: *Brain Mechanisms in Mental Retardation* (N. A. Buchwald and M. A. B. Brazier, eds.), pp. 401–420, Academic Press, New York.

Dodge, P. R., Mancall, E. L., Crawford, J. D., Knapp, J., and Paine, R. S., 1959, Hypoglycemia complicating treatment of phenylketonuria with a phenylalanine-deficient diet, *N. Engl. J. Med.* 260:1104–1111.

Edson, N. L., 1935, Ketogenesis-antiketogenesis. I. The influence of ammonium chloride on ketone-body formation in liver, *Biochem. J.* 29:2082–2094.

Edwards, D. J., and Blau, K., 1972, Aromatic acids derived from phenylalanine in the tissues of rats with experimentally induced phenylketonuria-like characteristics, *Biochem. J.* 130:495–503.

Efron, M. L., Kang, E. S., Visakorpi, J., and Fellers, F. X., 1969, Effect of elevated plasma phenylalanine levels on other amino acids in phenylketonuria and normal subjects, *J. Pediatr.* 74:399–405.

Embden, G., and Baldes, K., 1913, Über den Abbau des Phenylalanins im tierischen Organismus, *Biochem. Z.* 55:301–322.

Embden, G., Salomon, H., and Schmidt, F., 1906, Über Acentonbildung in der Leber, *Beitr. Chem. Physiol. Pathol.* 8:129–155.

Evered, D. F., 1956, The excretion of amino acids by the human, *Biochem. J.* 62:416–427.

Felig, P., 1975, Amino acid metabolism in man, *Ann. Rev. Biochem.* 44:933–955.

Felig, P., Wahren, J., and Ahlborg, G., 1973, Uptake of individual amino acids by the human brain, *Proc. Soc. Exp. Biol. Med.* 142:230–231.

Fellman, J. H., Vanbellinghen, P. J., Jones, R. T., and Koler, R. D., 1969, Soluble and mitochondrial forms of tyrosine aminotransferase. Relationship to tyrosinemia, *Biochemistry* 8:615–622.

Fellman, J. H., Fujita, T. S., and Roth, E. S., 1972, Substrate specificity of *p*-hydroxyphenylpyruvate hydroxylase, *Biochim. Biophys. Acta* 268:601–604.

Fernstrom, J. D., and Wurtman, R. J., 1971a, Brain serotonin content: physiological dependence on plasma tryptophan levels, *Science* 173:149–152.

Fernstrom, J. D., and Wurtman, R. J., 1971b, Brain serotonin content: increase following ingestion of carbohydrate diet, *Science* 174:1023.

Fincham, J. R. S., 1966, *Genetic Complementation*, W. A. Benjamin, New York.

Finkelstein, J. D., Mudd, H. S., Irreverre, F., and Laster, L., 1964, Homocystinuria due to cystathionine synthase, *Science* 146:785–787.

Fisher, D. B., and Kaufman, S., 1972a, The inhibition of phenylalanine and tyrosine hydroxylases by high oxygen levels, *J. Neurochem.* 19:1359–1365.

Fisher, D. B., and Kaufman, S., 1972b, Stimulation of rat liver phenylalanine hydroxylase by phospholipids, J. Biol. Chem. 247:2250–2252.

Fisher, D. B., and Kaufman, S., 1973a, Tetrahydropterin oxidation without hydroxylation catalyzed by rat liver phenylalanine hydroxylase, J. Biol. Chem. 248:4300–4304.

Fisher, D. B., and Kaufman, S., 1973b, The stimulation of rat liver phenylalanine hydroxylase by lysolecithin and α-chymotrypsin, J. Biol. Chem. 248:4345–4353.

Fisher, D. B., Kirkwood, R., and Kaufman, S., 1972, Rat liver phenylalanine hydroxylase, an iron enzyme, J. Biol. Chem. 247:5161–5167.

Floyd, J. F., Fajans, S. S., Conn, J. W., Knopf, R. F., and Rull, J., 1966, Stimulation of insulin secretion by amino acid, J. Clin. Invest. 45:1487–1502.

Fölling, A., 1934a, Utzkillese aus Fenylpyroduesyre i Urinieren som Stoffskifesanomali Forvindelse med Imvecilitet, Nord. Med. Tidskr. 8:1054.

Fölling, A., 1934b, Über Ausscheidung von Phenylbrenztraubensaüre in den Harn als Stoffwechselanomalie in Verbindung mit Imbezzillität, Z. Physiol. Chem. 227:169.

Fölling, A., Closs, K., and Gammes, T., 1938, Vorlaufige Schlüsselfolgerungen aus Belastungversucher mit Phenylalanin an Menschen und Tieren, Z. Physiol. Chem. 256:1.

Foote, J. L., Allen, R. J., and Agranoff, B. W., 1965, Fatty acids in esters and cerebrosides of human brain in phenylketonuria, J. Lipid Res. 6:518–524.

Friedman, P. A., and Kaufman, S., 1971, A study of the development of phenylalanine hydroxylase in fetuses of several mammalian species, Arch. Biochem. Biophys. 146:321–326.

Friedman, P. A., and Kaufman, S., 1973, Some characteristics of partially purified human liver phenylalanine hydroxylase, Biochim. Biophys. Acta 293:56–61.

Friedman, P. A., Kappelman, A. H., and Kaufman, S., 1972a, Partial purification and characterization of tryptophan hydroxylase from rabbit hindbrain, J. Biol. Chem. 247:4165–4173.

Friedman, P. A., Kaufman, S., and Kang, E. S., 1972b, Nature of the molecular defect in PKU and hyperphenylalaninemia, Nature (London) 240:157–159.

Friedman, P. A., Lloyd, T., and Kaufman, S., 1972c, Production of antibodies to rat liver phenylalanine hydroxylase: cross-reactivity with other pterin-dependent hydroxylases, Mol. Pharmacol. 8:501–510.

Friedman, P. A., Fisher, D. B., Kang, E. S., and Kaufman, S., 1973, Detection of hepatic 4-phenylalanine hydroxylase in classical phenylketonuria, Proc. Natl. Acad. Sci. U.S.A. 70:552–556.

Fuller, R. W., and Baker, J. C., 1974, Increased conversion of a phenylalanine load to tyrosine in tetraiodoglucagon-treated rats, Biochem. Biophys. Res. Commun. 58:945–950.

Fuller, R. W., Snoddy, H. D., and Bromer, W. W., 1972, Increase of hepatic L-phenylalanine:pyruvate aminotransferase by glucagon in rats, Mol. Pharmacol. 8:345–352.

Gaull, G. E., and Sturman, J. A., 1971, Vitamin B₆ dependency on homocystinuria, Br. Med. J. 3:532–533.

Gerstl, R. L., Malamud, N., Eng, L. F., and Hayman, R. B., 1967, Lipid alterations in human brains in phenylketonuria, Neurology 17:51–57.

Gibbs, N. K., and Woolf, L. I., 1959, Tests for phenylketonuria—results of a one year programme for its detection in infancy and among mental defectives, Br. Med. J. 2:532–535.

Gimenez, C., Valdivieso, F., and Mayor, F., 1974, Glycolysis in the brain and liver of rats with experimentally induced phenylketonuria, Biochem. Med. 11:81–86.

Giroud, A., Leblond, C. P., Ratsimanga, R., and Gero, E., 1938a, Le taux normal en acide ascorbique, Bull. Soc. Chim. Biol. 20:1079–1087.

Giroud, A., Rabinowicz, M., and Hartmann, E., 1938b, Le taux realise chex l'homme, Bull. Soc. Chim. Biol. 20:1097–1101.

Glazer, R. I., and Weber, G., 1971, The effects of phenylpyruvate and hyperphenylalaninemia on incorporation of [6-³H]glucose into macromolecules of slices of rat cerebral cortex, *J. Neurochem.* **18**:2371–2382.

Goodfriend, T. L., and Kaufman, S., 1961, Phenylalanine metabolism and folic acid antagonists, *J. Clin. Invest.* **40**:1743–1750.

Gould, R. M., and Dawson, R. M. C., 1976, Incorporation of newly formed lecithin into peripheral nerve myelin, *J. Cell. Biol.* **68**:480–496.

Graham-Smith, D. G., and Parfitt, A. G., 1970, Tryptophan transport across the synaptosomal membranes, *J. Neurochem.* **17**:1339–1353.

Gressner, A. M., 1974, Amino acid levels in liver and plasma of the rat during inhibition of transamination, *Biochem. Med.* **10**:199–207.

Grimm, U., Knapp., A., Schlenzka, K., and Reddemann, H., 1975, Phenylalaninhydroxylaseaktivität in der Leber als Parameter zur Unterscheidung der verschiedenen Formen der Hyperphenylalaninamien, *Clin. Chim. Acta* **58**:17–21.

Gross, I., and Warshaw, J. R., 1975, The influence of dietary deprivation on the enzymes of fatty acid synthesis in rat brain, *J. Neurochem.* **25**, 191–192.

Guroff, G., and Udenfriend, S., 1962, Studies on aromatic amino acid uptake by rat brain *in vivo*. Uptake of phenylalanine and of tryptophan; inhibition and stereoselectivity in the uptake of tyrosine by brain and muscle, *J. Biol. Chem.* **237**:803–806.

Guroff, G., King, W., and Udenfriend, S., 1961, The uptake of tyrosine by rat brain *in vitro, J. Biol. Chem.* **236**:1773–1777.

Guthrie, R., 1961, Blood screening for phenylketonuria, *J. Amer. Med. Assoc.* **178**:863.

Güttler, F., and Wamberg, E., 1972, Persistent hyperphenylalaninemia, *Acta. Paediatr. Scand.* **61**:321–328.

Hackney, I. M., Hanley, W. B., Davidson, W., and Linsao, L., 1968, Phenylketonuria: mental development, behavior, and termination of low phenylalanine diet, *J. Pediatr.* **72**:646–655.

Hanson, A., 1959, Action of phenylanine metabolites on glutamic acid decarboxylase and γ-aminobutyric acid α-ketoglutaric acid transaminase in brain, *Acta Chem. Scand.* **13**:1366–1374.

Harris, H., 1970, Heterozygotes, in: *The Principles of Human Biochemical Genetics* (A. Neuberger and E. L. Tatum, eds.), p. 173, North-Holland, Amsterdam.

Hartman, J. F., and Becker, R. A., 1973, Ultrastructural evidence against *in vivo* disaggregation of brain polyribosomes after administration of L-dopa or phenylalanine, *J. Neural Transm.* **34**:73–77.

Henson, C. P., and Cleland, W. W., 1964, Kinetic studies of glutamic oxaloacetic transaminase isozymes, *Biochemistry* **3**:338–345.

Himwich, H. E., and Fazekas, J. F., 1940, Cerebral metabolism in mongolian idiocy and phenylpyruvic oligophrenia, *Arch. Neurol. Psychiatry* **4**:1213–1218.

Hjalmarsson, O., Jagenburg, R., and Rodjer, S., 1971, Mild and severe PKU. Comparative studies in two infants, *Acta Paediatr. Scand.* **60**:11–16.

Hoffman, N. E., and Gooding, K. M., 1969, Gas chromatography of some urinary acid metabolites related to phenylketonuria, *Anal. Biochem.* **31**:471–479.

Hole, K., 1972*a*, Reduced 5-hydroxyindole synthesis reduces postnatal brain growth in rats, *Eur. J. Pharmacol.* **18**:361–362.

Hole, K., 1972*b*, Arousal defect in L-phenylalanine fed rats, *Dev. Psychobiol.* **5**:149–156.

Hole, K., 1972*c*, Behavior and brain growth in rats treated with *p*-chlorophenylalanine in the first weeks of life, *Dev. Psychobiol.* **5**:157–173.

Holtzman, N. A., Mellits, E. D., and Kallman, B. A., 1974*a*, Neonatal screening for phenylketonuria. II. Age dependence of initial phenylalanine in infants with PKU, *Pediatrics* **53**:353–357.

Holtzman, N. A., Meek, A. G., Mellits, E. D., and Kallman, C. H., 1974*b*, Neonatal screening for phenylketonuria. III. Altered sex ratio; extent and possible causes, *J. Pediatr.* **25**:175–181.

Hopper, S., and Segal, H. L., 1962, Kinetic studies of rat liver glutamic-alanine transaminase, *J. Biol. Chem.* **237**:3189–3195.

Horner, F. A., Streamer, C. W., Alejaudrino, L. L., Reed, L. H., and Ibbott, F., 1962, The termination of dietary treatment of phenylketonuria, *N. Eng. J. Med.* **266**:79–81.

Hsia, D. Y. Y., 1970, Phenylketonuria and its variants, *Prog. Med. Genet.* **7**:29–68.

Hsia, D. Y. Y., and Dobson, J., 1970, Altered sex ratio among phenylketonuric infants ascertained by screening the newborn, *Lancet* **1**:905–908.

Hsia, D. Y. Y., and Driscoll, K. W., 1956, Detection of the heterozygous carriers of phenylketonuria, *Lancet* **2**:1337–1338.

Hsia, D. Y. Y., Driscoll, K., Troll, W., and Knox, W. E., 1956, Detection by phenylalanine tolerance tests of heterozygous carriers of phenylketonuria, *Nature (London)* **178**:1239–1240.

Hsia, D. Y. Y., Knox, W. E., Quinn, K. V., and Paine, R. S., 1958, A one-year, controlled study of the effect of low-phenylalanine diet on phenylketonuria, *Pediatrics* **21**:178–202.

Huang, C. Y., and Kaufman, S., 1973, Studies on the mechanism of action of phenylalanine hydroxylase and its protein stimulator, *J. Biol. Chem.,* **248**:4242–4251.

Huang, C., Max, E. E., and Kaufman, S., 1973, Purification and characterization of phenylalanine hydroxylase stimulating protein from rat liver, *J. Biol. Chem.* **248**:4235–4241.

Huang, I., and Hsia, D. Y. Y., 1963, Studies on inhibition of 5-hydroxytryptophan decarboxylase by phenylalanine metabolites, *Proc. Soc, Exp. Biol. Med.* **112**:81–84.

Hudson, F. P., 1967, Termination of dietary treatment of phenylketonuria, *Arch. Dis. Child.* **42**:198–200.

Ikeda, M., Levitt, M., and Udenfriend, S., 1967, Phenylalanine as a substrate and inhibitor of tyrosine hydroxylase, *Arch. Biochem. Biophys.* **120**:420–427.

Jacoby, G. A., and LaDu, B. N., 1964, Studies on the specificity of tyrosine-α-ketoglutarate transaminase, *J. Biol. Chem.* **239**:419–424.

Jakubovic, A., 1971, Phenylalanine hydroxylating system in the human fetus at different developmental ages, *Biochim. Biophys. Acts* **237**:469–475.

Jéquier, E., 1968, Tryptophan hydroxylation in phenylketonuria, *Adv. Pharmacol.* **6B**:169–170.

Jéquier, E., Lovenberg, W., and Sjoerdsma, A., 1967, Tryptophan hydroxylase inhibition: the mechanism by which *p*-chlorophenylalanine depletes rat brain serotonin, *Mol. Pharmacol.* **3**:274–278.

Jervis, G. A., 1939, The genetics of phenylpyruvic oligophrenia, *J. Ment. Sci.* **85**:719–762.

Jervis, G. A., 1947, Studies on phenylpyruvic oligophrenia. The position of the metabolic error, *J. Biol. Chem.* **169**:651–656.

Jervis, G. A., 1953, Phenylpyruvic oligophrenia deficiency of phenylalanine-oxidizing system, *Proc. Soc. Exp. Biol. Med.* **82**:514–515.

Jervis, G. A., 1954, Phenylpyruvic oligophrenia (phenylketonuria), *Res. Publ. Assoc. Res. Nerv. Ment. Dis.* **33**:259–282.

Jervis, G. A., 1960, Detection of heterozygotes for phenylketonuria, *Clin. Chim. Acta* **5**:471–476.

Jervis, G. A., 1963, Pathogenesis of the mental defects, in: *Phenylketonuria* (F. L. Lyman, Ed.), pp. 101–113, Charles C. Thomas, Springfield, Illinois.

Johnson, R. C., and Shah, S. N., 1973, Effect of hyperphenylalaninemia on fatty acid composition of lipids of rat brain myelin, *J. Neurochem.* **21**:1225–1240.

Justice, P., O'Flynn, M. E., and Hsia, D. Y. Y., 1967, Phenylalanine hydroxylase activity in hyperphenylalaninemia, *Lancet* **1**:928–929.

Kang, E. S., Kaufman, S., and Gerald, P. S., 1970a, Clinical and biochemical observations of patients with atypical phenylketonuria, *Pediatrics* **45**:83–92.

Kang, E. S., Sollee, N., and Gerald, P. S., 1970b, Results of treatment and termination of the diet in phenylketonuria (PKU), *Pediatrics* **46**:881–890.

Kaufman, S., 1957, The enzymatic conversion of phenylalanine to tyrosine, *J. Biol. Chem.* **226**:511–524.

Kaufman, S., 1958a, A new cofactor required for the enzymatic conversion of phenylalanine to tyrosine, *J. Biol. Chem.* **230**:931–939.

Kaufman, S., 1958b, Phenylalanine hydroxylation cofactor in phenylketonuria, *Science* **128**:1506.

Kaufman, S., 1958c, Biochemistry in diagnosis, in: *Conference on Diagnosis in Mental Retardation*, pp. 125–144, The Training School at Vineland, New Jersey.

Kaufman, S., 1959, Studies on the mechanism of the enzymatic conversion of phenylalanine to tyrosine, *J. Biol. Chem.* **234**:2677–2682.

Kaufman, S., 1962, Aromatic hydroxylation in: *Oxygenases* (O. Hayaishi, ed.), pp. 129–179, Academic Press, New York.

Kaufman, S., 1963, The structure of phenylalanine hydroxylation cofactor, *Proc. Natl. Acad. Sci. U.S.A.* **50**:1085–1093.

Kaufman, S., 1964, Further studies on the structure of the primary oxidation product formed from tetrahydropteridines during phenylalanine hydroxylation, *J. Biol. Chem.* **239**:332–338.

Kaufman, S., 1967a, Metabolism of the phenylalanine hydroxylation cofactor, *J. Biol. Chem.* **242**:3934–3943.

Kaufman, S., 1967b, Pteridine cofactors, *Annu. Rev. Biochem.* **36**:171–184.

Kaufman, S., 1967c, Unanswered questions in the primary metabolic block in phenylketonuria, in: *Phenylketonuria and Allied Metabolic Diseases* (J. A. Anderson and K. F. Swaiman, eds.), pp. 205–213, U. S. Government Printing Office, Washington, D.C.

Kaufman, S., 1969, Phenylalanine hydroxylase of human liver: assay and some properties, *Arch. Biochem. Biophys.* **134**:249–252.

Kaufman, S., 1970, A protein that stimulates rat liver phenylalanine hydroxylase, *J. Biol. Chem.* **245**:4751–4759.

Kaufman, S., 1971, The phenylalanine hydroxylating system from mammalian liver, *Adv. Enzymol.* **35**:245–319.

Kaufman, S., and Fisher, D. B., 1970, Purification and some physical properties of phenylalanine hydroxylase from rat liver, *J. Biol. Chem.* **245**:4745–4750.

Kaufman, S., and Levenberg, B., 1959, Further studies on the phenylalanine hydroxylation cofactor, *J. Biol. Chem.* **234**:2683–2688.

Kaufman, S., and Max, E. E., 1971, Studies on the phenylalanine hydroxylating system in human liver and their relationship to pathogenesis of PKU and hyerphenylalaninemia, in: *Phenylketonuria and Some Other Inborn Errors of Amino Acid Metabolism* (H. Bickel, F. P. Hudson, and L. I. Woolf, eds.), pp. 13–19, Georg Thieme Verlag, Stuttgart.

Kaufman, S., and Fisher, D. B., 1974, Pterin-requiring aromatic amino acid hydroxylases, in: *Molecular Mechanisms of Oxygen Activation* (O. Hayaishi, ed.), pp. 285–369, Academic Press, New York.

Kaufman, S., Bridgers, W. F., Eisenberg, F., and Friedman, S., 1962, The source of oxygen in the phenylalanine hydroxylase and the dopamine-β-hydroxylase catalyzed reactions, *Biochem. Biophys. Res. Commun.* **9**:497–502.

Kaufman, S., Holtzman, N. A., Milstien, S., Butler, I. J., and Krumholz, A., 1975a, Phenylketonuria due to a deficiency of dihydropteridine reductase, *N. Engl. J. Med.* **293**:785–790.

Kaufman, S., Max, E. E., and Kang, E. S., 1975b, Phenylalanine hydroxylase activity in liver biopsies from hyperphenylalaninemia heterozygotes: deviation from proportionality with gene dosage, *Pediatr. Res.* **9**:632–634.

Kelley, W. N., Levy, R. I., Rosenbloom, F. M., Henderson, J. F., and Seegmiller, J. E., 1968, Adenine phosphoribosyl transferase deficiency: a previously undescribed genetic defect in man, *J. Clin. Invest.* **47**:2281–2289.

Kennaway, N. G., and Buist, N. R. M., 1971, Metabolic studies in a patient with hepatic cytosol tyrosine aminotransferase deficiency, *Pediatr. Res.* **5**:287–297.

Kenney, F. T., and Kretchmer, N., 1959, Hepatic metabolism of phenylalanine during development, *J. Clin. Invest.* **38**:2189–2196.

Kettler, R., Bartholini, G., Pletscher, A., 1974, *In vivo* enhancement of tyrosine hydroxylation in rat striatum by tetrahydrobiopterin, *Nature (London)* **249**:476–477.

Knopf, R. F., Conn, J. W., Fajans, S. S., Floyd, J. C., Guntsche, E. M., and Rull, J. A., 1965, Plasma growth hormone response to intravenous administration of amino acids, *J. Clin. Endocrinol. Metab.* **25**:1140–1144.

Knox, W. E., 1972, Phenylketonuria, in: *The Metabolic Basis of Inherited Disease* (J. B. Stanbury, J. B. Wyngaarden, and D. S. Fredrickson, eds.), pp. 266–295, McGraw-Hill, New York.

Knox, W. E., and Hsia, D. Y. Y., 1957, Pathogenic problems in phenylketonuria, *Amer. J. Med.* **22**:687–702.

Knox, W. E., and Messinger, E., 1958, The detection of the metabolic effect of the recessive gene for phenylketonuria, *Amer. J. Hum. Genet.* **10**:53–60.

Koch, R., Blaskovics, M., Wenz, E., Fishler, K., and Schaeffler, G., 1974, Phenylalaninemia and phenylketonuria, in: *Heritable Disorders of Amino Acid Metabolism* (W. L. Nyhan, ed.), pp. 109–140, John Wiley & Sons, New York.

Koe, B. K., and Weissman, A., 1966, *p*-Chlorophenylalanine: a specific depletor of brain serotonin, *J. Pharmacol. Exp. Ther.* **154**:499–516.

Kopelovich, L., Sweetman, L., and Nisselbaum, J. S., 1971, Kinetics of the inhibition of aspartate aminotransferase isozymes by DL-glyceraldehyde 3-phosphate, *Eur. J. Biochem.* **20**:351–362.

Korey, S. R., 1957, A possible mechanism in phenylketonuria, in: *Ross Pediatric Conference* (S. J. Onesti, ed.), pp. 34–36, Ross Laboratories, Columbus, Ohio.

Kornberg, H. L., 1966, Anaplerotic sequences and their role in metabolism, *Essays Biochem.* **2**:1–31.

Krebs, H. A., and deGasquet, P., 1964, Inhibition of gluconeogenesis by alpha-oxoacids, *Biochem. J.* **90**:149–154.

LaDu, B., 1967, Genetic variation in metabolic disorders, in: *Amino Acid Metabolism and Genetic Variation* (W. L. Nyhan, ed.), pp. 121–130, McGraw-Hill, New York.

LaDu, B., and Zannoni, V. G., 1967, Inhibition of phenylalanine hydroxylase in liver, in: *Phenylketonuria and Allied Diseases* (J. A. Anderson and K. F. Swaiman, eds.), pp. 193–202, U.S. Government Printing Office, Washington, D.C.

Land, J. M., and Clark, J. B., 1973, Effect of phenylpyruvate on enzymes involved in fatty acid synthesis in rat brain, *Biochem. J.* **134**:545–555.

Land, J. M., and Clark, J. B., 1974, Inhibition of pyruvate and β-hydroxybutyrate oxidation in rat brain mitochondria by phenylpyruvate and α-ketoisocaproate, *FEBS Lett.* **44**:348–351.

Laster, L., Spaeth, G. L., Mudd, H. S., and Finkelstein, S. D., 1965, Homocystinuria due to cystathionine synthase deficiency, *Ann. Intern. Med.* **63**:1117–1142.

Levine, S. Z., Gordon, H. H., and Marples, E., 1941a, A defect in the metabolism of tyrosine in premature infants. II. Spontaneous occurrence and eradication by vitamin C, *J. Clin. Invest.* **20**:209–219.

Levine, S. Z., Marples, E., and Gordon, H. H., 1941b, A defect in the metabolism of tyrosine and phenylalanine in premature infants. I. Identification and assay of intermediary products, *J. Clin. Invest.* **20**:199–207.

Light, I. J., Berry, H. K., and Sutherland, J. M., 1966, Aminoacidemia of prematurity, *Amer. J. Dis. Child.* **112**:229–236.

Lin, E. C. C., and Knox, W. E., 1957, Adaptation of the rat liver tyrosine α-ketoglutarate transaminase, *Biochim. Biophys. Acta* **26**:85–88.

Lin, E. C. C., and Knox, W. E., 1958, Specificity of the adaptive response of tyrosine-α-ketoglutarate transaminase in the rat, *J. Biol. Chem.* **233**:1186–1189.

Lin, E. C. C., Pitt, B. M., Civen, M., and Knox, W. E., 1958, The assay of aromatic amino acid transaminations and keto acid oxidation by the enol borate-tautomerase method, *J. Biol. Chem.* **233**:668–673.

Linneweh, F., and Ehrlich, M., 1962, Zur Pathogenese des Schwachsinns bei Phenylketonurie, *Klin. Wochenschr.* **40**:225–226.

Linneweh, F., and Socher, H., 1965, Über den Einfluss diatetischer Prophylaxie auf die Myelogenese bei der Leucinose, *Klin. Wochenschr.* **43**:926–930.

Linneweh, F., Ehrlich, M., Graul, E. H., and Hundeshagen, H., 1963, Über den Aminosäuren-Transport bei phenylketonurischer Oligophrenie, *Klin. Wochensch.* **41**:253–255.

Lipton, M. A., Gordon, R., Guroff, G., and Udenfriend, S., 1967, *p*-Chlorophenylalanine-induced chemical manifestations of phenylketonuria in rats, *Science* **156**:248–250.

Lo, G. S., Lee, S., Cruz, N. L., and Longenecker, J. B., 1970, Temporary induction of phenylketonuria-like characteristics in infant rats: effect on brain protein synthesis, *Nutr. Rep. Int.* **2**:59–72.

Loo, Y. H., 1967, Characterization of a new phenylalanine metabolite in phenylketonuria, *J. Neurochem.* **14**:813–821.

Loo, Y. H., and Mack, K., 1972, Effect of hyperphenylalaninemia on vitamin B_6 metabolism in developing rat brain, *J. Neurochem.* **19**:2377–2383.

Loo, Y. H., and Ritman, P., 1967, Phenylketonuria and vitamin B_6 function, *Nature (London)* **213**:914–916.

Lovenberg, W., Jéquier, E., and Sjoerdsma, A., 1968, Tryptophan hydroxylation in mammalian systems, *Adv. Pharmacol.* **6A**:21–35.

Lowden, J. A., and LaRamee, M. A., 1969, Hyperphenylalaninemia: the effect on cerebral amino acid levels during development, *Can. J. Biochem.* **47**:883–888.

Mabry, C. C., Denniston, J. C., Nelson, T. L., and Son, C. D., 1963, Maternal phenylketonuria. A cause of mental retardation in children without the metabolic defect, *N. Engl. J. Med.* **269**:1404–1408.

MacInnes, J. W., and Schlesinger, K., 1971, Effects of excess phenylalanine on *in vitro* and *in vivo* RNA and protein synthesis and polyribosome levels in brains of mice, *Brain Res.* **29**:101–110.

Makulu, D. R., Smith, E. F., Bertino, J. R., 1973, Lack of dihydrofolate reductase activity in brain tissue of mammalian species: possible implications, *J. Neurochem.* **21**:241–245.

Malamud, N., 1966, Neuropathology of phenylketonuria, *J. Neuropathol. Exp. Neurol.* **25**:254–268.

Mao, C. C., Guidotti, A., and Costa, E., 1974, Interactions between γ-aminobutyric acid and guanosine cyclic $3',5'$-monophosphate in rat cerebellum, *Mol. Pharmacol.* **10**:736–745.

McGee, M. M., Greengard, O., and Knox, W. E., 1972, The quantitative determination of phenylalanine hydroxylase in rat tissues. Its developmental formation in liver, *Biochem. J.* **127**:669–674.

McGeer, E. G., and McGeer, P. L., 1973, Amino acid hydroxylase inhibitors, *Metab. Inhibitors* **4**:45–105.

McGeer, E. G., and Tischler, B., 1959, Vitamin B_6 and mental deficiency. The effects of large doses of B_6 (pyridoxine) in phenylketonuria, *Can. J. Biochem. Physiol.* **37**:485–491.

McIlwain, H., 1966, *Biochemistry and the Central Nervous System,* 3rd Ed., J & A Churchill, London.

McKean, C. M., 1972, The effects of high phenylalanine concentration on serotonin and catecholamine metabolism in human brain, *Brain Res.* 47:469–476.

McKean, C. M., and Peterson, N. A., 1970, Glutamine in the phenylketonuric central nervous system, *N. Engl. J. Med.* 283:1364–1367.

McKean, C. M., Schanberg, S. M., and Giarman, N. J., 1962, A mechanism of the indole defect in experimental phenylketonuria, *Science* 137:604–605.

McKean, C. M., Boggs, D. E., and Peterson, N. A., 1968, The influence of high phenylalanine and tyrosine on the concentrations of essential amino acids in brain, *J. Neurochem.* 15:235–241.

McLean, A., Marwich, M. J., and Clayton, B. E., 1973, Enzymes involved in phenylalanine metabolism in the human foetus and child, *J. Clin. Pathol.* 26:678–683.

Meister, A., 1965, *Biochemistry of the Amino Acids,* 2nd Ed., Vol. II, p. 907, Academic Press, New York.

Meister, A., Udenfriend, S., and Bessman, S. P., 1956, Diminished phenylketonuria in phenylpyruvic oligophrenia after administration of L-glutamine, L-glutamate or L-asparagine, *J. Clin. Invest.* 35:619–626.

Menkes, J. H., 1966, Cerebral lipids in phenylketonuria, *Pediatrics* 37:967–978.

Menkes, J. H., 1968, Cerebral proteolipids in phenylketonuria, *Neurology* 18:1003–1008.

Menkes, J. H., and Avery, M. E., 1963, The metabolism of phenylalanine and tyrosine in the premature infant, *Bull. Johns Hopkins Hosp.* 113:301–319.

Menkes, J. H., and Solcher, H., 1967, Maple syrup urine disease: effects of diet therapy on cerebral lipids, *Arch. Neurol.* 16:486–491.

Miller, R. K., and Berndt, W. O., 1974, Characterization of neutral amino acid accumulation by human term placental slices, *Amer. J. Physiol.* 227:1236–1242.

Miller, J. E., and Litwack, G., 1971, Purification, properties, and identity of liver mitochondrial tyrosine aminotransferase, *J. Biol. Chem.* 246:3234–3240.

Miller, A. L., Hawkins, R. A., and Veech, R. L., 1973, Phenylketonuria: phenylalanine inhibits brain pyruvate kinase *in vivo, Science* 179:904–906.

Milstien, S., and Kaufman, S., 1975a, Studies on the phenylalanine hydroxylase system in liver slices, *J. Biol. Chem.* 250:4777–4781.

Milstien, S., and Kaufman, S., 1975b, Studies on the phenylalanine hydroxylase system *in vivo.* An *in vivo* assay based on the liberation of deuterium or tritium into the body water from ring-labeled L-phenylalanine, *J. Biol. Chem.* 250:4782–4785.

Mitoma, C., 1956, Studies on partially purified phenylalanine hydroxylase, *Arch. Biochem. Biophys.* 60:476–484.

Mitoma, C., Auld, R. M., and Udenfriend, S., 1957, On the nature of enzymatic defect in phenylpyruvic oligophrenia, *Proc. Soc. Exp. Biol. Med.* 94:634–635.

Miyamoto, M., and Fitzpatrick, T. B., 1957, Competitive inhibition of mammalian tyrosinase by phenylalanine and its relationship to hair pigmentation in phenylketonuria, *Nature (London)* 179:199–200.

Morales, D. R., and Greenberg, D. M., 1964, Purification and properties of dihydrofolate reductase of sheep liver, *Biochim. Biophys. Acta* 85:360–376.

Moss, A. R., and Schoenheimer, R., 1940, The conversion of phenylalanine to tyrosine in normal rats, *J. Biol. Chem.* 135:415–429.

Munro, H. N., 1970, Free amino acid pools and their role in regulation, in: *Mammalian Protein Metabolism* (H. N. Munro, ed.), p. 299, Academic Press, New York.

Nadler, H. L., and Hsia, D. Y. Y., 1961, Epinephrine metabolism in phenylketonuria, *Proc. Soc, Exp. Biol. Med.* 107:721–722.

Neame, K. D., 1961, Phenylalanine as inhibitor of transport of amino-acids in brain, *Nature (London)* 192:173–174.

Nelson, W. E., 1959, *Textbook of Pediatrics,* 7th Ed., pp. 50–61, W. B. Saunders Co., Philadelphia.

Neubauer, O., 1909, Über den Abbau der Aminosäuren im gesunden and kranken Organismus, *Dtsch Arch. Klin. Med.* **95**:211–256.

Neubatier, O., and Falta, W., 1904, Über das Schicksal einiger aromatisher Saüren bei der Alkapturie, *Z. Physiol. Chem.* **42**:81–109.

Nielsen, K. H., 1969, Rat liver phenylalanine hydroxylase. A method for the measurement of activity with particular reference to the distinctive features of the enzyme and the pteridine cofactor, *Eur. J. Biochem.* **7**:360–369.

Nisselbaum, J. S., and Bodansky, O., 1964, Immunochemical and kinetic properties of anionic and cationic glutamic-oxaloacetic transaminase separated from human heart and human liver, *J. Biol. Chem.* **239**:4232–4236.

Nordyke, E. L., and Roach, M. K., 1974, Effect of hyperphenylalaninemia on amino acid metabolism and compartmentation in neonatal rat brain, *Brain Res.* **67**:479–488.

Oates, J. A., Nirenberg, P. Z., Jepson, J. B., Sjoerdsma, A., and Udenfriend, S., 1963, Conversion of phenylalanine to phenylethylamine in patients with phenylketonuria, *Proc. Soc, Exp. Biol. Med.* **112**:1078–1081.

Obata, K., and Takeda, K., 1969, Release of γ-aminobutyric acid into the fourth ventricle induced by stimulation of the cats cerebellum, *J. Neurochem.* **16**:1043–1047.

O'Brien, D., and Ibbot, F. A., 1966, Effect of prolonged phenylalanine loading on the free amino acid and lipid content of the infant monkey brain, *Dev. Med. Child. Neurol.* **8**:724–728.

Odessey, R., and Goldberg, A. L., 1972, Oxidation of leucine by rat skeletal muscle, *Amer. J. Physiol.* **223**:1376–1383.

Okuno, E., Minatogawa, Y., Noguchi, T., and Kido, R., 1975, Purification and characterization of rat liver mitochondrial phenylalanine pyruvate aminotransferase, *Life Sci.* **17**:211–218.

Oldendorf, W. H., 1971, Brain uptake of radiolabeled amino acids, amines, and hexoses after arterial injection, *Amer. J. Physiol.* **221**:1629–1639.

Oldendorf, W. H., 1973*a*, Stereospecificity of blood–brain barrier permeability to amino acids, *Amer. J. Physiol.* **224**:967–969.

Oldendorf, W. H., 1973*b*, Saturation of blood–brain barrier transport of amino acids in phenylketonuria, *Arch. Neurol.* **28**:45–48.

Oldendorf, W. H., Sisson, W. B., and Silverstein, A., 1971, Brain uptake of selenomethionine Se 75. II. Reduced brain uptake of selenomethionine Se 75 in phenylketonuria. *Arch. Neurol.* **24**:524–528.

Paine, R. A., 1957, The variability in manifestations of untreated patients with phenylketonuria (phenylpyruvic aciduria), *Pediatrics* **20**:290–301.

Pare, C. M., Sandler, M., and Stacey, R. S., 1957, 5-Hydroxytryptamine deficiency in phenylketonuria, *Lancet* **1**:551–553.

Pare, C. M. B., Sandler, M., and Stacey, R. S., 1958, Decreased 5-hydroxytryptophan decarboxylase activity in phenylketonuria, *Lancet* **2**:1099–1101.

Partington, M. W., 1962, Variations in intelligence in phenylketonuria, *Can. Med. Assoc. J.* **86**:736–743.

Partington, M. W., and Vickery, S. K., 1974, Phenylketonemia in phenylketonuria, *Neuropaediatrie* **5**:125–137.

Patel, M. S., 1972, The effect of phenylpyruvate on pyruvate metabolism in rat brain, *Biochem. J.* **128**:677–684.

Patel, M. S., and Arinze, I. J., 1975, Phenylketonuria: metabolic alterations induced by phenylalanine and phenylpyruvate, *Amer. J. Clin. Nutr.* **28**:183–188.

Patel, A. J., and Balazs, R., 1970, Manifestation of metabolic compartmentation during the maturation of rat brain, *J. Neurochem.* **17**:955–971.

Patel, M. S., Grover, W. D., and Auerbach, V. H., 1973, Pyruvate metabolism by homogenates of human brain: effects of phenylpyruvate and implications for the etiology of the mental retardation in phenylketonuria, *J. Neurochem.* **20**:289–296.

Penrose, L., and Quastel, J. H., 1937, Metabolic studies in phenylketonuria, *Biochem. J.* **31**:266–274.

Perry, T. L., Hansen, S., Tischler, B., and Hestrin, M., 1964, Defective 5-hydroxylation of tryptophan in phenylketonuria, *Proc. Soc, Exp. Biol. Med.* **115**:118–123.

Perry, T. L., Hansen, S., Tischler, B., and Bunting, R., 1967a, Determination of heterozygosity for phenylketonuria on the amino acid analyzer, *Clin. Chim. Acta* **18**:51–56.

Perry, T. L., Tischler, B., Hansen, S., and MacDougall, L., 1967b, A simple test for heterozygosity for phenylketonuria, *Clin. Chim. Acta* **15**:47–50.

Perry, T. L., Hansen, S., Tischler, B., Bunting, R., and Diamond, S., 1970, Glutamine depletion in phenylketonuria, *N. Engl. J. Med.* **282**:761–766.

Perry, T. L., Sander, H. D., Hansen, S., Lesk, D., Klaster, M., and Gravlin, L., 1972, Free amino acids and related compounds in five regions of biopsied cat brain, *J. Neurochem.* **19**:2651–2656.

Pirrung, J., Gottesman, L., and Crandall, D. I., 1957, The metabolism of *p*-methoxyphenylalanine and *p*-methoxyphenylpyruvate, *J. Biol. Chem.* **229**:199–210.

Poser, C. M., and Van Bogaert, L., 1959, Neuropathologic observations in phenylketonuria, *Brain* **82**:1–9.

Prensky, A. L., and Moser, H. W., 1966, Brain lipids, proteolipids and free amino acids in maple syrup urine disease, *J. Neurochem.* **13**:863–874.

Prensky, A. L., Carr, S., and Moser, H. W., 1968, Development of myelin in inherited disorders of amino acid metabolism. A biochemical investigation, *Arch. Neurol.* **19**:552–558.

Rabinowitz, M., Olson, M. E., and Greenberg, D. M., 1954, Independent antagonism of amino acid incorporation into protein, *J. Biol. Chem.* **210**:837–849.

Raiha, N. C. R., 1973, Phenylalanine hydroxylase in human liver during development, *Pediatr. Res.* **7**:1–4.

Rampini, S., 1973, Die kongenitalen Störungen des Phenylalaninstoffwechsels, *Schweiz. Med. Wochenschr.* **103**:537–546.

Rampini. S., Anders, P. W., Curtius, H. C., and Marthaler, T., 1969, Detection of heterozygotes for· phenylketonuria by column chromatography and discriminatory analysis, *Pediatr. Res.* **3**:287–297.

Rauch, H., and Yost, M. T., 1963, Phenylalanine metabolism in dilute-lethal mice, *Genetics* **48**:1487–1495.

Rees, K. R., 1955, cited in Bickel, H., Boscott, R. J., and Gerrard, J., Observations on the biochemical error in phenylketonuria and its dietary control, in: *Biochemistry of the Developing Nervous System* (H. Waelsch, ed.), p. 417, Academic Press, New York.

Rembold, H., 1964, Untersuchungen über den Stoffwechsel des Biopterins und über die polarographische Charakterisierung von Pteridinen, in: *Pteridine Chemistry* (W. Pfleiderer and E. C. Taylor, eds.), pp. 465–483, Pergamon Press, Oxford.

Rembold, H., and Buff, K., 1972, Tetrahydrobiopterin, a cofactor in mitochondrial electron transfer, *Eur. J. Biochem.* **28**:579–585.

Renson, J., Weissbach, H., and Udenfriend, S., 1962, Hydroxylation of tryptophan by phenylalanine hydroxylase, *J. Biol. Chem.* **237**:2261–2264.

Rey, F., Pellie, C., Sivy, M., Blandin-Savoja, F., Rey, J., and Frezal, J., 1974, Influence of age on *ortho*-hydroxyphenylacetic acid excretion in phenylketonuria and its genetic variants, *Pediatr. Res.* **8**:540–545.

Roberts, S., and Morelos, B. S., 1965, Regulation of cerebral metabolism of amino acids. IV. Influence of amino acid levels on leucine uptake, utilization and incorporation into protein *in vivo*, *J. Neurochem.* **12**:373–387.

Roberts, S., and Morelos, B. S., 1976, Role of ribonuclease action in phenylalanine-induced disaggregation of rat cerebral polyribosomes, *J. Neurochem.* **26**:387–400.

Rosenblatt, D., and Scriver, C. R., 1968, Heterogeneity in genetic control of phenylalanine metabolism in man, *Nature (London)* **218**:677–678.

Ryan, W. L., and Carver, M. J., 1966, Free amino acids of human foetal and adult liver, *Nature (London)* **212**:292–293.

Saugstad, L. F., 1972, Birth weights in children with phenylketonuria and in their siblings, *Lancet* **1**:809–813.

Sauberlich, H. E., 1961, Studies on the toxicity and antagonism of amino acids for weanling rats, *J. Nutr.* **75**:61–72.

Schreier, K., and Flaig, H., 1956, Urinary excretion of indole-pyruvic acid in normal conditions and in Fölling's disease, *Klin. Wochenschr.* **34**:1213.

Scrimgeour, K. G., and Cheema, S., 1971, Quinonoid dihydropterin reductase, *Ann. N.Y. Acad. Sci.* **186**:115–118.

Shah, S. N., Peterson, N. A., and McKean, C. M., 1969, Inhibition of sterol synthesis *in vitro* by metabolites of phenylalanine, *Biochim. Biophys. Acta* **187**:236–242.

Shah, S. N., Peterson, N. A., and McKean, C. M., 1970, Cerebral lipid metabolism in experimental hyperphenylalaninemia: incorporation of ^{14}C-labeled glucose into total lipids, *J. Neurochem.* **17**:279–284.

Shah, S. N., Peterson, N. A., and McKean, C. M., 1972a, Impaired myelin formation in experimental hyperphenylalaninemia, *J. Neurochem.* **19**:479–485.

Shah, S. N., Peterson, N. A., and McKean, C. M., 1972b, Lipid composition of human cerebral white matter and myelin in phenylketonuria, *J. Neurochem.* **19**:2369–2376.

Sherwin, C. P., and Kennard, K. S., 1919, Toxicity of phenylacetic acid, *J. Biol. Chem.* **40**:259–264.

Shiman, R., Akino, M., and Kaufman, S., 1971, Solubilization and partial purification of tyrosine hydroxylase from bovine adrenal medulla, *J. Biol. Chem.* **246**:1330–1340.

Siegel, F. L., Aoki, K., and Colwell, R. E., 1971, Polyribosome disaggregation and cell-free protein synthesis in preparations from cerebral cortex of hyperphenylalaninemic rats, *J. Neurochem.* **18**:537–547.

Smith, M. E., 1974, Labeling of lipids by radioactive amino acids in the central nervous system, *J. Neurochem.* **23**:435–438.

Smith, I., and Wolff, O. H., 1974, Natural history of phenylketonuria and influence of early treatment, *Lancet* **2**:540–544.

Smith, I., Clayton, B. E., and Wolff, O., 1975a, A variant of phenylketonuria, *Lancet* **1**:328–329.

Smith, I., Clayton, B. E., and Wolff, O. H., 1975b, A new variant of phenylketonuria with progressive neurological illness unresponsive to phenylalanine restriction, *Lancet* **1**:1108–1111.

Snyderman, S. E., Norton, P., and Holt, L. E., Jr., 1955, Effect of tyrosine administration in phenylketonuria, *Fed. Proc. Fed. Amer. Soc. Exp. Biol.* **14**:450–451.

Sokoloff, L., 1960, Metabolism of the central nervous system *in vivo*, in: *Handbook of Physiology–Neurophysiology* (J. Field, H. W. Magoun, and V. E. Hall, eds.), pp. 1843–1864, American Physiological Society, Washington, D.C.

Spydervold, O. S., Zaheer-Baquer, N., McLean, P., and Greenbaum, A. L., 1974, The effect of quinolinic acid on the content and distribution of hepatic metabolites, *Arch. Biochem. Biophys.* **164**:590–601.

Stave, U., and Armstrong, M. D., 1973, Tissue free amino acid concentrations in perinatal rabbits, *Biol. Neonate* **22**:374–387.

Stein, W. H., and Moore, S., 1954, The free amino of human blood and plasma, *J. Biol. Chem.* **211**:915–926.

Stein, W. H., Bearn, A. G., and Moore, S., 1954, The amino acid content of the blood and urine in Wilson's disease, *J. Clin. Invest.* **33**:410–419.

Stephenson, J. B. P., and McBean, M. S., 1967, Diagnosis of phenylketonuria (phenylalanine hydroxylase deficiency, temporary and permanent), *Br. Med. J.* **3**:579–581.

Storm, C. B., and Kaufman, S., 1968, The effect of variation of cofactor and substrate structure on the action of phenylalanine hydroxylase, *Biochem. Biophys. Res. Commun.* 32:788–793.

Swaiman, K. F., Hosfield, W. B., and Lemieux, B., 1968, Elevated plasma phenylalanine concentration and lysine incorporation into ribosomal protein of developing brain, *J. Neurochem* 15:687–690.

Taniguchi, K., and Armstrong, M. D., 1963, The enzymatic formation of *o*-hydroxyphenylacetic acid, *J. Biol. Chem.* 238:4091–4097.

Taniguchi, K., Kappe, T., and Armstrong, M. D., 1964, Further studies on phenylpyruvate oxidase. Occurrence of side chain rearrangement and comparison with *p*-hydroxyphenylpyruvate oxidase, *J. Biol. Chem.* 239:3389–3395.

Tashian, R. E., 1961, Inhibition of brain glutamic acid decarboxylase by phenylalanine, valine, and leucine derivatives: a suggestion concerning the etiology of the neurological defect in phenylketonuria and branched-chain ketonuria, *Metabolism* 10:393–402.

Tischler, B., and McGeer, E. G., 1958, Vitamin B_6 in mental deficiency: xanthurenic acid excretion in phenylketonurics, *Can. Med. Assoc. J.* 78:954–955.

Tong, J. H., and Kaufman, S., 1975, Tryptophan hydroxylase: purification and some properties of the enzyme from rabbit hindbrain, *J. Biol. Chem.* 250:4152–4158.

Tourian, A., Goddard, J., and Puck, T. T., 1969, Phenylalanine hydroxylase activity in mammalian cells, *J. Cell. Physiol.* 73:159–170.

Treiman, D. M., and Tourian, A., 1973, Phenylalanine hydroxylase in dilute lethal mice, *Biochim. Biophys. Acta* 313:163–169.

Udenfriend, S., 1966, Tyrosine hydroxylase, *Pharmacol. Rev.* 18:43–51.

Udenfriend, S., 1967, The primary enzymatic defect in phenylketonuria and how it may influence the central nervous system, in: *Phenylketonuria and Allied Metabolic Diseases* (J. A. Anderson, and K. F. Swaiman, eds.), pp. 1–8, U.S. Dept of Health, Education and Welfare, Washington, D.C.

Udenfriend, S., and Bessman, S. P., 1953, The hydroxylation of phenylalanine and antipyrine in phenylpyruvic oligophrenia, *J. Biol. Chem.* 203:961–966.

Udenfriend, S., and Cooper, J. R., 1952, The enzymatic conversion of phenylalanine to tyrosine, *J. Biol. Chem.* 194:503–511.

Umezawa, K., Sakamoto, Y., and Ichihara, K., 1959, The metabolism of *p*-methylphenylalanine and *p*-methoxyphenylalanine, *J. Biochem.* 46:941–944.

Vandeman, P. R., 1963, Termination of dietary treatment for phenylketonuria, *Amer. J. Dis. Child.* 100:492–495.

Van den Berg, C. J., 1970, Compartmentation of glutamate metabolism in the developing brain: experiments with labelled glucose, acetate, phenylalanine, tyrosine and proline, *J. Neurochem.* 17:973–983.

Van den Berg, C. J., 1971, Utilization of substrates for energy production by the growing brain, *Psychiatr. Neurol. Neurochir.* 74:427–431.

Vollmin, J. A., Bosshard, H. R., Muller, M., Rampini, S., and Curtius, H. C., 1971, Determination of urinary aromatic acids by gas chromatography, *Z. Klin. Chem. Klin. Biochem.* 9:402–404.

Volpe, J. J., Lyles, T. O., Roncari, D. A. K., and Vagelos, P. R., 1973, Fatty acid synthetase of developing brain and liver. Content, synthesis and degradation during development, *J. Biol. Chem.* 248:2502–2513.

Wadman, S. K., Van der Heiden, C., and Van Sprang, F. J., 1971a, Abnormal tyrosine and phenylalanine metabolism in patients with tyrosyluria and phenylketonuria; gas-liquid chromatographic analysis of urinary metabolites, *Clin. Chim. Acta* 34:277–287.

Wadman, S. K., Van Sprang, F. J., Van der Heiden, C., and Ketting, D., 1971b, Quantitation of urinary phenylalanine metabolites in phenylketonuria (PKU), in: *Phenylketonuria and Some Other Inborn Errors of Amino Acid Metabolism* (H. Bickel, F. B. Hudson, and L. I.

Woolf, eds.), pp. 65–72, Georg Thieme Verlag, Stuttgart.

Wallace, H. W., Moldave, K., and Mesiter, A., 1957, Studies on conversion of phenylalanine to tyrosine in phenylpyruvic oligophrenia, *Proc. Soc. Exp. Biol. Med.* **94**:632–633.

Wapnir, R. A., Hawkins, R. L., and Stevenson, J. H., 1971, Ontogenesis of phenylalanine and tryptophan hydroxylation in rat brain and liver, *Biol. Neonate* **118**:85.

Watt, D. D., and Martin, P. R., 1969, Phenylalanine antimetabolite effect on development. I. Behavioral effects of D, L-4-chlorophenylalanine in the young rat, *Life Sci.* **8**:1211–1222.

Weber, G., Glazer, R. I., and Ross, R. A., 1970, Regulation of human and rat brain metabolism: inhibitory action of phenylalanine and phenylpyruvate on glycolysis, protein, lipid, DNA and RNA metabolism, in: *Advances in Enzyme Regulation* (G. Weber, ed.), pp. 13–36, Pergamon Press, New York.

Weil-Malherbe, H., 1955, Blood adrenaline and intelligence, in: *Biochemistry of the Developing Nervous System* (H. Waelsch, ed.) pp. 458–465, Academic Press, New York.

Williamson, J. R., and Corkey, B. E., 1969, Assay of intermediates of the citric acid cycle and related compounds by fluorometric methods, in: *Methods in Enzymology* (J. M. Lowenstein, ed.), p. 445, Academic Press, New York.

Wood, J. L., and Cooley, S. L., 1954, Substitution of α-keto acids for five amino acids essential for growth of the rat, *Proc. Soc. Exp. Biol. Med.* **85**:409–411.

Woods, M. N., and McCormick, D. B., 1964, Effects of dietary phenylalanine on activity of phenylalanine hydroxylase from rat liver, *Proc. Soc. Exp. Biol. Med.* **116**:427–430.

Woolley, D. W., and Van der Hoeven, T., 1965, Serotonin deficiency in infancy as a cause of a mental defect in experimental phenylketonuria, *Int. J. Neuropsychiatry* **1**:529–544.

Woolf, L. I., 1951, Excretion of conjugated phenylacetic acid in phenylketonuria, *Biochem. J.* **49**:ix–x.

Woolf, L. I., and Vulliamy, D. G., 1951, Phenylketonuria with a study of the effect upon it of glutamic acid, *Arch. Dis. Childhood* **26**:487–494.

Woolf, L. I., Griffiths, R., and Moncrieff, A., 1955, Treatment of phenylketonuria with a diet low in phenylalanine, *Br. J. Med.* **1**:57–64.

Woolf, L. I., Cranston, W. I., and Goodwin, B. L., 1967, Genetics of phenylketonuria, *Nature (London)* **213**:882–885.

Woolf, L. I., Goodwin, B. L., Cranston, W. I., Wade, D. N., Woolf, F., Hudson, F. B., and McBean, M. S., 1968, A third allele at the phenylalanine-hydroxylase locus in mild phenylketonuria, *Lancet* **1**:114–117.

Woolf, L. I., Jakubovic, A., Woolf, F., and Bory, P., 1970, Metabolism of phenylalanine in mice homozygous for the gene "dilute lethal," *Biochem. J.* **119**:895–903.

Yuwiler, A., and Louttit, R. T., 1961, Effects of phenylalanine diet on brain serotonin in the rat, *Science* **134**:831–832.

Zachmann, M., Cleveland, W. W., Sandberg, D. H., and Nyhan, W. L., 1966, Concentrations of amino acids in plasma and muscle, *Amer. J. Dis. Child.* **112**:283–289.

Zamecnik, P. C., and Keller, E. B., 1954, Relationship between phosphate energy donors and incorporation of labeled amino acids into proteins, *J. Biol. Chem.* **209**:337–353.

Zannoni, V. G., and Moraru, E., 1969, Mechanism of phenylalanine hydroxylation and "phenylketonuria" in dilute-lethal mice, *FEBS Symp.* **19**:347–354.

Zannoni, V. G., Weber, W. W., van Valen, P., Rubin, A., Berstein, R., and LaDu, B. N., 1966, Phenylalanine metabolism and "phenylketonuria" in dilute-lethal mice, *Genetics* **54**:1391–1399.

Zelnicek, E., and Slama, J., 1971, Phenylpyruvate and o-hydroxyphenylacetate in phenylketonuric urine, *Clin. Chim. Acta.* **35**:496–497.

TRYPTOPHAN IN THE CENTRAL NERVOUS SYSTEM: Regulation and Significance

S. N. YOUNG AND T. L. SOURKES

Laboratory of Neurochemistry
Department of Psychiatry
McGill University
Montreal, Quebec, Canada

1. INTRODUCTION

Tryptophan is an indispensable amino acid in animal nutrition. Without it young animals cease to grow, and adult animals can no longer maintain nitrogen equilibrium. Hence, dietary intake is ultimately the starting point for any consideration of the regulation of the content of this amino acid in the CNS. Dietary deficiency of tryptophan results quickly in a decrease in the content of this amino acid in brain (Gessa *et al.*, 1974; Dickerson and Pao, 1975). This depletion also accounts for the known decrease in tissue 5-hydroxytryptamine (5-HT, serotonin), including that of the brain (Eber and Lembeck, 1958; Gal and Drewes, 1962; Culley *et al.*, 1963). Some species (e.g., pig and rat) can use D-tryptophan for part of their nutritional requirement (Baker *et al.*, 1971) presumably converting it by way of oxidative deamination followed by asymmetric transamination to the L-amino acid. The rat can use the α-keto analogue of tryptophan for growth, but this compound is not as effective as the amino acid itself (Pond *et al.*, 1964).

That portion of the dietary tryptophan that is liberated in the intestinal tract by proteolysis, and which is not drawn off by the intestinal flora for baterial protein synthesis or catabolized by tryptophanase, is absorbed and carried to the liver in the hepatic portal blood. The transport of tryptophan across the mucosal membrane is an active process occurring against a concentration gradient, as demonstrated by Spencer and Samiy (1960) and Cohen and Huang (1964).

In the liver, tryptophan is metabolized to protein or is catabolized primarily by tryptophan pyrrolase (L-tryptophan-2,3-dioxygenase, EC 1.13.1.12). Both processes are readily influenced by physiological changes and by certain drugs. Although incorporation of tryptophan into protein in the liver and other tissues is quantitatively the most important pathway for this amino acid, much more is known about the factors involved in the very sensitive regulation of tryptophan pyrrolase than for protein synthesis. Tryptophan contained in proteins is eventually released during protein catabolism. In contrast to this process, metabolism to N'-formylkynurenine is an irreversible pathway. The fluxes through these two pathways in the liver interact, and the factor that affects both of them is the content of tryptophan in the liver. This is itself influenced by the flux of the amino acid into and from the liver, thus adding to the complexity of factors in the periphery that make tryptophan available to the blood and, hence, to the brain. The concentration of tryptophan in venous blood plasma of four species is shown in Table 1. The level is of the order of 30–70 μmol/liter in the fasting animal.

Tryptophan is carried in the plasma largely bound in a loose form at sites on serum albumin; these sites may be shared with free fatty acids and

TABLE 1. Tryptophan Concentration of Blood Plasma (μmol/liter)

| Species | Tryptophan | | Number of determinations | References |
	Total	NAB[d]		
Man	67.6 ± 4.9[a]	11.3 ± 1.5	29	Young *et al.* (1975)
Dog	48.9 ± 3.6		17	Geddes *et al.* (1973)
Pig	34.3 ± 2.9	4.4 ± 0.6	8	Curzon *et al.* (1973)
Rat	78.4 ± 0.1	9.3 ± 1.3	10	Moir (1974)
Rat (fed)[b]	138.1 ± 0.9	13.7 ± 0.5	20	Gessa and Tagliamonte (1974)
Rat (fasted)[c]	88.2 ± 0.8	22.5 ± 0.6	20	Gessa and Tagliamonte (1974)

[a] Mean ± S.E.
[b] Allowed food for 2 hr before being killed.
[c] Fasted for 24 hr.
[d] Non-albumin-bound.

certain drugs. It is the only amino acid that is transported in this manner, and considerable research has been carried out recently to determine whether it is the free or albumin-bound tryptophan that determines delivery of the amino acid to the tissues.

It has become clear in recent years that both the content of tryptophan in the CNS and the concentration of tryptophan in the fluid that bathes it varies sensitively with conditions at the periphery. A discussion of this subject is a major goal of the present chapter.

In a normal animal, the content of tryptophan does not vary much among the various regions of the CNS (Table 2). This mean value is somewhat higher than in the CSF for the same species (Table 3). Utilization of tryptophan in the CNS, as in the liver, involves a reversible component, i.e., synthesis into protein, and several irreversible pathways, of which the most important functionally is the formation of 5HT.

TABLE 2. Tryptophan in the Central Nervous System

Species	Region	Tryptophan content (nmol/g fresh weight)	Number of determinations	References
Man	N. caudatus	59.3 ± 25.5	9	Birkmayer et al. (1974)
	Putamen	79.3 ± 16.2	9	
	Pallidum	75.9 ± 15.2	9	
	S. nigra oral	80.3 ± 21.1	9	
	S. nigra caudal	120.5 ± 10.3	9	
	N. amygdalae	81.3 ± 15.2	9	
	Gyri cinguli	74.9 ± 19.6	9	
	Raphe	79.3 ± 27.9	9	
	N. ruber	109.7 ± 15.2	9	
Dog	N. caudatus	25.0 ± 3.4	5	Eccleston et al. (1968)
	Hypothalamus	37.2 ± 4.4	5	
	Thalamus	38.2 ± 3.9		
Pig	N. caudatus	20.6 ± 1.5	8	Curzon et al. (1973)
	Hypothalamus	23.0 ± 1.5	8	
	Thalamus	18.6 ± 1.0	8	
	Cortex	20.1 ± 1.47	8	
Rat	Brain	27.9 ± 0.3[a]	20	Gessa and Tagliamonte
	Brain	17.6 ± 0.2[b]	20	(1974)
Rat	N. caudatus	11.3	6	Knott and Curzon (1974)
	Hypothalamus	23.0	6	
	Cortex	8.8	6	
	Midbrain	8.3	6	
	Hindbrain	8.8	6	
Rat	Raphe nuclei	22.5 ± 2.0	7	Deguchi et al. (1973)
	Rest of brain stem	8.3 ± 0.5	6	

[a] Animals were fasted for 24 hr.
[b] Animals were fed during 2 hr before being killed.

TABLE 3. Tryptophan Concentration of CSF (nmol/ml)

| Species | Ventricular | CSF compartments | | | References |
		Cisternal	Mixed[a]	Lumbar	
Man	4.2 ± 0.3[b] (8)	3.4 ± 0.6 (5)	2.3 ± 0.2 (14)	2.5 ± 0.1 (57)	Young *et al.* (1974a)
Man			2.2 ± 0.3 (3)		Eccleston *et al.* (1970)
Man				2.6 ± 0.8 (26)	Ashcroft *et al.* (1973a)
Dog	5.4 ± 0.4 (19)	5.4 ± 0.5 (21)			Geddes *et al.* (1973)
Rat (fed)		2.1 ± 0.2 (7)			Young *et al.* (1976a)
Rat (starved)[c]		2.5 ± 0.1 (8)			
Rat		3.4 (9)			Modigh (1975)
Rat		1.0 ± 0.2 (7)			Moir (1974)

[a] The CSF obtained after injection of air in pneumoencephalography.
[b] Mean ± S.E. (number of determinations in parentheses).
[c] 18 hr without food.

The content of tryptophan in brain depends not only on the availability of tryptophan in the blood but also on the uptake of this amino acid at the blood–brain barrier and at the cell membrane. The specific processes whereby tryptophan crosses the membranes in the CNS have been under careful study in recent years. Transport into brain has been studied at the *in vitro* level with both slices and synaptosomes, and at the *in vivo* level through animal and clinical investigation. Efflux processes occur, but they are less well understood. This efflux presumably takes place from brain to blood, from spinal cord to blood, from CNS to CSF, and from CSF to blood. The origin of tryptophan in CSF, whether from adjacent nervous tissue or from the blood, is also not clear.

2. PERIPHERAL METABOLISM OF TRYPTOPHAN

An adult mammal in nitrogen balance maintains a corresponding balance with respect to all the essential amino acids, including tryptophan. However, over short periods of time the content of free tryptophan in an animal will vary depending on the rates of dietary intake, excretion, and metabolism of tryptophan. The content of tryptophan in brain also depends on the distribution of the free tryptophan within the animal. Thus, uptake of tryptophan by peripheral tissues will play a role in regulating brain tryptophan.

After a protein meal, the concentrations of amino acids in portal plasma are elevated but the increases are much smaller in venous plasma. This is due to uptake of amino acids by the liver and other tissues. The liver disposes of these amino acids fairly rapidly by catabolism and by protein synthesis, while much larger amounts of the amino acids are held for longer periods of time in the free form in muscle (Christensen, 1964). These processes, which tryptophan has in common with the other amino acids, have not been investigated in relation to the brain but obviously must be involved in the regulation of the content of cerebral amino acids.

Tryptophan seems to play an important role in the incorporation of amino acids into protein since it is the least abundant amino acid in the pool available for protein synthesis (Munro, 1968). Perhaps because of this an increase in the content of tryptophan in liver causes an increase in ribosome aggregation (Sidransky et al., 1971). This effect presumably speeds the incorporation of free tryptophan into protein after a meal. Because of the small pool size of free tryptophan, a diet deficient in this amino acid will result in sufficient incorporation of tryptophan into protein in the liver and other tissues to deplete free tryptophan stores. Thus, when rats that have been trained to eat their normal daily meal in a period of 2 hr are given a tryptophan-deficient diet, there is a marked fall in serum tryptophan (Biggio et al., 1974; Gessa et al., 1974). In addition, in brain, tryptophan and 5-HT decline 85% and 58%, respectively. This effect persists for more than 24 hr. Other factors that stimulate protein synthesis might also be expected to decrease brain tryptophan; thus, it is of interest that administration of growth hormone to hypophysectomized rats decreases tryptophan and 5-hydroxyindoleacetic acid (5-HIAA) in brain (Cocchi et al., 1975).

Although tryptophan incorporation into protein can decrease brain tryptophan, this effect is only temporary. When the protein is broken down, the component amino acids again become available. This is illustrated by the situation in starvation, when brain tryptophan is elevated (Curzon et al., 1972; Perez-Cruet et al., 1972). Measurement of tryptophan in the portal vein, hepatic vein, and abdominal aorta of starved rats has shown that there is a net release from the liver and the portal-drained tissues (Bloxam et al., 1974); release from the liver is greater. There is a fall in the tryptophan content of the liver accompanying the output from that organ; this is probably caused by altered transport at the cell membrane. Thus in this situation, the extra tryptophan in the brain is supplied by breakdown of protein in the liver and other tissues (Table 4).

Except for protein synthesis, the main metabolic routes of tryptophan are irreversible. Like most amino acids, tryptophan is catabolized to

TABLE 4. Changes in the Level of Tryptophan of Rat Plasma, Liver, and Brain after Starvation for 24 Hours[a]

	Number of rats	Fed	Starved	Change (%)
Hepatic vein	16	119.0 ± 3.8	117.2 ± 3.5	−2
Portal vein	16	113.7 ± 3.1	101.8 ± 2.9	−10
Hepatic vein–portal vein[b]	16	5.3 ± 2.3	15.3 ± 1.9	+189
Liver	8	44.3 ± 1.2	33.2 ± 0.9	−25
Brain	8	16.3 ± 0.8	21.2 ± 1.0	+30

[a] Values in nmol/ml for plasma and nmol/g for tissue are given as mean ± S.E. Data are taken from Bloxam *et al.* (1974).
[b] This represents the tryptophan released by the liver.

tricarboxylic acid intermediates, but unlike most of them tryptophan can be converted to other functionally important compounds as well. Quantitatively, the most important pathway of tryptophan catabolism is the "kynurenine" pathway. This pathway leads to the formation of NAD. In man, approximately 1 mg niacin is formed for each 60 mg of tryptophan in the diet (Horwitt *et al.*, 1956), but the flux through the kynurenine pathway must be much larger than this since many other metabolites of this pathway are found in the urine (Leklem, 1971).

The first enzyme in the kynurenine pathway is tryptophan pyrrolase, which is found primarily in the liver. Tryptophan pyrrolase is induced by both glucocorticoids and tryptophan (Knox, 1966). It has been suggested that this induction does not lead to an increase in tryptophan breakdown, as would be expected from elevated enzyme levels, but that changes in the content of tryptophan in liver do alter the rate of tryptophan catabolism, even under circumstances where the enzyme activity is unchanged (Kim and Miller, 1969). There are indeed circumstances in which the activity of tryptophan pyrrolase, measured *in vitro,* does not reflect the breakdown of tryptophan by tryptophan pyrrolase *in vivo* (Young *et al.,* 1974b). However, tryptophan breakdown *in vivo* does increase after cortisol administration (Young and Sourkes, 1975). This result has been demonstrated by following tryptophan breakdown as the rate of expiration of $^{14}CO_2$ after injection of rats with [ring-2-^{14}C]tryptophan. The tryptophan pyrrolase reaction is the rate-limiting step in this process (Young *et al.,* 1974b). The increase in tryptophan catabolism after cortisol is not as great as the increase in tryptophan pyrrolase activity detected *in vitro.* This is due to a decline in the content of tryptophan in liver probably brought about by its high rate of catabolism, which will decrease the degree of saturation of the enzyme with its substrate (Young and Sourkes, 1975). These results indicate that increases in enzyme activity definitely will increase the rate of tryptophan breakdown, although saturation of the enzyme with its substrate is also an important factor.

As induction of tryptophan pyrrolase can increase tryptophan catabolism, there exists the possibility that it might deplete the pool of free tryptophan sufficiently to decrease levels of this amino acid in brain. This problem has been studied mainly for effects on brain 5-HT, and the measurement of tryptophan has often been omitted. However, it seems likely that effects on brain 5-HT would be mediated through changes in the content of brain tryptophan (see Section 4). Brain 5-HT levels decline after induction of rat tryptophan pyrrolase by either hydrocortisone (Green and Curzon, 1968; Yuwiler *et al.*, 1971) or corticosterone (Scapagnini *et al.*, 1969). The content of brain tryptophan also decreases under these conditions (Green *et al.*, 1975a). Some investigators have not observed a decline in the content of brain 5-HT (Benkert and Matussek, 1970; Hillier *et al.*, 1975), but this discrepancy may be partly explained by the different dose and time used in the second study (Green *et al.*, 1975b).

A much larger and longer-lasting induction of tryptophan pyrrolase is obtained with the substrate analogue DL-α-methyltryptophan (αMeTrp). Injection of αMeTrp into rats causes a severe depletion of tryptophan in the liver, blood, and brain (Sourkes *et al.*, 1970) and 5-HT in the brain, although the presence of some α-methyl 5-HT gives artificially high apparent values of 5-HT (Roberge *et al.*, 1972).

There is further evidence that is consistent with the idea that the decrease in the levels of brain tryptophan and 5-HT are mediated by the rise in tryptophan pyrrolase. In the Mongolian gerbil *(Meriones unguiculatus)* tryptophan pyrrolase is induced by αMeTrp but not by cortisol. Whereas αMeTrp causes a decrease of the content of tryptophan and 5-HT in gerbil brain, cortisol does not affect them (Green *et al.*, 1975a). In rats over a certain age (about 100 days), hydrocortisone also no longer induces tryptophan pyrrolase nor does it lower brain 5-HT and 5-HIAA (Green and Curzon, 1975). It is interesting that an injection of cortisol into neonatal rats results in premature induction of tryptophan pyrrolase, with a concomitant reduction of the content of 5-HT in the brain (Yuwiler and Geller, 1973). These data, taken together, provide strong evidence that an increase in the rate of tryptophan catabolism in the liver alters tryptophan metabolism in the brain.

This system has also been studied by applying the stress of immobilization to rats in order to induce tryptophan pyrrolase through release of endogenous corticosterone. But this situation is more complex. Immobilization of rats for 3 hr, too short a time for changes in the brain to be effected through the tryptophan pyrrolase mechanism, causes a slight increase of the content of brain tryptophan and 5-HIAA, even in adrenalectomized animals. The content of 5-HT remains unchanged under these circumstances, but after 5 hr of immobilization, levels of 5HT decline (Curzon *et al.*, 1972). This latter effect may be partially mediated by a rise in pyrrolase, as 5-HT fell less in adrenalectomized rats (Curzon and Green, 1969). The increase of the content of brain 5-HIAA after 5 hr of immobilization may be related to increased firing of 5-HT neurons (Aghajanian *et al.*, 1967).

Any effect that tryptophan pyrrolase induction has on the content of brain tryptophan may be modified by the metabolic status of the animal. It has been shown that administration of glucose before glucocorticoids diminishes both the increase in activity of tryptophan pyrrolase and the fall of brain 5-HT (Yuwiler *et al.*, 1971).

The decline in the content of brain tryptophan and 5-HT after induction of tryptophan pyrrolase in liver is probably due, at least partly, to diversion of tryptophan away from the brain into the kynurenine pathway. However, this may not be the only mechanism. If increased catabolism of tryptophan in the liver can divert tryptophan away from the brain, it might be expected that repeated induction of tryptophan pyrrolase would lower the content of brain 5-HT even further. However, when rats were injected with cortisone daily for 5 days, the levels of brain 5-HT were unchanged 3 hr after the last injection (Shah *et al.*, 1968). However, even under these circumstances, induction of pyrrolase did prevent the increase in the content of brain 5-HT after a tryptophan load. It is likely that the increase in 5-HT is blocked by an increase in tryptophan catabolism, since tryptophan loads cause a very large increase in tryptophan catabolism in the liver even after the independent induction of tryptophan pyrrolase (Young and Sourkes, 1975). In this situation, the liver is protecting the brain against the tryptophan load. However, there is another mechanism that may be operating in situations where no tryptophan load is given.

When kynurenine, 3-hydroxykynurenine, and 3-hydroxyanthranilic acid (the first major metabolites of tryptophan along the kynurenine pathway) are injected into rats in quite small doses, there is a fall in brain 5-HT (Green and Curzon, 1970). Although kynurenine and 3-hydroxy-kynurenine (at 1 mM concentration) inhibit tryptophan transport into brain slices (Green and Curzon, 1970; Kiely and Sourkes, 1972), it is unlikely that this effect is important at physiological concentrations of these substances. 3-Hydroxyanthranilic acid, which also causes the lowering of brain 5-HT levels, does not inhibit tryptophan transport. The mechanism of action of these tryptophan metabolites on tryptophan metabolism in brain is unknown. However, when tryptophan pyrrolase is induced, it is possible that the increased concentration of metabolites along the kynurenine pathway is partly responsible for the decline in brain tryptophan content.

Another factor to be considered in interpreting these results is the possible interaction between the kynurenine pathway and protein synthesis. Induction of tryptophan pyrrolase with either αMeTrp or cortisol lowers the tryptophan content in liver as well as affecting that of the brain (Sourkes *et al.*, 1970; Young and Sourkes, 1975). It is known that the decline of the tryptophan content in liver after αMeTrp is accompanied by a decrease in protein synthesis and an increase in protein catabolism (Sourkes, 1971). Thus, protein catabolism may partly offset the decline in the free tryptophan pool owing to excess metabolism along the kynurenine pathway. Before we can understand fully how peripheral tryptophan metabolism affects brain tryptophan, we must understand the factors that influence the rate of tryptophan catabolism in the liver, the rate of protein synthesis and degradation, and the size of the free tryptophan pools in peripheral tissues.

Protein synthesis and the kynurenine pathway are probably the only routes of tryptophan metabolism with sufficient capacity to influence brain tryptophan levels. Other pathways, such as 5-HT formation in the intestine, decarboxylation of tryptophan to tryptamine, and transamination, are not normally quantitatively important. However, when the plasma tryptophan concentration is elevated after a massive dose of tryptophan (1 g/kg), the administration of a peripherally acting inhibitor of aromatic amino acid decarboxylase will raise that concentration even further (David *et al.*, 1974). Thus, decarboxylation to tryptamine may play a role in protecting the brain from very large tryptophan loads.

3. TRYPTOPHAN TRANSPORT INTO CEREBRAL TISSUES

3.1. Factors in Plasma Affecting Uptake of Tryptophan into the Brain

Peripheral tryptophan metabolism can affect the availability of tryptophan to the brain, but any changes in the brain must be mediated through changes in the plasma. The rate of uptake of tryptophan in the brain will depend on the concentration of tryptophan in plasma, factors in the plasma that affect uptake, and the activity of the actual uptake system. All these are known to vary. As discussed in the previous section the control of the tryptophan concentration in plasma is incompletely understood. The factors in the plasma that affect uptake have been the subject of a controversy centering around the role of albumin and of the large neutral amino acids in influencing the content of tryptophan in brain.

While the other amino acids in plasma are in free solution, most of the tryptophan in plasma is bound to albumin (McMenamy *et al.*, 1957; cf. Table 1). There is approximately one tryptophan binding site per albumin molecule (McMenamy and Oncley, 1958). The proportion of tryptophan in plasma that is in free solution, i.e., non-albumin-bound (NAB), depends not only on the albumin concentration but on that of the nonesterified fatty acids (NEFA). NEFA bind to albumin at the same site as tryptophan and thereby displace it (McMenamy, 1964). The tryptophan binding site on albumin corresponds to the primary binding site for medium-chain NEFA and the secondary site for long-chain NEFA (Cunningham *et al.*, 1975). Thus the addition of NEFA to plasma *in vitro* causes NAB tryptophan to increase. Moreover, the percentage of NAB tryptophan in human plasma is positively and significantly correlated with the NEFA concentration (Curzon *et al.*, 1974). A nonlinear relationship between NEFA and NAB tryptophan has been reported in rat plasma. There may be other factors affecting the binding of tryptophan to albumin in the rat, as no relationship between NEFA and NAB tryptophan was seen in rats injured by limb ischaemia (Stoner *et al.*, 1975).

A physiological compound that is known to displace tryptophan is bilirubin, and jaundice is associated with an increase in NAB tryptophan (McArthur *et al.*, 1971*a*). In addition, a variety of pharmacological agents displace tryptophan from albumin; these agents include antirheumatic drugs, such as salicylates and indomethacin, as well as probenecid, chlorpromazine, diphenylhydantoin, tolbutamide, the benzodiazepines, and clofibrate (McArthur *et al.*, 1971*b*; Lewander and Sjöström, 1973; Spano *et al.*, 1974; Iwata *et al.*, 1975; Müller and Wollert, 1975). The clinical significance of the action of some of those drugs in displacing tryptophan from albumin binding sites has been discussed recently (Paoletti *et al.*, 1975).

A variety of methods have been used for measuring NAB tryptophan in plasma and in albumin solutions. These methods include equilibrium dialysis and various ultrafiltration methods. However, the binding of tryptophan to albumin varies with temperature and pH (McMenamy, 1964), so that any useful method must keep the temperature constant, preferably at 37°, and must maintain the CO_2 content, and, thus, the pH of plasma within the physiological range. Unfortunately, not all methods in current use are sufficiently rigorous in these respects (Flentge *et al.*, 1974). Another factor to be considered in methods, such as equilibrium dialysis, which takes several hours, is a possible increase in NEFA owing to the action of lipases present in the plasma on endogenous triglycerides.

When probenecid and salicylate are administered to rats, the tryptophan level increases in brain and falls in plasma. As both drugs decrease the binding of tryptophan to albumin, these data indicate that it is the NAB tryptophan, and not the total plasma tryptophan, that controls the content of tryptophan in the brain (Tagliamonte *et al.*, 1971*a*). Consistent with this analysis is the finding that both these drugs, as well as clofibrate, increase NAB tryptophan and brain tryptophan *in vivo* in similar proportions (Gessa and Tagliamonte, 1974).

Other data also point to the control of the levels of brain tryptophan by the concentration of NAB tryptophan in plasma. In the newborn rat, most of the tryptophan in serum is in the free form and salicylate does not increase the plasma NAB tryptophan, nor does it increase the content of brain tryptophan. The high concentration of NAB tryptophan in these rats may be partly responsible for their high content of brain tryptophan (Bourgoin *et al.*, 1974). Additional evidence comes from loading experiments. After a tryptophan load, the rise in brain tryptophan is more in keeping with the rise in NAB tryptophan than with the total plasma concentration of this amino acid (Tagliamonte *et al.*, 1973). Factors that influence plasma NEFA, and therefore affect the concentration of NAB tryptophan, influence brain tryptophan similarly. For example, starvation and lipolytic drugs such as heparin, caffeine, Dopa, and isoprenaline increase plasma NEFA, NAB tryptophan, and brain tryptophan. Nicotinic acid can partially block the effect of starvation through its antilipolytic effect (Knott and Curzon, 1972; Curzon and Knott, 1974).

The increase in tryptophan in the rat brain after 3 hr of immobilization is accompanied by an increase in plasma NAB tryptophan and is probably associated with an increase in sympathetic nervous activity, and thus an increase in plasma NEFA (Curzon *et al.*, 1972). Less severe stresses also increase plasma NAB tryptophan. Thus, if group-housed fasted rats are removed from cages at 2-min intervals and killed, the continuing disturbance causes a progressive increase in plasma NEFA and NAB tryptophan, while the total plasma tryptophan falls (Curzon and Knott, 1975). The changes are blocked by nicotinic acid and propranolol. The effects on serum tryptophan are not accompanied by any change in the brain tryptophan content. The lack of change in the brain is probably due to the rapidity and short duration of changes in the plasma. These data indicate that experiments on the relationship between plasma NAB tryptophan and brain tryptophan must be designed with great care.

The hypothesis that the concentration of plasma NAB tryptophan influences the content of brain tryptophan has not gone without challenge. After carbohydrate or insulin administration to rats, there is a decrease in plasma NEFA and NAB tryptophan and an increase in total plasma and brain tryptophan (Madras *et al.*, 1973, 1974). Also, the diurnal variation of rat brain tryptophan follows the diurnal variation of the total plasma tryptophan (Fernstrom and Wurtman, 1974). Thus, in these two situations it is the total and not the plasma NAB tryptophan that is correlated with the brain tryptophan. However, in another study of circadian changes, there was no consistent relationship between brain tryptophan and either plasma NAB or total tryptophan (Morgan *et al.,* 1975).

It is possible that the total plasma tryptophan concentration could control the brain content if its uptake system has a higher affinity for tryptophan than albumin does. However, diffusion of tryptophan into the brain could only depend on the plasma NAB tryptophan. If diffusion plays an appreciable role in the transport of tryptophan into the brain and if the brain transport system does have a higher affinity for tryptophan than albumin, then both the total and NAB plasma tryptophan could play a role in regulating brain tryptophan. So the results mentioned are not necessarily at variance. However, to show that the total plasma amino acid concentration can influence the brain content, it will be necessary to demonstrate that the brain is capable of removing tryptophan from plasma when it is bound to albumin. Recent results indicate that albumin can inhibit tryptophan uptake at the level of the blood–brain barrier (Etienne *et al.,* 1976) as measured by the technique of Oldendorf (1971). These data indicate that the plasma NAB tryptophan must have some role in regulating brain tryptophan.

In the studies mentioned, the level of tryptophan was measured in the plasma and brain. The results do not allow for any changes in the activity of the uptake system that might be independent of changes in plasma tryptophan concentration. It is known that the rate of uptake into brain cells can vary. Thus, tryptophan transport into brain slices shows a diurnal variation (Héry, *et al.,* 1972) and can be influenced by pharmacological agents such as dibutyryl cyclic AMP (Tagliamonte *et al.,* 1972*b*). It is likely that the controversy surrounding the role of albumin in controlling the content of brain tryptophan will be resolved when measurements of brain and plasma tryptophan are made in conjunction with measurements of tryptophan uptake into the brain.

Albumin is not the only component of plasma that can influence the content of brain tryptophan; other amino acids can also have an effect. In studies using brain slices (Blasberg and Lajtha, 1965) and animals treated with amino acids (Guroff and Udenfriend, 1962) data have shown that there are a small number of specific transport systems, each of which acts on a group of amino acids. Tryptophan shares a transport system with other large neutral amino acids, including leucine, isoleucine, valine, phenylalanine, and tyrosine. These amino acids can inhibit tryptophan transport across the blood–brain barrier (Oldendorf, 1971) and across the brain cell membrane (Kiely and Sourkes, 1972). Accordingly, when rats ingest protein-containing food, there is no change in the content of brain tryptophan despite a large increase in the concentration of plasma tryptophan (Fernstrom and Wurtman 1972), for the protein-containing food also elevates the concentration of the amino acids in plasma that compete with tryptophan for uptake into the brain. The level of brain tryptophan is correlated with the ratio of the concentration of plasma tryptophan to the sum of the concentrations of tyrosine, phenylalanine, leucine, isoleucine, and valine in plasma.

3.2. The Uptake System of the Brain

Transport of the amino acids from plasma into brain cells occurs in two stages. The first step is at the level of the blood–brain barrier, which is localized in the endothelium of brain capillaries. Transfer at this stage may be mediated by a cycle in which the amino acid reacts with glutathione to give the γ-glutamyl amino acid; the derivative is later broken down to release the free amino acid (Orlowski *et al.*, 1974). Amino acids are transported across the capillary walls into the CNS extracellular fluid. The second transport step is from the extracellular fluid across the cell membrane into the cell. Both steps have been studied with respect to tryptophan.

The most extensive studies have been made on brain tissue, in particular, on slices. Slices will, of course, contain a heterogenous population of cells, and to overcome this problem, both cultured glial cells and synaptosomes have been studied. Although synaptosomes are all derived from neurons, there will be heterogeneity with respect to the neurotransmitter they contain. Only a very small portion of any synaptosomal preparation will synthesize 5-HT from tryptophan. However, despite these difficulties, much useful information has been obtained on transport of tryptophan into brain tissue. In fact, more is known about transport at the cell membrane than at the blood–brain barrier.

Transport of tryptophan into tissue from the cerebral cortex of the rat against a concentration gradient utilizing energy-requiring processes was first described by Joanny, *et al.* (1968) and Barbosa *et al.* (1970). Since then, the kinetics of tryptophan transport into cerebral cortex slices have been analyzed in detail. These studies have considered both the rate of uptake and the equilibrium attained against a concentration gradient. The relationship between the rate of transport of labeled tryptophan from the medium bathing the brain tissue into the intracellular water conforms to the hyperbolic function (Kiely and Sourkes, 1972):

$$v = aS^b \tag{1}$$

where v is the rate of uptake, S is the initial concentration of tryptophan in the incubation medium, and a and b are constants.

The hyperbolic nature of Eq. (1) suggests that there is a multiple type of mechanism for tryptophan transport into brain slices. In fact, the relation between v and S also satisfactorily fits Eq. (2) (Vahvelainen and Oja, 1972):

$$v = V_m S/(S + K_t) + K_d S \qquad (2)$$

where v is the velocity of influx; V_m is the maximal velocity of saturable transport; K_t is the transport constant, equivalent to a Michaelis–Menten constant, K_m; K_d is the diffusion constant; and S is the concentration of the amino acid in the medium. Values for the kinetic constants in Eq. (2) for immature and adult rat brain cortex slices are presented in Table 5. Vahvelainen and Oja (1972) employed a correction for extracellular water (in which some of the substrate tryptophan would be dissolved and, hence, could give rise to an overestimate of the true amount transported to the interior of the cells). They used inulin as the marker, but they determined the "inulin space" after only 5 min incubation; this is insufficient time for inulin to attain equilibrium, so that the short period underestimates the extracellular space and overestimates intracellular volume.

Kiely and Sourkes derived a third relationship between v and S that is very similar to Eq. (2), but includes a correction factor for efflux of tryptophan from the tissues:

$$v = V_{max} S/(S + K_m) + K_d S - c \qquad (3)$$

where V_{max} and K_m correspond to V_m and K_t in Eq. (2), and c is the correction factor for efflux. This form of the equation has also been adduced for transport of amino acids in the isolated perfused dog brain (Gilboe and Betz, 1971).

Both Eqs. (2) and (3) are readily interpretable in terms of transport mechanisms, the first term in each corresponding to Michaelis–Menten kinetics involving binding of substrate to an intermediate carrier and the second term corresponding to diffusion. The negative constant in Eq. 3 empirically corrects for efflux of tryptophan accumulated within the cells as a result of uptake. The specific experimental conditions in the work of Kiely and Sourkes (1972), such as the relatively long incubation period (15 min), would favor efflux. In fact, the rate of this process has been estimated as 0.90 mmol/liter during the incubation period; this is equivalent to 60 μmol tryptophan lost from each liter of intracellular water per minute.

In more recent work in the authors' laboratory, undertaken with a very short incubation period (2.5 min) to minimize efflux and to ensure the best estimate of initial uptake rates, a satisfactory fit of data has been obtained to Eq. (2). At the same time, it has been shown (Denizeau et al., 1976) that the graphical methods used to evaluate the constants in Eqs. (2) and (3) (Kiely and Sourkes, 1972) give somewhat inaccurate estimates. However, these estimates can be used as the starting point for an iterative calculation of more accurate constants. The values obtained by this method for various regions of the brain are set out in Table 5.

TABLE 5. Kinetic Constants of Transport of Tryptophan in Rat Brain

Tissue and brain region	V_{max}	K_m (mM)	K_d	References
Brain slices				
Cortex, adult	38 nmol g⁻¹min⁻¹	0.30	32 ml g⁻¹min⁻¹	Vahvelainen and Oja (1972)
7-days old	17 nmol g⁻¹min⁻¹	0.14	35 ml g⁻¹min⁻¹	
adult[a]	0.402 nmol liters⁻¹min⁻¹	0.829	0.063 min⁻¹	Kiely and Sourkes (1972)
Hypothalamus	0.18 nmol liters⁻¹min⁻¹	0.07	0.37 min⁻¹	Denizeau and Sourkes (1977)
Corpus striatum	0.13 nmol liters⁻¹min⁻¹	0.06	0.54 min⁻¹	
Cerebellum	0.16 nmol liters⁻¹min⁻¹	0.11	0.39 min⁻¹	
Raphe nuclei	0.29 nmol liters⁻¹min⁻¹	0.30	0.08 min⁻¹	
Synaptosomal preparations (P₂ fraction)				
Brain (except cerebellum), adult	45 nmol g⁻¹min⁻¹	1.0	—[b]	Grahame-Smith and Parfitt (1970)
1-day old[c]	62 nmol g⁻¹min⁻¹	0.19		Knapp and Mandell (1973)
Striatum, adult		0.03		
adult[d]	38 nmol g⁻¹min⁻¹d	0.02		Belin et al. (1974)

[a] "Liters" refers to intracellular water (noninulin space).
[b] No evidence for a multiple transport system.
[c] Calculations were made on the basis of data in the reference.
[d] Based on an assumed content of 7.5 g protein/100 g fresh weight gray matter.

It is also possible to obtain a good fit of the points to Eq. (4):

$$v = \frac{V'S}{S + K'_t} + \frac{V''S}{S + K'_t} \tag{4}$$

This represents a multiple system comprising two carriers, with high and low affinities, respectively, for the substrate, and has been frequently adopted by investigators in the transport field. For example, in regard to tryptophan transport in nervous tissue, Baumann *et al.* (1974) have employed Eq. (4) in the kinetic analysis of their data for cultured C6 glial cells, trypsin-disaggregated cells of the cerebral cortex, and synaptosomal preparations. They have calculated K_t values for their cortical preparations of about 5.8×10^{-6} M and 1.0^{-3} M (high and low K_m values, respectively), and 7.0×10^{-6} M and 8×10^{-4} M for the glial cells. Knapp and Mandell (1973) have also assumed that this two-carrier system is operating in respect to tryptophan transport into synaptosomes isolated from rat striatum.

It is important to remember the dictum of Lineweaver and Burk (1934) in respect to kinetic estimates, which states that conformity of fit to a particular equation is not necessarily proof that the phenomenon under study actually works in the fashion described by the formula. The analysis of systems described by Eq. (1) has usually rested with the method of Lineweaver and Burk, in which the reciprocals of v and S are plotted against one another. For either Eq. (2) or Eq. (4), this method gives a curved line. When the line is extrapolated to the v^{-1} axis, it cuts the ordinate above zero for Eq. (4) but passes through zero in the case of Eq. (2). These two possibilities are not easily distinguished when fitting a line to experimentally determined points. The decision between the two models of transport must be based upon specific experimental tests rather than upon graphical ones. In the case of tryptophan, uptake experiments have provided a basis for rejecting the two-carrier system for tryptophan uptake into slices (Denizeau *et al.*, 1976) and for recognizing that uptake of this amino acid occurs through a combination of energy-requiring transport and diffusion, as already demonstrated for Ehrlich ascites tumor cells (Jaquez, 1961) and suggested for human brain tumors (Snodgrass and Iversen, 1974). No tryptophan-binding carrier has been detected yet in nervous tissue, but Wiley (1970) has described such a protein in *Neurospora crassa*.

The values of V_m and K_t in Eq. (2) are larger in the cortical slices cut from brain of adult rats than from brain of 7-day-old animals (Vahvelainen and Oja, 1972), but K_d shows little difference. The differences between the two age groups are greater for phenylalanine and tyrosine than for tryptophan, histidine, and leucine.

The time for tryptophan uptake to attain equilibrium at 37°C has been determined *in vitro* with a number of brain preparations. With cerebral cortical slices from rat incubated with 10^{-3} M tryptophan, equilibrium is reached in 60 min. After this time period, losses of tryptophan occur and these losses are attributed to disintegrative processes (Kiely and Sourkes, 1972). Equilibrium is attained after 20 min for the P_2 fraction of rat brain containing synaptosomes incubated with 5×10^{-4} M tryptophan (Grahame-Smith and Parfitt, 1970), but only 2 min with cultured glial cells in a medium containing 6×10^{-6} M or 10^{-3} M tryptophan (Baumann et al., 1974). Forn and his colleagues (Forn et al., 1972) have claimed that brain stem slices from rat accumulate more tryptophan when they are incubated in a medium containing dibutyryl cyclic AMP. As their experiments made use of a 90-min incubation period, it is possible that the effect is not on tryptophan transport. Cyclic AMP may merely have maintained integrity of cellular membranes for a longer period than would otherwise have been possible.

The uptake of tryptophan into brain tissue is inhibited by certain amino acids. These inhibitions have been demonstrated with brain slices and synaptosomal preparations (Green and Curzon, 1970; Grahame-Smith and Parfitt, 1970; Kiely and Sourkes, 1972; Vahvelainen and Oja, 1975; Denizeau and Soukes, 1976). Inhibitory amino acids include phenylalanine, p-chlorophenylalanine, tyrosine, dopa (3,4-dihydroxyphenylalanine), α-methyltryptophan, kynurenine, leucine, isoleucine, and valine. Competing amino acids are thought to do so because they share the same postulated transport carrier as tryptophan itself, the carrier that handles the "large neutral amino acids." However, a detailed kinetic study of inhibition of uptake of some of these amino acids into brain cortex slices by phenylalanine indicates that there is more than one mechanism for the transport of these amino acids into brain cells. If there was one mechanism, the transport constant K_t for an amino acid would be the same as the inhibition constant K_i. Whereas the K_i for phenylalanine when inhibiting tyrosine uptake is the same as its K_t, the K_i is different when phenylalanine inhibits tryptophan or leucine (Vahvelainen and Oja, 1975).

Some investigators have obtained inhibition with methionine, histidine, glutamic acid, and 3-hydroxykynurenine, but others have not been able to verify this. Noncompeting amino acids include lysine, arginine, glutamine, gamma-aminobutyric acid, glycine, alanine, threonine, proline, 4-methyltryptophan, and D-tryptophan. Among metabolities of tryptophan that do not affect its uptake are N'-formylkynurenine, and the following acids: anthranilic, 3-hydroxyanthranilic, quinolinic, and xanthurenic (Green and Curzon, 1970; Kiely and Sourkes, 1972). The 4- and 5-hydroxytryptophans appear to favor the uptake of tryptophan (Kiely and Sourkes, 1972).

Efflux of tryptophan from brain tissue has been studied in vitro only to a limited extent. In P_2 fractions (Grahame-Smith and Parfitt, 1970), but not in brain slices (Denizeau and Sourkes, 1976), preincubation of the material with cold tryptophan or with several other amino acids will increase both the rate of uptake and the rate of efflux of labeled tryptophan.

Uptake of tryptophan at the level of the blood–brain barrier has been studied by the technique of Oldendorf (1971). In this method a solution containing ^{14}C-labeled tryptophan and tritiated water is injected into the common carotid artery of a rat with sufficient pressure to stop blood flow temporarily. Thus, the injected solution passes through the brain as a bolus, which mixes very little with the blood. After 15 s, a time long enough for the bolus to be washed out of the brain microcirculation by the blood flow, the rat is decapitated. The ratio of ^{14}C to ^{3}H in the injected solution is compared with this ratio in the brain. Thus, the measure of uptake, referred to as the brain uptake index (BUI or I_b) and expressed as a percentage, is a ratio of ratios.

Uptake is compared with tritiated water as a freely diffusible standard and yet the BUI can be as high as 30% with concentrations of tryptophan around the physiological concentration (Pardridge and Oldendorf, 1975). Phenylalanine, leucine, and tyrosine have even higher BUIs. Thus, there must be very rapid exchange of amino acids between blood and brain in the capillaries. In this situation, efflux could be very important in controlling the brain content of amino acids and is in need of study in the living animal.

Uptake of amino acids by the brain in live animals has also been studied by Baños *et al.* (1973). In their method they first inject intravenously the amino acid they are to study to determine its kinetics of absorption and removal. From this they calculate the rate of injection of a solution of the amino acid that will ensure a constant plasma concentration and specific activity. They then carry out their experiments in animals in which a steady-state concentration of the amino acid is maintained over a period of at least 3 min. The net uptake measured under these conditions presumably will depend partly on uptake by brain cells as well as movement across the blood–brain barrier, and partly on efflux.

Seta *et al.* (1973) have studied uptake by brain when amino acids are infused intravenously over a period of hours. They found that exchange of amino acids between plasma and brain is rapid but incomplete. They suggested that there are rapidly and slowly exchangeable compartments in brain.

Uptake rates determined *in vivo* are given in Table 6. Rates of uptake have been calculated from the brain uptake index by Pardridge and Oldendorf (1975). Values for the same amino acids as taken up by cortex slices of adult rats are also shown for comparison. The first and last columns for *in vitro* and *in vivo* data, respectively, list estimated values of a theoretical maximal velocity, attained at very high concentrations of the amino acid. These values for tryptophan agree very well, but the others do not show the same concordance. When the data for the two modes are compared with respect to uptake rates that could be expected to obtain at physiological concentrations of the respective amino acids, there is good agreement for both tryptophan and phenylalanine. Although there are discrepancies for

TABLE 6. Influx of Amino Acids into the Brain of Rats: Comparison of Rates Observed *in Vitro* and *in Vivo* (nmol g^{-1}min^{-1})

Amino acid	In vitro[a]		In vivo	
	V_{max}	Comparison[b]	Steady-state method[c]	Bolus injection[d]
Tryptophan	38[e]	3.0	2.5	33[e]
Phenylalanine	101	16.1	14.0	30
Tyrosine	101	20.7	6.2	46
Histidine	145	13.9	4.9	38
Leucine	62	15.0	6.5	33

[a] Data from Vahvelainen and Oja (1972). Slices of cerebral cortex of adult rats were used.
[b] In order to compare the results of Vahvelainen and Oja (1972) obtained *in vitro* with those of Baños *et al.* (1973) derived from physiological experiments, the rate of uptake, *v*, was calculated for a concentration of 10^{-4} *M* of the amino acids in the incubation medium. In the case of tryptophan, because of binding to albumin, the calculation took into account only the concentration of free tryptophan in the plasma, estimated as 2×10^{-5} *M*.
[c] Data from Baños *et al.* (1973). The influx rates have been standardized to a plasma amino acid concentration of 10^{-4} *M*.
[d] Data from Pardridge and Oldendorf (1975). Values shown are estimates of maximal velocity, assuming a plateau is reached with large concentrations of tryptophan in the bolus.
[e] Maximal velocities were estimated by kinetic–graphical methods.

the other three amino acids measured, the rank orders are in reasonable agreement, with tyrosine and leucine having greater uptake rates than histidine (Table 6).

There are other similarities between uptake of amino acids by brain tissue and by brain capillaries. Thus, study of tryptophan influx *in vivo* by the Oldendorf method shows that BUI is highest for the low concentrations of the amino acid, and as that concentration increases the BUI falls rapidly at first and more slowly later on (Pardridge and Oldendorf, 1975). Kinetic analysis of the results provides good evidence for a diffusional component in the uptake of this amino acid as well as active transport.

In tissue slices, tryptophan uptake is inhibited by the "large neutral" amino acids that share a common carrier with tryptophan. Oldendorf (1971) has shown that at the blood–brain barrier inhibition can also occur with methionine and histidine and has suggested that the blood–brain barrier has only two carriers for amino acid uptake, one for basic amino acids and one for the rest. Competition for tryptophan uptake by other amino acids has been studied in physiological situations by the infusion technique.

In hyperinsulinemia there is increased uptake of amino acids into skeletal muscle, which leads to a reduction in their concentrations in the blood. For example, free tryptophan in the serum may decline considerably, and the branched-chain amino acids even more. Under these conditions, the entry of tryptophan into the brain is greatly increased (Daniel *et al.*, 1975). This could arise from reduced competition by other amino acids, but studies of animals in the steady state with respect to labeled amino acids in plasma show that the competition is really evident only when the competing amino acids are quite elevated in the plasma (Baños *et al.*, 1974*a*). Thus, the level of phenylalanine in blood must be raised fourfold in order to reduce the entry of tryptophan into the brain by 50%, and other amino acids must be increased even more (Baños *et al.*, 1974*b*). Daniel *et al.* (1975) therefore suggest an alternative explanation for the effect of high insulin levels on tryptophan influx into the brain, namely, an increase in the activity of the effective transport system. However, there is no specific evidence for such an adaptive increase.

The blood–brain barrier is not completely stereospecific. Yuwiler (1973) has obtained a BUI by the Oldendorf technique for D-tryptophan of 3.49, i.e., about one-tenth that for the L-amino acid. Analogous studies of intestinal absorption of tryptophan by rat and hamster show that the L-amino acid is taken up about four to five times faster than D-tryptophan (Bosin *et al.*, 1974).

Passage of tryptophan into the CSF from brain and blood, or from CSF into brain, has received very little attention. There is some evidence for exchange of certain amino acids between CSF in the ventricles and the surrounding parenchymal tissues of the brain (Levin *et al.*, 1966*b*). The accumulation of labeled tryptophan in different brain regions 7 min after its intraventricular injection in the rat shows some diurnal variations, the concentrations attained being lower during dark hours than light, at least in the hypothalamus and brain stem (Héry *et al.*, 1972). It is not clear whether this is a light/dark effect or some intrinsic rhythm. In the case of glycine, transfer from the CSF to blood occurs by a saturable process in the rat (Franklin *et al.*, 1975).

Movement of tryptophan from the CSF has been studied by Geddes *et al.*, (1973), who showed that the clearance rate in a recirculatory ventricu-locisternal perfusion system, applied to the dog, is independent of the concentration of tryptophan in the inflow fluid. They concluded from their work that the amino acid is removed by a nonsaturable mechanism and by bulk absorption of the CSF, without intervention of active transport or a significant rate of metabolism of tryptophan (cf. Lajtha and Toth, 1962). This agrees with the conclusion drawn about the clearance of other amino acids from the CSF of cat (Levin *et al.*, 1966*a*), but in recent years the role of the choroid plexus in active transport of amino acids has become generally recognized (Katzman and Pappius, 1973). Saturable transport has been demonstrated with cat choroid plexus *in vitro* (Lorenzo and Cutler, 1969) and *in vivo* (Cutler, 1970; Lorenzo, 1974). Lysine, for example, is removed by an active transport mechanism and a diffusional system

(Cutler, 1970), corresponding to the terms of Eq. (2). Beyond the ventricles, the arachnoid membrane and pial vessels may assume the transport function of the choroid plexus and ependymal tissue (Lorenzo, 1974; Levin et al., 1974).

4.   CEREBRAL TRYPTOPHAN

The significance of tryptophan in the brain stems not only from its indispensable role in protein synthesis but also from the nature of its metabolic products. Of the latter, the most important is 5-HT. The other pathways are quantitatively smaller and it is not known what role, if any, the metabolites formed along these pathways play in the functioning of the brain. There has been some speculation about possible interactions between the various pathways of tryptophan metabolism in the brain, but there is little experimental evidence on this. These pathways will be discussed in this section.

4.1.   Protein Synthesis

As mentioned in section 3.2, there is a rapid exchange of amino acids between plasma and brain; and tracer infusion studies indicate that the major portion of each amino acid is in this dynamic state (Seta et al., 1973). Thus, the half-life of many cerebral amino acids is less than 30 min (Lajtha, 1964).

The high rate of protein turnover relative to the pool size of the free amino acids probably also contributes to this short half-life. It has been estimated from the rate of turnover of protein in brain that the amount of amino acids that are incorporated and released from protein within 30 min equals the total cerebral pool of most free amino acids (Lajtha, 1974). Thus, the free amino acid pool in brain exchanges rapidly with both free amino acids in plasma and cerebral protein. This is a property that tryptophan has in common with other amino acids.

Tryptophan is the least abundant amino acid in the pool available for protein synthesis (cf. Section 2.2) and it is involved in the regulation of ribosomal aggregation in the liver. Whether tryptophan plays a similar role in the regulation of protein synthesis in the brain is not clear. When 7-day-old rats are given a single injection of phenylalanine, there is a reduction of brain tryptophan and a disaggregation of brain polyribosomes (Aoki and Siegel, 1970). The disaggregation of polyribosomes under these conditions follows the time course of decrease of brain tryptophan and not the increase of phenylalanine. However, owing to technical difficulties, it has not been possible to demonstrate that tryptophan administration can reverse the effect of phenylalanine on brain ribosomes (Siegel *et al.*, 1971). This effect is no longer seen at 4 weeks of age.

Two findings indicate that the content of tryptophan does not limit protein synthesis in the adult brain. The first of these has to do with the content of tryptophyl-tRNA in the brain. These nucleic acids are present in immature mouse brain in concentrations only slightly greater than in adult animals (Johnson and Chou, 1973), although the rate of cerebral protein synthesis is much greater in the immature mice (Johnson and Luttges, 1966). The second finding is that the tryptophan-activating enzyme has a particularly low K_m for tryptophan, as studied in the brain of the adult water-buffalo (Liu *et al.*, 1973).

Alterations in ribosome aggregation in amino acid loading experiments do not necessarily reflect the situation *in vivo* even in immature animals. Thus, administration of phenylalanine to immature rats can activate ribonuclease activity in the brain and it is possible that the ribosome disaggregation seen under these conditions occurs due to increased ribonuclease activity after the death of the animal (Roberts, 1974). The problem of regulation of protein synthesis in the brain is a difficult one. Amino acid imbalances in general seem to affect it. Also, the results of studies of incorporation of amino acids into protein can vary with the route of administration. Thus, after intravenous administration of phenylalanine, the incorporation of amino acids from the blood into protein is decreased, but intracisternal administration suggests that protein synthesis in the brain may actually be accentuated (Roberts, 1974). Recently, these problems have been discussed in greater detail than is possible here (Barondes, 1974; Oja *et al.*, 1974; Roberts, 1974). At the moment it seems unlikely that tryptophan plays a special role in regulating brain protein synthesis. However, it may be possible that some of the effects of tryptophan on the brain and behavior are mediated through changes in the rate of synthesis of brain protein rather than through changes in 5-HT formation.

The data given refer to ribosome aggregation and protein synthesis in whole brains or brain slices. In whole brain, only a small proportion of tryptophan is metabolized to 5-HT and most of the rest goes into proteins. However, in 5-HT neurons, which comprise only a very small portion of brain, 5-HT synthesis is presumably relatively much more important. The pineal gland, which synthesizes 5-HT very actively as an intermediate in the pathway to melatonin, uses 100 times more tryptophan for 5-HT synthesis than for protein formation (Wurtman *et al.*, 1969). In neurons containing biogenic amines, most of the amine synthesis occurs in nerve endings (Hillarp *et al.*, 1966) and only a small proportion in the cell body, where most of the protein synthesis occurs. Some protein synthesis occurs in mitochondria, including those located in nerve endings, and in view of this there might be competition between tryptophan hydroxylase (E.C. 1.14.16.4) and tRNA for substrate, although this remains speculative (Barondes, 1974).

The content of tryptophan in the brain may influence protein synthesis through direct mechanisms, but it may also have an indirect influence through 5-HT formation. Administration of L-dopa or L-5-HTP to rats caused disaggregation of brain polysomes and suppression of protein synthesis in whole brain (Weiss *et al.*, 1974). This effect is mediated by the corresponding amines dopamine and 5-HT. The amines may work through receptors, as the appropriate receptor antagonists can block the ribosomal disaggregation (Weiss *et al.*, 1975). Tryptophan could have a similar effect, but the increase of brain 5-HT that it provokes is much more localized than that after 5-HTP, so that it would have a more limited effect, if any.

Tryptophan is incorporated into peptides as well as proteins in brain. The content of peptides with NH_2-terminal tryptophan is as high as 5.6 μg/g (tryptophyl-alanine equivalents) in whole rat brain and their content is as much as seven times this in the pituitary (Edvinsson *et al.*, 1973). Tryptophan plays a crucial role in the hypothalamic luteinizing-hormone-releasing hormone (LHRH), a decapeptide, for the tryptophan in the 3-position is essential for biological activity (Yanaihara *et al.*, 1973).

4.2. 5-Hydroxytryptamine

5-HT is thought to be a neurotransmitter in the mammalian CNS. The cell bodies of the putative 5-HT neurons are present in the raphe nuclei of the brain stem (Dahlström and Fuxe, 1964). Axons project from these cell bodies down into the spinal cord and up into the mid- and forebrain (Fuxe, 1965; Andén *et al.*, 1966). Electrical stimulation of raphe neurons *in vivo* produces a decrease in the content of 5-HT in the forebrain and an increase in its metabolite 5-HIAA (Aghajanian *et al.*, 1967). These data indicate that electrical stimulation of the cell bodies causes release of 5-HT at the synapses. Single-unit recordings from 5-HT neurons in the raphe nuclei have been made, and the firing rate proves to be correlated with raphe cell fluorescence, detected histochemically, and the content of 5-HT (Aghajanian, 1972). These findings support the idea that 5-HT is a neurotransmitter, but the evidence is not conclusive (for discussion, see Krnjevic, 1974).

Biosynthesis of 5-HT in the CNS occurs solely within 5-HT neurons. This is due to the fact that the rate-limiting enzyme in 5-HT formation, tryptophan hydroxylase, is localized within those neurons (Kuhar *et al.*, 1972; Joh *et al.*, 1975). A large body of evidence indicates that tryptophan hydroxylase is not normally saturated with its substrate, tryptophan, and that changes in the content of tryptophan in brain affect the rate of 5-HT synthesis. Although this is by no means the only factor affecting the rate of 5-HT synthesis, a detailed discussion of control of 5HT turnover is not within the scope of this review, and we will consider only the influence of tryptophan on this process. Other factors regulating 5-HT have recently been discussed elsewhere (Costa and Meek, 1974; Hamon and Glowinski, 1974; Lovenberg and Victor, 1974; Mandell *et al.*, 1974; Gál, 1975).

Two types of evidence show that tryptophan hydroxylase in brain is unsaturated. The first comes from *in vivo* studies which indicate that changes in the content of brain tryptophan can influence the rate of 5-HT synthesis. The second is derived from the kinetic studies of tryptophan hydroxylase which suggest that the K_m of the enzyme with respect to tryptophan is high compared with the probable tryptophan content in 5-HT neurons. The first indication that tryptophan hydroxylase is unsaturated was the demonstration that tryptophan administration to rats pretreated with a monoamine oxidase inhibitor (MAOI) increased the brain 5-HT content (Hess and Doepfner, 1961). Later studies (Ashcroft *et al.*, 1965) showed that, in the absence of a MAOI, large loads of tryptophan increased both 5-HT and 5-HIAA concentrations in the brain although 5-hydroxy-tryptophan (5-HTP) remained undetectable. This finding suggested both that tryptophan hydroxylase is rate-limiting in the formation of 5-HT and that it is unsaturated with respect to tryptophan. Doses of tryptophan as low as 12.5 mg/kg (less than 5% of the normal daily dietary intake) can raise the brain 5-HT content significantly (Fernstrom and Wurtman, 1971). Investigations on the accumulation of 5-HT in rat brain after inhibition of monoamine oxidase (Grahame-Smith, 1971) and on the accumulation of 5-HTP in mouse brain after inhibition of aromatic amino acid decarboxylase (Carlsson and Lindqvist, 1972) give similar results. With increasing trypto-phan doses, the accumulation of 5-HT or 5-HTP, which is used as an index of the rate of tryptophan hydroxylation, reaches a plateau at a brain tryptophan content of about 360 nmol/g. At this brain tryptophan content, tryptophan hydroxylation occurs at about 2.5 times the normal rate; this suggests that normally the enzyme is slightly less than half-saturated. The control of 5-HT synthesis by tryptophan in the brain is seen not only in tryptophan-loading experiments. Many drugs that affect the brain trypto-phan content of rats influence the content of 5-HT or 5-HIAA in the same direction, while other drugs that do not influence brain indoleamines do not affect brain tryptophan (Tagliamonte *et al.*, 1971*b*). Other conditions in which there are parallel variations of brain tryptophan and 5-hydroxyin-doles include the increases after immobilization and food deprivation (Cur-zon *et al.*, 1972) and the decreases after induction of tryptophan pyrrolase with α-methyltryptophan (Sourkes, 1971) and cortisol (Green *et al.*, 1975*a*).

Changes in indoleamine metabolism caused by changes in the tryptophan content of tissues are not uniform throughout the brain (Moir and Eccleston, 1968). In particular, in the rat hypothalamus, which contains relatively large amounts of tryptophan and 5-HT, changes in indoleamine metabolism are smaller than in other areas perhaps because of the low 5-HT turnover rate found in this region (Knott and Curzon, 1974; Curzon and Marsden, 1975).

Because tryptophan hydroxylase is located solely within 5-HT neurons (Kuhar *et al.*, 1972), the increase of 5-HT after tryptophan administration will occur only in 5-HT neurons. Thus, after a tryptophan load, the pattern of increase of indoleamines in the brain regions resembles that found under physiological conditions (Moir and Eccleston, 1968) and there is a selective enhancement of histochemical fluorescence (owing to 5-HT) of raphe neurons (Aghajanian and Asher, 1971). The latter property can be used in single-cell recording experiments to mark raphe neurons by iontophoretic injection of tryptophan (Aghajanian and Haigler, 1974).

The large body of evidence described indicates that the brain tryptophan content limits the rate of 5-HT formation, and *in vitro* evidence is consistent with this. The K_m for rabbit-hindbrain tryptophan hydroxylase is 50 μM with tetrahydrobiopterin (BH$_4$) as cofactor (Friedman *et al.*, 1972). The normal brain tryptophan content is equivalent to about 30 μM, so that at this concentration tryptophan hydroxylase would be just under half saturated with its substrate. This is in good agreement with the data obtained *in vivo*, but one must recognize the assumptions made in the assessment of the *in vitro* results. Tryptophan hydroxylase shows different kinetic properties, including different K_m values, with different pteridine cofactors (Tong and Kaufman, 1975). Although it is commonly assumed that tetrahydrobiopterin is the natural cofactor, there is no conclusive evidence that it is. The K_m of the particulate enzyme (presumably in synaptosomes), which can be determined without exogenous pteridine, is the same as that found with BH$_4$ (Ichiyama *et al.*, 1970). However, with tetrahydrobiopterin (but not other cofactors) added *in vitro*, tryptophan displays substrate inhibition at concentrations above 0.3mM, while *in vivo* no substrate inhibition is seen with brain tryptophan contents equivalent to 0.7 mM (Grahame-Smith, 1971). The discovery of a factor in brain with pteridine like properties, capable of activating tryptophan hydroxylase (Gál and Roggeveen, 1973), casts further doubts on tetrahydrobiopterin as the natural cofactor.

Although the K_m for tryptophan with the natural cofactor is uncertain, the content of tryptophan in tissues is even more uncertain. To extrapolate from the tryptophan content of whole brain to the content of tryptophan in free solution in the synaptosomes of 5-HT neurons leaves a lot to faith. Thus, the agreement of the *in vivo* and *in vitro* data on the degree of saturation of tryptophan hydroxylase with its substrate may owe more to chance than to a valid analysis of data obtained *in vitro.*

4.3. Metabolites of 5-Hydroxytryptamine

Although the most important pathways of tryptophan metabolism in brain are to protein and 5-HT, a variety of other compounds are formed from this amino acid. Some of these are formed by further metabolism of 5-HT and some through other pathways. 5-HT is metabolized primarily by monoamine oxidase (MAO) (E.C. 1.4.3.4) to 5-hydroxyindoleacetaldehyde. This terminates the neuronal action of 5-HT. The aldehyde has not been detected in brain because it is further metabolized to 5-hydroxytryptophol and 5-HIAA. Kinetic studies have shown that the aldehyde is much more reactive toward the aldehyde dehydrogenase (aldehyde: NAD oxidoreductase, EC 1.2.1.3) than to the aldehyde reductase (alcohol: NADP oxidoreductase, EC 1.1.1.2) (Turner *et al.,* 1974; Duncan and Sourkes, 1974). This explains why 5-HT is metabolized almost completely to 5-HIAA.

Conjugation is an alternative route of 5-HT inactivation and is readily detected if MAO is inhibited. In rats pretreated with pargyline, about 8% of the ^{14}C-labeled 5-HT found in brain after intravenous ^{14}C-labeled 5-HTP is present as 5-HT-*O*-sulphate (Gál, 1972).

Metabolism of 5-HT in the brain may not only terminate 5-HT action but also lead to compounds with a functional significance of their own. In the pineal gland, which is not considered part of the CNS, tryptophan and 5-HT are precursors of melatonin (*N*-acetyl-5-methoxytryptamine). The presence in rat hypothalamus of 5-methoxytryptamine (Green *et al.,* 1973) and melatonin (Koslow, 1974) has been demonstrated with mass spectrometry. These compounds are present even in pinealectomized animals and may be formed in the brain. There is an enzyme in rat brain capable of acetylating 5-methoxytryptamine (Paul *et al.,* 1975) but it is not known how the *O*-methylation occurs.

The brain is capable of *N*-methylating indoleamines as well as *O*-methylating them. Intracisternally injected [14C]tryptamine leads to the formation of the *N*-methyl and *N,N*-dimethyl derivatives in rat brain (Saavedra and Axelrod, 1972*a*). *In vitro* this enzyme requires *S*-adenosyl-methionine, as does the enzyme in human brain which methylates 5-HT to bufotenine (*N,N*-dimethyl 5-HT) (Mandell and Morgan, 1971). These methylated derivatives are of great interest because of their psychoactive properties. However, it remains to be shown that they are definitely formed in brain under physiological or pathological circumstances. At the moment, to postulate a role for brain tryptophan in controlling their formation would be purely speculative.

4.4. Decarboxylation to Tryptamine

Present evidence suggests that aromatic amino acid decarboxylase (E.C. 4.1.1.28) in the CNS is present in two locations, monoaminergic neurons and capillaries (Andén et al., 1972). The enzyme is most active toward dopa and 5-HTP but will metabolize other amino acids also. The K_m for tryptophan (1.4×10^{-2} M) is much greater than that for 5-HTP (5.4×10^{-6} M) (Ichiyama et al., 1970). Because the K_m for tryptophan is so high, the concentration of tryptamine in brain should be much lower than that of 5-HT. The actual concentration found depends on the method used. A radiochemical derivative assay (Snodgrass and Horn, 1973) and an isotopic-enzymatic assay (Saavedra and Axelrod, 1972b) give a brain tryptamine content of 100–300 pmol/g while a mass-spectrometric method, which may be more specific, gives a value of 3 pmol/g (Philips et al., 1974) (compared with the 5-HT content of brain, which is about 3 nmol/g). The regional distribution of tryptamine roughly approximates that of 5-HT, although it is relatively high in the caudate nucleus (Philips et al., 1974), a region rich in dopamine. This indicates that tryptamine may be formed by the decarboxylase of both catecholamine and indoleamine neurons. In support of this suggestion, destruction of both catecholamine neurons (with 6-hydroxydopamine) and of 5-HT neurons (by lesions of the raphe nuclei) leads to a decline of rat brain tryptamine (of 24% and 29%, respectively) (Marsden and Curzon, 1974). A significant portion of the tryptamine in brain is associated with the synaptosomal fraction (Boulton and Baker, 1975).

MAO inhibitors, the tryptophan hydroxylase inhibitor parachlorophenylalanine (PCP), and large doses of tryptophan can all cause an increase of brain tryptamine (Snodgrass and Horn, 1973; Saavedra and Axelrod, 1973; Marsden and Curzon, 1974). Tryptophan and MAO inhibitors presumably act by increasing the rate of synthesis and decreasing the rate of breakdown of tryptamine. The mechanism of action of PCP is unknown, but it is possible that inhibition of tryptophan hydroxylase increases the content of tryptophan in 5-HT neurons. Alternatively, 5-HTP, which decreases after PCP, may competitively inhibit decarboxylation of tryptophan.

After small changes in brain tryptophan, there is no net change in the content of brain tryptamine (Snodgrass and Horn, 1973; Marsden and Curzon, 1974). It is possible that even with a small increase in the rate of synthesis of tryptamine, in these circumstances there would not be an increase in the actual tryptamine content because of its very high turnover in brain. Thus, the half-life in brain of exogenous tryptamine is only 52 min even after inhibition of MAO and is less in untreated animals (Wu and Boulton, 1973).

The presence of tryptamine in monoamine neurons suggests a functional role, but there is no evidence of what this role may be.

4.5. Transamination

Rat brain contains an aromatic amino acid transaminase that is active toward tryptophan and 5-HTP, among other compounds (Fonnum *et al.,* 1964); α-ketoglutarate, pyruvate, and phenylpyruvate are capable of acting as receptors for the amino group (Lees and Weiner, 1975). The K_m for tryptophan is of the order of 1 mM (Lees and Weiner, 1975); hence, there is probably little metabolism of tryptophan through this pathway. However, it is possible that indoleacetic acid and 5-HIAA could be formed in brain through transamination and decarboxylation rather than by way of tryptamine and 5-HT. Since transaminases are reversible, it is also possible that tryptophan could be formed in brain from its keto acid. However, it is unlikely that either of these processes occurs to an appreciable extent under physiological circumstances. The functional significance of brain transaminase is not known.

4.6. Pyrrolase

Brain homogenates are capable of converting tryptophan to kynurenine (Gál *et al.,* 1966), but the enzymes capable of performing this reaction have not been studied until recently. Tryptophan pyrrolase in the liver has a much greater activity than the brain enzyme toward L-tryptophan (Gál, 1974). However, the brain enzyme has a broader range of specificity, for it is active toward 5-HTP (Tsuda *et al.,* 1972), D-tryptophan, and melatonin (Hirata *et al.,* 1974). The brain enzyme was found in one case (Tsuda *et al.,* 1972) to require FAD for maximal activity. In another case, the properties of the enzyme were similar to the intestinal enzyme (Yamamoto and Hayaishi, 1967) in that it required methylene blue, ascorbate, and the superoxide anion for maximal activity (Hirata *et al.,* 1974). The intestinal pyrrolase, like the liver pyrrolase, is a heme enzyme but the liver enzyme requires oxygen for activity, not superoxide (Hirata and Hayaishi, 1971).

The presence of a tryptophan pyrrolase, or, rather indoleamine pyrrolase, in brain may have considerable functional significance. After intracisternal injection of [^{14}C]melatonin, the only metabolite detectable in rat brain is N-acetyl-5-methoxykynurenine formed by the action of pyrrolase (Hirata *et al.*, 1974). Other indoleamines may also be metabolized through this pathway. 5-Hydroxykynurenamine, which could be formed by the action of pyrrolase on 5-HT, has been detected in mouse brain (Makino *et al.*, 1962) and can produce contractions of dog and human cerebral arteries through 5-HT receptors (Toda, 1975).

The presence of an indoleamine pyrrolase in brain greatly increases the possible number of tryptophan metabolites that may be present in brain. The results described indicate that some of these metabolites could have functional significance and that for some indoleamines in brain, degradation by pyrrolase may be an important pathway. Much work remains to be done in this relatively new field.

4.7. Other Pathways

It is known that tryptophan in brain can be metabolized into protein and peptides, hydroxylated on the pathway to 5-HT, decarboxylated, transaminated, and degraded by pyrrolase. These metabolic fates include many of the known biological reactions of tryptophan but there still remain other pathways of tryptophan metabolism in the brain that could occur. For example, some amino acids, such as aspartic acid, are N-acetylated in brain as is 5-methoxytryptamine (to give melatonin). Could the same occur for tryptophan? Also, tryptophan is the source of ommochromes, which are among the most widely distributed pigments in nature. One pigment found in the brain is melanin . Tryptophan can be incorporated into melanin in a mouse melanoma (DeAntoni *et al.*, 1974), and it has been shown that in extracts of human brain 5-HT can be incorporated into melanin (Rodgers and Curzon, 1975). However, there is no evidence *in vivo* that tryptophan or its metabolites can be incorporated into melanin in brain.

5. CSF TRYPTOPHAN

The transport of tryptophan between brain, blood, and CSF was discussed in Section 3.2. Since transport can occur among all three compartments, it is pertinent to ask whether the concentration of CSF tryptophan reflects the level of this amino acid in the brain or in plasma. This is a question of importance in clinical studies, because CSF is more accessible than brain in humans. The evidence available suggests that CSF tryptophan is controlled by the same factors that influence brain tryptophan and that the concentration of this amino acid in the CSF, about 4 nmol/ml in the ventricular fluid (Table 3), reflects its brain content.

The administration of tryptophan to humans increases the concentration of 5-HIAA in lumbar CSF, and the CSF concentrations of tryptophan and 5-HIAA are positively and significantly correlated (Ashcroft *et al.,* 1973*b*). This is evidence that, under these circumstances, the CSF tryptophan does reflect the brain concentration, at least within the compartment(s) where 5-HT is formed and metabolized. Consistent with this is the finding that in rats given a diet supplemented with large amounts of tryptophan, there is a correlation between the increase in the brain and CSF concentrations of that compound (Modigh, 1975). This relationship is seen in rats under more physiological circumstances as well. The changes of brain tryptophan owing to diurnal variation, 24-hr starvation, and glucose feedings are reflected by similar changes in CSF tryptophan (Young *et al.,* 1976*a*). However, the CSF values show greater variability than the brain values. This variability might explain why CSF tryptophan and 5-HIAA concentrations are correlated in human lumbar CSF after a tryptophan load (Ashcroft *et al.,* 1973*b*), but not without the load (Ashcroft *et al.,* 1973*a;* Young *et al.,* 1974*a*).

The data obtained with experimental animals indicate that CSF tryptophan taken during clinical studies probably reflects the brain tryptophan but, because of the variability of the CSF results, measurements would have greater predictive value for groups of patients than for particular individuals.

In current work on experimental animals, the CSF is taken from the cisterna magna and is thus in close contact with the brain. However, in clinical studies, lumbar CSF tryptophan is often used. That lumbar CSF is not in contact with the brain probably is not important. Although the tryptophan metabolite 5-HIAA shows a steep concentration gradient from ventricles, through cisterna magna, to the lumbar region (Sourkes, 1973), tryptophan does not. Thus, in lumbar CSF samples taken during pneumoencephalography, the concentration of tryptophan in the CSF before injection of air was the same as that after injection of air (Young *et al.,* 1974*a*). The sample taken after air injection contains CSF from higher compartments as well as the lumbar subarachnoid space. In one patient with a block of CSF flow in the thoracic region, the tryptophan concentration was the same in cisternal and lumbar CSF (Young *et al.,* 1973). These data suggest that tryptophan is uniformly distributed throughout the CSF. There is an apparent contradiction in the fact that the concentration of tryptophan in ventricular CSF is significantly higher than that in lumbar CSF (Young *et al.,* 1974*a*), but more recent results indicate that the elevated ventricular CSF tryptophan is probably associated with the pathological conditions necessitating the surgery during which the samples were taken (Young *et al.,* 1976*b*).

Clinical studies so far reveal that CSF tryptophan and thus brain tryptophan in the human are influenced by the same types of factor that affect brain tryptophan in the rat. Tryptophan loading produces an increase in lumbar CSF tryptophan that starts at 2 hr and reaches a maximum at 8 hr (Eccleston *et al.*, 1970). The lag in the increase in CSF tryptophan after a load may indicate that the CSF responds to changes more slowly than the CNS. However, this is certainly not so in other circumstances, since in two patients there were parallel variations of serum NAB tryptophan and ventricular CSF tryptophan over a 24-hr period (Young *et al.*, 1976*b*). This study of ventricular CSF supports the idea that it is the NAB tryptophan and not the total plasma tryptophan that controls the brain content. In these two cases, variation in the serum concentration of the large neutral amino acids that compete for tryptophan uptake by brain did not influence the CSF tryptophan. However, in another study lumbar CSF was taken pre- and postprandially from patients. The change in CSF tryptophan in this study could be explained by the change in the ratio of NAB tryptophan to the large neutral amino acids in plasma (Perez-Cruet *et al.*, 1974).

Another situation in which clinical and experimental studies are in agreement concerns the effect of liver damage. In dogs, during acute hepatic coma, the contents of most amino acids in brain rise; the levels of the aromatic amino acids, i.e., histidine, phenylalanine, tyrosine, and tryptophan, increased to the greatest extent (Mattson *et al.*, 1970). In humans with cirrhosis of the liver, lumbar CSF tryptophan is greatly elevated although patients in coma show no greater increase than those not in coma (Young *et al.*, 1975).

6. TRYPTOPHAN IN CLINICAL STATES

6.1. Introduction

A role for tryptophan in physiological states and in mental diseases has been discussed for many years. In 1961, Pollin *et al.* (1961) showed that tryptophan given with a monoamine-oxidase inhibitor to schizophrenic patients causes "mild to marked changes characterized primarily by mood elevation, increased involvement and extroversion, an early and transitory phase of somnolence, and more active deep tendon reflexes." Since that time many groups of investigators have studied the effects of the amino acid in mental patients. Most recently, Greenwood and her colleagues have carefully assessed the acute psychological effects of tryptophan administered intravenously (Greenwood *et al.*, 1974) and orally (Greenwood *et al.*, 1975) to normal adults. Although subjects rated themselves as drowsy under the influence of tryptophan, and the electroencephalogram correspondingly registered an increase of slow-wave activity, none went to sleep. Contrary to some other reports about the effects of tryptophan, there was no euphoria. These papers should be consulted for a detailed coverage of the literature on the psychopharmacological actions of tryptophan.

6.2. Sleep

The drowsiness caused by tryptophan may be clinically useful and it has been suggested that tryptophan is a natural hypnotic (Hartmann, 1974). Several studies have investigated the effect of tryptophan administration on sleep in man. Although the results are not in agreement as to the exact effects on the stages of sleep, the effects were generally found to be beneficial. Normal humans had a reduction by half in sleep latency (Hartmann et al., 1974) and patients with insomnia showed increased total sleep and decreased intermittent and early-morning wakefulness (Wyatt et al., 1970). In a study of tryptophan in affective disorders, tryptophan significantly increased sleep time in manic patients but not in unipolar or bipolar depressed patients (Murphy et al., 1974). Tryptophan may be connected with sleep physiologically; in a study with 6 young women, there was a significant correlation between plasma NAB tryptophan and the amount of REM sleep (Chen et al., 1974). These results generally favor Jouvet's concept of the importance of 5-HT in maintenance of the sleep state in experimental animals (Jouvet and Pujol, 1974).

6.3. Schizophrenia

Tryptophan has been tested on a few occasions as a potential therapeutic agent in schizophrenic patients. Such studies have been prompted variously by reports of the sedative action of tryptophan or by claims that it has some activating effect as mentioned above. A current hypothesis states that there is a protein of the α_2-globulin fraction of plasma that favors the uptake of tryptophan into the brain and that this protein is increased in schizophrenia (Frohman et al., 1969). With such a mechanism, a more rapid uptake of this amino acid would occur, and the overproduction of 5-HT or some other methylated indolic compound might be responsible for the mental symptoms. Warner et al. (1972) argue that cortical slices from rat brain incubated with plasma from schizophrenic patients take up more tryptophan from the incubation medium than slices bathed in plasma of controls; however, their data show that the increase is under 6% and is not statistically significant. Because some of the work of Frohman's group made use of chicken erythrocytes as the model tissue for testing plasma factors, Guchait *et al.* (1975) examined the effect of human plasma on the uptake of tryptophan by these cells. They noted a stimulation of uptake by plasma, whether it came from normal persons or from schizophrenic patients; hospitalization itself did not play a role. Himwich has also emphasized the probable role of abnormal indole (tryptophan) metabolism in schizophrenia (Himwich, 1971). This abnormality may result from a disturbance of regulatory mechanisms that operate at the intestinal or hepatic level (Payne *et al.*, 1974), but these are not yet understood.

Although Bender and Bamji (1974) found that schizophrenic patients had only half as much tryptophan in plasma as normal subjects did, Domino and Krause (1974*a,b*) have been unable in repeated studies to establish any such difference (Table 7). But in their study of the kinetics of plasma tryptophan following tryptophan loading, Domino and Krause noted that the amino acid reaches a peak earlier in the schizophrenic patients and is cleared from the plasma more slowly. The half-life of tryptophan in the plasma was the same in both groups (Domino and Krause, 1974*a*).

In connection with the measurement of tryptophan levels in schizophrenic patients, some of whom were receiving neuroleptic drugs, the results with these substances in experimental animals are of interest. The *acute* administration of chlorpromazine or haloperidol lowers the total tryptophan concentration in rat plasma (Tagliamonte *et al.*, 1971*b;* Bender, 1975), but raises the level of the NAB tryptophan (Bender, 1975). There is no change in the content of this amino acid in brain (Tagliamonte *et al.*, 1971*a*) even with the increased free tryptophan in plasma, but there is a greater rate of uptake of tryptophan into the brain (Bender, 1975). Thus, there must be greater utilization of tryptophan under these conditions or a greater flux of the amino acid from the brain. With *repeated* administration of chlorpromazine over an 11-day period, the effect wears off, so that both total and free serum tryptophan are within the control range (Bender, 1975). Nevertheless, the uptake of a radioactive dose of tryptophan is still greater than normal (Bender, 1975).

6.4. Mental Depression

6.4.1. Tryptophan Metabolism

Mentally depressed patients sometimes show changes in the plasma levels of tryptophan. Two examples are summarized in Table 7. According to Coppen *et al.* (1972*a*) the tryptophan concentration in CSF is low in mental depressives, as compared with a control group, but this has not been confirmed by Ashcroft *et al.* (1973*a*). It has been claimed that there

TABLE 7. Plasma Tryptophan Concentrations in Schizophrenia and Mental Depression

| Tryptophan (nmol/ml) | | | | |
| Control subjects | | Psychiatric patients[a] | | References |
Total	NAB	Total	NAB	
A. Schizophrenic patients				
83.7 ± 3.1[b] (21)	8.3 ± 0.3	45.1 ± 3.9** (10)	6.4 ± 0.5*	Bender and Bamji (1974)
73.9 ± 2.9 (18)	9.3 ± 1.0	71.0 ± 2.0 (26)	8.8 ± 0.5	Domino and Krause (1974a)
433.9 ± 31.3[c] (10)	83.2 ± 13.2	373.2 ± 15.7 (19)	56.8 ± 4.9	Domino and Krause (1974a)
79.8 ± 4.9 (5)		78.8 ± 2.9[d] (27)		Domino and Krause (2974b)
B. Depressed patients				
44.1 − 78.4[e]		12.2 − 19.6[e]		Lehmann (1972)
56.8 ± 3.4[f] (6)	5.9 ± 0.5	52.4 ± 3.4 (6)	3.4 ± 0.5***	Coppen et al. (1973)

[a] Probability of chance difference from controls: * < 0.02, ** < 0.01, *** < 0.001. For other differences between patients and controls, $P > 0.05$.
[b] Adult males, 27–50 years old; fasted overnight. Schizophrenic patients were receiving large doses of chlorpromazine and orphenadrine, neither one of which, according to the authors, affects tryptophan binding to serum albumin.
[c] Peak concentrations attained 1–2 hr after oral dose of tryptophan, 32 mg/kg.
[d] The patients in this study were chronic schizophrenics.
[e] The patients in this study were suffering from carcinoid syndrome.
[f] These values are for the six depressed patients after their recovery (test/retest).

must be an increased tryptophan metabolism in depressed patients because they show elevated rates of excretion of the metabolites kynurenine and 3-hydroxykynurenine (Cazzullo *et al.*, 1966; Rubin, 1967; Curzon and Bridges, 1970). This, too, has been disputed (Frazer *et al.*, 1973). By means of kinetic data applied to a bicompartmental model, Shaw *et al.* (1975) have derived information about the size of tryptophan pools and the fluxes of those pools. The two compartments in the simplified model are the free and bound tryptophan in plasma and interstitial fluid and the tryptophan in the cellular pool of amino acids. The concentrations of tryptophan in the pools are $65 \times 10^{-6} M$ and $130 \times 10^{-6} M$, respectively.

These two pools did not differ in size between depressed or manic patients, or those who had recovered, but this group of patients with affective disorder had a lower mean pool size than in controls. Flux of tryptophan from the body fluids into the cellular pool and the flux from that pool were not altered when patients recovered from their illness. The group of depressed patients and those who had recovered, taken together, had a significantly smaller flux of the amino acid from plasma to cellular pool than either controls or manic patients. Moreover, losses of tryptophan from the cellular pool occurred at a significantly slower rate in this group than in the controls. The fluxes for these two pools in the controls were 3.1 and 1.6 μmol liters^{-1} min^{-1}. The flux from the cellular pool to the plasma and interstitial fluid was about 1.5μ mol liters^{-1} min^{-1}. It is difficult to reconcile these measurements with the more conventionally derived ones previously mentioned, but perhaps this will be feasible when the modeling can separate out the components of the cellular pool so that tryptophan fluxes can be accounted for through the CNS and liver.

In Shaw's work it was noted that the two pools were significantly greater in size in males than in females.

6.4.2. Therapeutic Use of Tryptophan in Affective Disorders

On the basis of the suggestion that mental depression stems from a deficiency of serotonin in the brain (Coppen, 1967; Lapin and Oxenkrug, 1969), tryptophan has been tested in patients with this disease. The postulated decrease in 5-HT might come about in part under the influence of exaggerated levels of cortisol in the plasma, causing the activity of tryptophan pyrrolase of the liver to be increased (Curzon, 1969; Lapin and Oxenkrug, 1969). Elevated adrenocortical activity is common in depression (Rubin and Mandell, 1966) and cortisol induces pyrrolase in man (Altman and Greengard, 1966) as in other species. It can be readily shown by animal experiments that the administration of adrenocortical hormone results in a decrease of brain 5-HT and that this is mediated by a rise in tryptophan pyrrolase activity (Section 2).

The results of some of the studies of tryptophan in mental depression are set out in Table 8. The clinical evidence that tryptophan has antidepressant action is weak (Carroll, 1971), but it probably has some effect in maniacal states. The possibility that tryptophan exerts an action in affective states through formation of an amine cannot be summarily dismissed on the basis of trials carried out thus far, for the circumstances of these trials have not necessarily been the most favorable from a metabolic point of view. For example, the tendency toward overactivity of the adrenal cortex in depression could contribute to an elevated pyrrolase level, which would be counterproductive to the administration of tryptophan. The amino acid load itself favors an increase in pyrrolase activity, and it is interesting that in many of the trials the amino acid was given according to a regimen and in an amount that would favor peaking of pyrrolase activity rather than the maintenance, as far as possible with this kind of treatment, of fairly even levels of the enzymic activity. The authors have suggested, in fact, that the doubtful efficacy of tryptophan as an antidepressant medication may be due to its rapid catabolism, and that combination of tryptophan with a pyrrolase inhibitor, such as nicotinamide, could restrict the action of pyrrolase sufficiently to prolong and increase the rise in brain 5-HT and even eliminate undesirable side effects owing to formation of certain tryptophan metabolites in excess (Young and Sourkes, 1974). MacSweeney (1975) has obtained some preliminary evidence that the combination of tryptophan and nicotinamide is superior to unilateral electroshock therapy (given twice weekly) in the treatment of unipolar depression.

The data in Table 8 indicate that lower doses of tryptophan may be more effective in alleviating depression than higher doses. Thus, in studies on depressed patients in whom tryptophan was ineffective, the mean dose given was 8.7 ± 0.9 (S.E.) g for seven studies. (Where a range of doses was given the middle of the range was used in calculating the mean.) This value is 44% higher ($P < 0.05$) than the mean of 6.1 ± 0.6 for the six studies in which some effect was found. This might indicate that induction of tryptophan pyrrolase by tryptophan causes catabolism of much of the load at higher doses.

The efficacy of tryptophan supplements in restoring normal mental functions to a patient with carcinoidosis has been reported by Lehmann (1966). He has concluded that the 5-HT-secreting tumor was using up so much of the daily supply of dietary tryptophan that the brain was being deprived of this essential amino acid. Appropriate supplementation of the patient with sufficient tryptophan brought her from a stuporous akinetic state, through a depressive–catatonic phase associated with muscular spasticity, and eventually to a brief hypomanic phase before apparent mental recovery. This case of carcinoid tumor is somewhat unusual in provoking

TABLE 8. Some Studies of L-Tryptophan in Affective Disorders

Diagnosis	Number of patients	Dose of tryptophan (g)	Duration of treatment (days)	Specific clinical effect	Side effects	References[a]
Depression	41	5–7[b]	28	Equal to ECT		1
	12	7	21	No effect (less than ECT)	Nausea, lightheadedness, visual blurring	2
	38	6	28	Equal to imipramine	Nausea, lightheadedness, minimal changes in body temperature pattern	3
	8	2–8	8–18	Some improvement		4
	5	9	10	No effect		5
	12	9	28	Equal to imipramine		6
	10	6	14	No effect		7
	17	3–6	42	Equal to imipramine		8
	22	6–8	28	No effect (less than ECT)		9
	16	9.6 ± 0.4	20 ± 2	No effect		10
	22	6	21	Almost as good as imipramine	Less than imipramine	11
	4	11–16	42	No effect		12
	9	9	14 ± 1	No effect		13
Manic state	10	6	14	Slightly superior to chlorpromazine	Constipation, blurred vision, drowsiness, palpitations, paraesthesia seen in some patients during first two weeks, seldom thereafter	14
	10	9.6 ±0.4	20 ± 2	Partial antimanic effect		10
Bipolar depression	8	8	16	No effect		15
	8	9.6 ± 0.4	20 ± 2	Partial antidepressant effect		10
	2	11–16	42	No effect		12
	3	9	14 ± 1	No effect		13

[a] References: (1) Coppen et al. (1967), (2) Carroll et al. (1970), (3) Broadhurst (1970), (4) Bowers (1970), (5) Dunner and Goodwin (1972), (6) Coppen et al. (1972c), (7) Gayford et al. (1973), (8) Kline and Shah (1973), (9) Herrington et al. (1974), (10) Murphy et al. (1974), (11) Jensen et al. (1975), (12) Mendels et al. (1975), (13) Dunner and Fieve (1975), (14) Prange et al. (1974), (15) Bunney et al. (1971).
[b] The racemic amino acid was used.

such profound mental symptoms, although other less striking cases have been reported in which mental symptoms accompanying the carcinoid syndrome responded to tryptophan administration (Lehmann, 1972).

Another clinical use of tryptophan is in the depression that sometimes is observed in Parkinson's disease or accompanies the use of dopa in its treatment. The natural disease is characterized by the profound loss of dopamine from the basal ganglia and some associated structures of the brain, but there is also a decrease of norepinephrine and 5-HT. It is this last feature that has prompted trials of tryptophan, on the grounds that administration of dopa makes up for only part of the deficiency. Generally, tryptophan has little or no beneficial action in Parkinson's disease on its own, but in combination with dopa there may be a significant advantage over dopa alone (Coppen *et al.*, 1972*b*).

In one study, tryptophan was measured in brain areas of Parkinsonian patients who had been on dopa. The patients who exhibited psychotic symptoms in response to the dopa treatment had a lower content of tryptophan in all brain areas compared with those patients who did not develop psychoses (Birkmayer *et al.*, 1974; Table 9). Thus, the mental symptoms caused by dopa may be associated with lowered brain 5-HT. In experimental animals administration of L-tryptophan can normalize brain 5-HT in dopa-treated rats (Fahn *et al.*, 1975), and in patients, tryptophan has been used successfully to alleviate the mental symptoms that develop during dopa therapy (Birkmayer and Neumayer, 1972; Lehmann, 1973; Miller and Nieburg, 1974; Gehlen and Müller, 1974).

On this topic the question has been raised as to whether the use of large amounts of dopa, an amino acid that competes with tryptophan for the same transport system in the brain (and presumably at membranes of other organs as well), interferes with the movement of tryptophan across membranse. *In vitro* experiments with slices of rat brain indicate that there is a rather weak inhibition of tryptophan uptake by dopa, when the latter is present at a concentration corresponding to its peak level in plasma in clinical usage, i.e., about 4 μg/ml, or approximately 2×10^{-5} M, and tryptophan is present at 5×10^{-5} M, i.e., about twice the normal concentration of free tryptophan in plasma (Kiely and Sourkes, 1972). Nevertheless, *in vivo* it is possible to lower the content of tryptophan in rat brain in acute experiments by giving large doses of dopa, greater than those used in the treatment of Parkinson's disease [Table 9(a)]. The results are not entirely uniform, for Curzon and Knott (1974) obtained a decrease in total plasma tryptophan, an increase in the free tryptophan in plasma, and a corresponding increase in the content of the amino acid in brain. However, their analyses were done 2 hr after dopa administration, i.e., later than in the other experiments summarized in Table 9(b). These investigations point out

TABLE 9(a). Effect of L-Dopa on Tryptophan Content of Serum, Plasma, Brain, and Liver

Species	Treatment with dopa (other conditions)	Source	Tryptophan content		P	References
			Controls	Dopa		
Man[a]	0.6 g, p.o. (Case 4) 4 hr	Serum (nmol/ml)	780	390		Lehmann (1973)
	0.2 g, p.o. (Case 5) 1 hr	Serum (nmol/ml)	690	440		
	Same, but 2 hr (Case 6)	Serum (nmol/ml)	980	390		
	Individualized therapy for Parkinson's disease (Cases 1–3)	Serum (nmol/ml)	590–930	190–490[b]		
Rat	0.5 g/kg, 2 hr	Plasma, total (nmol/ml)	129 ± 3	83 ± 2	< 0.02	Curzon and Knott (1974)
		Plasma, NAB	9.8 ± 1.2	24.02 ± 2.4[c]	< 0.001	
		Brain (nmol/g)	12.7 ± 0.8	22.0 ± 1.0	< 0.001	
Rat	0.2 g/kg, i.p., 1 hr	Brain (nmol/g)	43.1 ± 2.5	29.9 ± 0.9[d]	< 0.001	Karobath et al. (1971)
Rat	0.2 g/kg, i.p., 10 min[e]	Brain (nmol/g)	13.7 ± 1.1	9.8 ± 0.6	<0.01	Algeri and Cerletti (1974)
Rat	0.25 g/kg,[f] 1 hr	Brain (nmol/g)	18.1 ± 2.3	13.2 ± 2.6	< 0.0[g]	Fahn et al. (1975)
Rat	0.5 or 1.0 g/kg, s.c., daily for 8 days	Liver (nmol/g)	30 ± 2	30 ± 1[h]	> 0.05	Liu et al. (1974)

[a] In these cases serum tryptophan was determined under conditions of tryptophan-loading on two occasions: before administration of a dose of L-dopa (control values) and the specified time after it.

[b] Patients were receiving L-dopa for the treatment of Parkinson's disease. Cases 1 and 2 had subnormal tryptophan concentrations before treatment with dopa.

[c] There was a highly significant increase in the concentration of unesterified fatty acid in the plasma at this time.

[d] Content of brain tyrosine and serotonin was also significantly lowered.

[e] The L-dopa caused a lowering of the content of tryptophan at 20 and 30 min also, but the effect was not statistically significant.

[f] All rats received carbidopa 25 mg/kg to inhibit peripheral decarboxylase. Injections were made intraperitoneally.

[g] There was also a significant decrease in brain tryptophan at 0.5 and 2 hr.

[h] Determined 2 hr after the last injection of L-dopa. At this time the contents of threonine and tyrosine were significantly increased.

TABLE 9(b).   Tryptophan in Brain Regions of Parkinsonian Patients Treated with
L-Dopa (nmol/g)[a]

	No psychotic symptoms	Number of patients	Psychotic symptoms	Number of patients
Nucl. caudat.	73 ± 14	6	24	1
Putamen	69 ± 15	6	24	1
Pallidum	69 ± 15	6	34 ± 15	3
S. nigra oral	83 ± 15	6	29 ± 5	3
S. nigra caudal	108 ± 20	6	29 ± 20	3
N. amygdalae	83 ± 15	6	10,25	2
Gyri cinguli	83 ± 15	6	15	1
Raphe	78 ± 15	6	29 ± 15	5
N. ruber	117 ± 10	6	54 ± 24	5

[a] All patients without psychotic symptoms had been treated with L-dopa. Those with psychotic symptoms had also been treated with L-tryptophan. No medication had been given for 2 days before death. Patients without psychotic symptoms had the same brain tryptophan content as controls (see Table 2). Data are taken from Birkmayer *et al.* (1974).

that by 2 hr, an excess of neurotransmitter has been formed from the dopa and released to the adipose tissue, among other places, where it causes lipolysis and an increase in the concentration of free fatty acid in the plasma, with the consequences for tryptophan metabolism that have already been described.

The decline in brain tryptophan in some Parkinsonian patients is probably due to a decrease in tryptophan absorption in the intestine caused by chronic dopa treatment (Lehmann, 1973) rather than by competition for entry into the brain.

6.5.   Electroconvulsive Therapy

Electroshock administered to humans or rats does not alter the total concentration of tryptophan in the plasma (Table 10), but there is an immediate and significant increase in the plasma NAB tryptophan. This persists for some time and seems to be a function of lipolysis; the free fatty acids released from adipose stores into the circulation compete successfully against tryptophan for binding sites on serum albumin. Electroshock experiments in the rat show that the plasma NAB tryptophan increases in this species also, and that it is associated with a very large increase in the content of tryptophan in brain (Table 10).

TABLE 10. Effect of Electroshock on Tryptophan Content of Tissues

Treatment	Tissue	Before	After	Time	References
Depressed patients on ECT	Plasma, total (nmol/ml)	41.1 ± 2.0 (18)	39.2 ± 2.9 (18)	1 min	Stelmasiak and Curzon (1974)
			42.6 ± 3.4 (18)	30 min	
	Plasma, NAB (nmol/ml)	5.4 ± 0.5 (18)	6.9 ± 0.5 (18)	1 min	
			5.9 ± 0.5 (18)	30 min	
Rats given single ECT	Plasma, total (nmol/ml)	137 ± 5 (150)	142 ± 5 (21)	1 hr	Tagliamonte et al. (1972a)
	Brain (nmol/g)	22.5 ± 5(150)	42.1 ± 1.5 (21)	1 hr	

6.6. Lithium Treatment

In the compartmental analysis carried out by Shaw and his colleagues (Shaw *et al.*, 1975), eight patients treated for a 3-week period with lithium had a mean serum lithium concentration of 0.9 ± 0.6 mM and a reduced concentration of tryptophan in the pool made up of plasma and interstitial fluid. Treatment of depression with tricyclic antidepressant drugs for a similar period of time led to a marked reduction in the concentration of tryptophan in the cellular pool of amino acids (Shaw *et al.*, 1975).

In contrast to the reduction in plasma tryptophan as reported for man, the parenteral administration of lithium salts to rats in amounts of 3–4 mEq/kg daily leads to a significant increase in the plasma and CSF concentrations of tryptophan within 5 days (Table 11). This does not occur with lithium salt added to the diet, but both routes of administration of the salt provoke large increases in the content of the amino acid in brain (Table 11). The effect of lithium on the disposition of tryptophan in the brain has also been examined by measuring the rate of disappearance of radioactive tryptophan from that organ after rats receive a standard dose by the intravenous route (Schubert, 1973). In these experiments, lithium-treated animals have a higher specific activity of the brain tryptophan immediately after the end of the infusion, and they lose tryptophan from the brain at a slower rate than controls do. Furthermore, the rats receiving lithium accumulate more 5-HT and 5-HIAA in brain and spinal cord than their controls. Lithium has been reported to stimulate tryptophan uptake *in vitro* by synaptosomes (Knapp and Mandell, 1975) but not by brain slices (Kiely, 1974).

6.7. Hepatic Coma

Liver disease is sometimes complicated by neurological and psychiatric symptoms that presage the development of coma. In some of the cases with chronic hepatic cirrhosis there is expansion of the collateral circulation between the intestine and the systematic vessels, so that blood coming from the gut bypasses the liver. This condition can be approached experimentally by surgical means, such as through portocaval anastomosis. In the "natural" disease, portal–systematic encephalopathy, there are changes in certain important blood constituents but none of these has been unequivocally associated with the severity of the clinical state. Nevertheless, the rise in blood ammonia is the most characteristic and represents a serious danger signal in the progress of the disease. Among the metabolic changes engendered by major liver dysfunction, there are disturbances of tryptophan metabolism. In cases of chronic hepatic encephalopathy, the concentration of tryptophan in human CSF is greatly elevated (Müting, 1962; Young *et*

TABLE 11. Effect of Lithium on Tryptophan Content of Plasma, Brain, and CSF

Treatment	Number	Plasma, serum, or CSF (nmol/ml)	P	Brain (nmol/g)	P	References
Controls	150	136 ± 3		22.5 ± 0.3		Tagliamonte et al. (1971b)
Lithium[a]	26	170 ± 4	< 0.001	35.8 ± 1.4	<0.001	
Controls	8	62.2 ± 6.9		9.3 ± 0.5		Schubert (1973)
Lithium[b]	8	66.6 ± 5.9[c]	> 0.05	12.2 ± 0.5	>0.01	
Controls	5	98.4 ± 3.4		14.2 ± 0.4		Iwata et al. (1974)
Lithium[d]	5	142 ± 4	< 0.05	18.4 ± 0.5	<0.05	
Controls	7	2.0 ± 0.2[e]		25.5 ± 0.7		Young et al. (1976a)
Lithium[a]	7	2.5 ± 0.2[e]	~ 0.5	31.3 ± 1.9	<0.05	

[a] 1.62 mEq/kg Li i.p. twice daily, for 5 days in the experiments of Tagliamonte et al. (1971b), 6 days in those of Young et al. (1976).

[b] Each kg of diet contained 40 mEq Li. The experiment lasted for 7 days.

[c] Neither total nor NAB serum tryptophan was significantly different from the control values.

[d] 2 mEq/kg was injected i.p. twice daily for 5 days.

[e] CSF, nmol/ml.

al., 1974*a*); this probably reflects an increase in brain tryptophan. Although CSF 5-HIAA is elevated (Knell *et al.*, 1972), its rise after probenecid administration is not quantitatively different from that seen in controls (Lal *et al.*, 1974). Thus, 5-HT turnover in the CNS does not seem to be abnormal. The large rise in CSF tryptophan is not explained by changes in the disposition of tryptophan in the serum; the total plasma tryptophan is normal and serum NAB tryptophan is only slightly elevated. The situation is different in fulminant hepatic failure where plasma NAB tryptophan was found to be elevated eight-fold (Knell *et al.*, 1974). Data are given in Table 12.

In the experimental analogue of chronic liver disease in the rat there are increases in both free and total serum tryptophan, as well as in skeletal muscle and various regions of the brain (Table 12). The content of tryptophan in liver is unchanged. The rate of uptake into the brain of intracisternally injected radioactive tryptophan is doubled (Baldessarini and Fischer, 1973).

Rapidly developing changes are brought about by acute devascularization of the liver. In pigs, this operation causes the content of tryptophan in the brain to increase significantly. This is consonant with the observed increase in the free tryptophan of plasma, which runs parallel to the rising concentration of free fatty acids (Curzon *et al.*, 1973; Table 12); the increase occurs despite a decrease in total plasma tryptophan. In acutely operated animals, the change in concentration in plasma of amino acids competing in transport with tryptophan apparently does not influence the content of tryptophan in the brain (Buxton *et al.*, 1974).

Knott and Curzon (1975) have used another model of acute hepatic disease. This one depends on the injection of carbon tetrachloride into rats to induce hepatic necrosis. Within 48 hr of this treatment, the animals have concentrations of free tryptophan and of free fatty acids in the plasma that are 2.5 times the control levels. At the same time, the total tryptophan concentration is unaltered from control values. The content of tryptophan in the liver, brain, and muscle is increased substantially, but this does not occur in the kidney. The plasma and brain changes are accompanied by increases in brain 5-HT and 5-HIAA.

Munro *et al.* (1975) have pointed out that in hepatic cirrhosis or portocaval shunt, the plasma concentrations of the branched-chain amino acids are subnormal. Owing to the hyperinsulinemia that accompanies the disturbance in liver function, this group of amino acids is removed from blood at an excessively high rate for catabolism. The deficiency of these amino acids in the plasma would then reduce the amount of competition that tryptophan encounters in membrane transport and would favor a greater rate of uptake of tryptophan into the brain. This would have as its consequences a greater rate of 5-HT formation, which, these investigators argue, could contribute to some of the neurological and mental accompani-

TABLE 12. CNS Tryptophan and Altered Hepatic Function

Investigation	Measurement	Control	Altered hepatic function
Rat, 2 months after portocaval anastomosis (Baldessarini and Fischer, 1973)	Brain		
	Whole organ (nmol/g)	24.0 ± 2.0	65.1 ± 9.8 (7)
	Uptake of [³H] tryptophan (nCi/g)	479.9 ± 44.1 (12)	940.2 ± 83.3 (7)
Rat, 3 weeks after portocaval anastomosis (Curzon et al., 1975)	Plasma (nmol/ml)		
	Total	73.5 ± 3.4 (10)	98.4 ± 4.4 (10)
	NAB	3.9 ± 0.5 (10)	8.3 ± 1.0 (10)
	Muscle (nmol/g)	24.0 ± 1.0 (10)	34.3 ± 1.5 (12)
	Liver	46.0 ± 1.5 (10)	47.0 ± 2.0 (11)
	Brain		
	Midbrain	20.6 ± 1.0 (12)	69.5 ± 5.4 (11)
	Hippocampus	26.0 ± 1.0 (12)	91.6 ± 5.9 (10)
	Pons and medulla	17.6 ± 1.0 (11)	67.6 ± 5.9 (10)
	Hypothalamus	36.2 ± 1.5 (5)	98.4 ± 8.8 (6)
Pig, 4–7.5 hr after hepatic devascularization (Curzon et al., 1973)	Plasma (nmol/ml)		
	Total	38.2 ± 3.4 (8)	28.9 ± 4.4 (4)
	NAB	5.4 ± 1.0 (8)	16.7 ± 1.0 (4)
	Brain (nmol/g)		
	Hypothalamus	23.0 ± 1.5 (8)	42.6 ± 4.9 (9)
	Thalamus	18.6 ± 1.0 (8)	36.2 ± 5.9 (9)
	Caudate nucleus	20.6 ± 2.0 (8)	33.8 ± 2.5 (7)
	Cerebral cortex	20.1 ± 1.5 (8)	33.8 ± 1.5 (7)
Patients with hepatic encephalopathy and coma (Hirayama, 1971)	Plasma, total (nmol/ml)	73.5 ± 4.9 (5)	117.5 ± 4.9 (13)
Patients with hepatic cirrhosis (Young et al., 1975)	Plasma (nmol/ml)		
	Total	67.6 ± 4.9 (29)	63.2 ± 6.4 (12)
	NAB	11.3 ± 1.5 (29)	16.7 ± 2.0 (12)
	CSF	2.8 ± 0.2 (29)	10.8 ± 2.0 (12)
Patients with fulminant hepatic failure (Knell et al., 1974)	Plasma (nmol/ml)		
	Total	52.4 ± 5.4 (23)	58.3 ± 19.6 (9)
	NAB	3.9 ± 1.0 (23)	34.3 ± 20.1 (9)

ments of severe cirrhosis, including coma. They advocate a clinical trial of branched-chain amino acids in the management of hepatic coma.

The competition for uptake is not completely eliminated, for the concentrations of phenylalanine and tyrosine are increased in the plasma in cirrhosis, and this could theoretically affect tryptophan transport. But apart from these considerations, in patients with hepatic cirrhosis it has been found that the tryptophan concentration of the CSF is increased five-fold (1.5–9 times a control mean value), and this suggests that the content of tryptophan in brain in this disease, not yet directly determined, is elevated just as in the animal models (Table 12). Moreover, if increased content of tryptophan in brain contributes to the genesis of coma in hepatic cirrhosis, it might be expected that this value would be greater in patients in coma. This comparison has also been made for CSF tryptophan and the results show that there is no difference in its concentration in patients with cirrhosis in coma as against those not in coma (Young *et al.,* 1975; Lal *et al.,* 1975). However, this does not preclude elevated brain tryptophan and 5-HT being among several factors that cause an imbalance between neuronal systems in the brain and thus cause coma.

6.8.   Phenylketonuria and Hartnup Disease

There is increased excretion of indoles in phenylketonuria. This stems from competition at the level of intestinal absorption between phenylalanine, whose concentration is elevated in the body fluids, and tryptophan (Spencer and Samiy, 1960). In the clinical manifestation of this competition in phenylketonuria (Drummond *et al.,* 1966; Yarbro and Anderson, 1966), there is a greater opportunity for bacterial action on tryptophan, the metabolic products then being absorbed and excreted in the urine. However, there is also competition at the point of transfer of tryptophan into the brain, and this process is considered sufficiently active to account for a relatively poor uptake of tryptophan and the consequent reduction in the brain's content of 5-HT (Yuwiler *et al.,* 1965).

A similar type of competition at the sites of absorption of amino acids in the intestine occurs in Hartnup disease (Rosenberg and Scriver, 1969). In this case the deficient uptake of tryptophan puts a strain on the amount of nicotinic acid available for bodily needs, and a pellagralike condition may develop, including even mental manifestations.

7.   REFERENCES

Aghajanian, G. K., 1972, Influence of drugs on the firing of serotonin-containing neurons in brain, *Fed. Proc. Fed. Amer. Soc. Exp. Biol.* **31**:91–96.

Aghajanian, G. K., and Asher, I. M., 1971, Histochemical fluorescence of raphe neurons: selective enhancement by tryptophan, *Science* 172:1159–1161.

Aghajanian, G. K., and Haigler, H. J., 1974, L-Tryptophan as a selective histochemical marker for serotoninergic neurons in single-cell recording studies, *Brain Res.* 81:364–372.

Aghajanian, G. K., Rosecrans, J. A., and Sheard, M. H., 1967, Serotonin: release in the forebrain by stimulation of midbrain raphe, *Science* 156:401–403.

Algeri, S., and Cerletti, C., 1974, Effects of L-dopa administration on the serotonergic system in rat brain: correlation between levels of L-dopa accumulated in the brain and depletion of serotonin and tryptophan, *Eur. J. Pharmacol.* 27:191–197.

Altman, K., and Greengard, O., 1966, Tryptophan pyrrolase induced in human liver by hydrocortisone: effect on excretion of kynurenine, *Science* 151:332–333.

Andén, N.-E., Dahlström, A., Fuxe, K., Larsson, K., Olson, L., and Ungerstedt, U., 1966, Ascending monoamine neurons to the telencephalon and diencephalon, *Acta Physiol. Scand.* 67:313–326.

Andén, N.-E., Engel, Y., and Rubenson,A., 1972, Central decarboxylation and uptake of L-dopa, *Naunyn-Schmiedebergs Arch. Pharmacol.* 273:11–26.

Aoki, K., and Siegel, F. L., 1970, Hyperphenylalaninemia: disaggregation of brain polyribosomes in young rats, *Science* 168:129–130.

Ashcroft, G. W., Eccleston, D., and Crawford, T. B. B., 1965, 5-Hydroxyindole metabolism in rat brain. A study of intermediate metabolism using the technique of tryptophan loading; I: methods, *J. Neurochem.* 12:489–492.

Ashcroft, G. W., Blackburn, I. M., Eccleston, D., Glen, A. I. M., Hartley, W., Kinloch, N. E., Lonergan, M., Murrary, L. G., and Pullar, I. A., 1973a, Changes on recovery in the concentrations of tryptophan and the biogenic amine metabolites in the cerebrospinal fluid of patients with affective illness, *Psychol. Med.* 3:319–325.

Ashcroft, G. W., Crawford, T. B. B., Cundall, R. L., Davidson, D. L., Dobson, J., Dow, R. C., Eccleston, D., Loose, R. W., and Pullar, I. A., 1973b, 5-Hydroxytryptamine metabolism in affective illness: the effect of tryptophan administration, *Psychol. Med.* 3:326–332.

Baker, D. H., Allen, N. K., Boomgaardt, J., Graber, G., and Norton, H. W., 1971, Quantitative aspects of D- and L-tryptophan utilization by the young pig, *J. Anim. Sci.* 33:42–46.

Baldessarini, R. J., and Fischer, J. E., 1973, Serotonin metabolism in rat brain after surgical diversion of the portal venous circulation, *Nature (London) New Biol.* 245:25–27.

Baños, G., Daniel, P. M., Moorhouse, S. R., and Pratt, O. E., 1973, The influx of amino acids into the brain of the rat *in vivo:* the essential compared with some non-essential amino acids, *Proc. R. Soc. London* 183:59–70.

Baños, G., Daniel, P. M., Moorhouse, S. R., and Pratt, O. E., 1974a, Inhibition of entry of some amino acids into the brain with observations on mental retardation in the aminoacidurias, *Psychol. Med.* 4:262–269.

Baños, G., Daniel, P. M., and Pratt, O. E., 1974b, Saturation of a shared mechanism which transports L-arginine and L-lysine into the brain of the living rat, *J. Physiol. (London)* 236:29–41.

Barbosa, E., Joanny, P., and Corriol, J., 1970, Accumulation active du tryptophane dans le cortex cérébral isolé du rat, *C. R. Séances Soc. Biol. Paris* 164:345–350.

Barondes, S. H., 1974, Do tryptophan concentrations limit protein synthesis at specific sites in the brain? in: *Aromatic Amino Acids in the Brain, Ciba Found. Symp. 22 (New ser.)* (G. E. W. Wolstenholme and D. W. Fitzsimons, eds.), pp. 265–281, Elsevier, Amsterdam.

Baumann, A., Bourgoin, S., Benda, P., Glowinski, J., and Hamon, M., 1974, Characteristics of tryptophan accumulation by glial cells, *Brain Res.* 66:253–263.

Belin, M.-F., Chouvet, G., and Pujol, J.-F., 1974, Transport synaptosomal du tryptophane et de la tyrosine cérébrale. Stimulation de la vitesse de capture après réserpine ou inhibition de la monoamine oxydase, *Biochem. Pharmacol.* 23:587–597.

Bender, A., 1975, Serum-amino acids and brain tryptophan uptake, *Lancet* 2:278–279.

Bender, D. A., and Bamji, A. N., 1974, Serum tryptophan binding in chlorpromazine-treated chronic schizophrenics, *J. Neurochem.* **22**:805–809.

Benkert, O., and Matussek, N., 1970, Influence of hydrocortisone and glucagon on liver tyrosine transaminase and on brain tyrosine, norepinephrine and serotonin, *Nature (London)* **228**:73–75.

Biggio, G., Fadda, F., Fanni, P., Tagliamonte, A., and Gessa, G. L., 1974, Rapid depletion of serum tryptophan, brain tryptophan, serotonin and 5-hydroxyindoleacetic acid by a trypto-phan-free diet, *Life Sci.* **14**:1321–1329.

Birkmayer, W., and Neumayer, E., 1972, Die Behandlung der Dopa-Psychosen mit L-Tryptopan, *Nervenaezt* **43**:76–78.

Birkmayer, W., Danielczyk, W., Neumayer, E., and Riederer, P., 1974, Nucleus ruber and L-dopa psychosis: biochemical post-mortem findings, *J. Neural Transm.* **35**:93–116.

Blasberg, R., and Lajtha, A., 1965, Substrate specificity of steady-state amino acid transport in mouse brain slices, *Arch. Biochem. Biophys.* **112**:361–377.

Bloxam, D. L., Warren, W. H., and White, P. J., 1974, Involvement of the liver in the regulation of tryptophan availability. Possible role in the responses of liver and brain to starvation, *Life Sci.* **15**:1443–1455.

Bosin, T. R., Hathaway, D. R., and Maickel, R. P., 1974, Intestinal absorption of tryptophan stereoisomers and analogs, *Arch. Intern. Pharmacodyn. Thér.* **212**:32–35.

Boulton, A. A., and Baker, G. B., 1975, The subcellular distribution of β-phenylethylamine, p-tyramine and tryptamine in rat brain, *J. Neurochem.* **25**:477–481.

Bourgoin, S., Faive-Bauman, A., Benda, P., Glowinski, J., and Hamon, M., 1974, Plasma tryptophan and 5-HT metabolism in the CNS of the newborn rat, *J. Neurochem.* **23**:319–327.

Bowers, M. B., 1970, Cerebrospinal fluid 5-hydroxyindoles and behaviour after L-tryptophan and pyridoxine administration to psychiatric patients, *Neuropharmacology* **9**:599–604.

Broadhurst, A. D., 1970, L-Tryptophan versus E.C.T., *Lancet* **1**:1392–1393.

Bunney, W. E., Brodie, K. H., Murphy, D. L., and Goodwin, F. K., 1971, Studies of alpha-methyl-para-tyrosine, L-dopa, and L-tryptophan in depression and mania, *Amer. J. Psychiatry* **127**:872–881.

Buxton, B. H., Stewart, D. A., Murray-Lion, I. M., Curzon, G., and Williams, R., 1974, Plasma amino acids in experimental acute hepatic failure and their relationship to brain tryptophan, *Clin. Sci. Mol. Med.* **46**:559–562.

Carlsson, A., and Lindqvist, M., 1972, The effect of L-tryptophan and some psychotropic drugs on the formation of 5-hydroxytryptophan in the mouse brain *in vivo, J. Neural Transm.* **33**:23–43.

Carroll, B. J., 1971, Monoamine precursors in the treatment of depression, *Clin. Pharmacol. Ther.* **12**:743–761.

Carroll, B. J., Mowbray, R. M., and Davies, B., 1970, Sequential comparison of L-tryptophan with E.C.T. in severe depression, *Lancet* **1**:967–969.

Cazzullo, C. L., Mangoni, H., and Mascherpa, A., 1966, Tryptophan metabolism in affective psychoses, *Br. J. Psychiatry* **112**:157–162.

Chen, C. N., Kalucy, R. S., Hartmann, M. K., Lacey, J. H., Crisp, A. H., Bailey, J. E., Eccleston, E. G., and Coppen, A., 1974, Plasma tryptophan and sleep, *Br. Med. J.* **4**:564–566.

Christensen, H. N., 1964, Free amino acids and peptides in tissues, in: *Mammalian Protein Metabolism,* Vol. I (H. N. Munro and J. B. Allison, eds.), pp. 105–123, Academic Press, New York.

Cocchi, D., di Giulio, A., Groppetti, A., Mantegazza, P., Muller, E. E., and Spano, P. F., 1975, Hormonal inputs and brain tryptophan metabolism: the effect of growth hormone, *Experientia* **31**:384–386.

Cohen, L. L., and Huang, K. C., 1964, Intestinal transport of tryptophan and its derivatives, *Amer. J. Physiol.* **206**:647.

Coppen, A., 1967, The biochemistry of affective disorders, *Br. J. Psychiatry* **113**:1237–1264.

Coppen, A., 1972, Indoleamines and affective disorders, *J. Psychiatr. Res.* **9**:163–171.

Coppen, A., Shaw, D. M., Herzberg, B., and Maggs, R., 1967, Tryptophan in the treatment of depression, *Lancet* **2**:1178–1180.

Coppen, A., Brooksbank, B. W. L., and Peet, M., 1972a, Tryptophan concentration in the cerebrospinal fluid of depressive patients, *Lancet* **1**:1393.

Coppen, A., Metcalfe, M., Carroll, J. D., and Morris, J. G. L., 1972b, Levodopa and L-tryptophan therapy in parkinsonism, *Lancet* **1**:654–658.

Coppen, A., Whybrow, P. C., Noguera, R., Maggs, R., and Prange, A. J., 1972c, The comparative antidepressant value of L-tryptophan and imipramine with and without attempted potentiation by liothyronine, *Arch. Gen. Psychiatry* **26**:234–241.

Coppen, A., Eccleston, E. G., and Peet, M., 1973, Total and free tryptophan concentration in the plasma of depressive patients, *Lancet* **2**:60–63.

Costa, E. and Meek, J. L., 1974, Regulation of biosynthesis of catecholamines and serotonin in the CNS, *Annu. Rev. Pharmacol.* **14**:491–511.

Culley, W. J., Saunders, R. N., Merts, E. T., and Jolly, D. H., 1963, Low tryptophan diet will decrease serotonin content of tissues, *Proc. Soc. Exp. Biol. Med.* **113**:645–648.

Cunningham, V. J., Hay, L., and Stoner, H. B., 1975, The binding of L-tryptophan to serum albumins in the presence of non-esterified fatty acids, *Biochem. J.* **146**:653–658.

Curzon, G., 1969, Tryptophan pyrrolase: a biochemical factor in depressive illness?, *Br. J. Psychiatry* **115**:1367–1374.

Curzon, G., and Bridges, P. K., 1970, Tryptophan metabolism in depression, *J. Neurol. Neurosurg. Psychiatry* **33**:698–704.

Curzon, G., and Green, A. R., 1969, Effects of immobilization on rat liver tryptophan pyrrolase and brain 5-hydroxytryptamine metabolism, *Br. J. Pharmacol.* **37**:689–697.

Curzon, G., and Knott, P. J., 1974, Effects on plasma and brain tryptophan in the rat of drugs and hormones that influence the concentration of unesterified fatty acid in the plasma, *Br. J. Pharmacol.* **50**:197–204.

Curzon, G., and Knott, P. J., 1975, Rapid effects of environmental disturbance on rat plasma unesterified fatty acid and tryptophan concentrations and their prevention by antilipolytic drugs, *Br. J. Pharmacol.* **54**:389–396.

Curzon, G., and Marsden, C. A., 1975, Metabolism of a tryptophan load in the hypothalamus and other brain regions, *J. Neurochem.* **25**:251–256.

Curzon, G., Joseph, M. H., and Knott, P. J., 1972, Effects of immobilization and food deprivation on rat brain tryptophan metabolism, *J. Neurochem.* **19**:1967–1974.

Curzon, G., Kantamaneni, B. D., Winch, J., Rojas-Bueno, A., Murray-Lyon, I. M., and Williams, R., 1973, Plasma and brain tryptophan changes in experimental acute hepatic failure, *J. Neurochem.* **21**:137–145.

Curzon, G., Friedel, J., Katamaneni, B. D., Greenwood, M. H., and Lader, M. H., 1974, Unesterified fatty acids and the binding of tryptophan in human plasma, *Clin. Sci. Mol. Med.* **47**:415–424.

Curzon, G., Kantamaneni, B. D., Fernando, J. C., Woods, M. S., and Cavanagh, J. B., 1975, Effects of chronic portocaval anastomosis on brain tryptophan, tyrosine and 5-hydroxytryptamine, *J. Neurochem.* **24**:1065–1070.

Cutler, R. W. P., 1970, Transport of lysine from cerebrospinal fluid of the cat, *J. Neurochem.* **17**:1017–1027.

Dahlström, A., and Fuxe, K., 1964, Evidence for the existence of monoamine-containing neurons in the central nervous system. 1. Demonstration of monoamines in the cell bodies of brain stem neurons, *Acta Physiol. Scand.* **62**, supplement 232.

Daniel, P. M., Love, E. R., Moorhouse, S. R., and Pratt, O. E., 1975, Amino acids, insulin and hepatic coma, *Lancet* **2**:179–180.

David, J. C., Dairman, W., and Udenfriend, S., 1974, On the importance of decarboxylation in the metabolism of phenylalanine, tyrosine and tryptophan, *Arch. Biochem. Biophys.* **160**:561–568.

DeAntoni, A., Allegri, G., Costa, C., and Bordin, F., 1974, Melanogenesis from tryptophan: biogenetic experiments with Harding-Passey mouse melanoma, *Experientia* **30**:600–601.

Deguchi, T., Sinha, A. K., and Barchas, J. D., 1973, Biosynthesis of serotonin in raphe nuclei of rat brain: effect of p-chlorophenylalanine, *J. Neurochem.* **20**:1329–1336.

Denizeau, F., and Sourkes, T. L., 1977, Regional transport of tryptophan in rat brain, *J. Neurochem.* in press.

Denizeau, F., Wyse, J. R., and Sourkes, T. L., 1976, Two models of multiple transport systems; computation of parameters, *J. Theoret. Biol.,* in press.

Dickerson, J. W., and Pao, S.-K., 1975, The effect of a low protein diet and exogenous insulin on brain tryptophan and its metabolites in the weanling rat, *J. Neurochem.* **25**:559–564.

Domino, E. F., and Krause, R. R., 1974*a*, Free and bound serum tryptophan in drug-free normal controls and chronic schizophrenic patients, *Biol. Psychiatry* **8**:265–279.

Domino, E. F., and Krause, R. R., 1974*b*, Plasma tryptophan tolerance curves in drug-free normal controls, schizophrenic patients and prisoner volunteers, *J. Psychiatr. Res.* **10**:247–261.

Drummond, K. N., Michael, A. F., and Good, R. A., 1966, Tryptophan metabolism in a patient with phenylketonuria and scleroderma: a proposed explanation of the indole defect in phenylketonuria, *Can. Med. Assoc. J.* **94**:834–838.

Duncan, R. J. S., and Sourkes, T. L., 1974, Some enzymic aspects of the production of oxidized or reduced metabolites of catecholamines and 5-hydroxytryptamine by brain tissues, *J. Neurochem.* **22**:663–669.

Dunner, D. L., and Goodwin, F. K., 1972, Effect of L-tryptophan on brain serotonin metabolism in depressed patients, *Arch. Gen. Psychiatry* **26**:364–366.

Dunner, D. L., and Fieve, R. R., 1975, Affective disorders: studies with amine precursors, *Amer. J. Psychiatry* **132**:180–183.

Eber, O., and Lembeck, F., 1958, Hydroxytryptamine content of the rat intestine at different tryptophan levels in nutrition, *Pflügers Arch. Gesamte Physiol. Menschen Tiere* **265**:563–566.

Eccleston, D., Ashcroft, G. W., Moir, A. T. B., Parker-Rhodes, A., Lutz, W., and O'Mahoney, D. P., 1968, A comparison of 5-hydroxyindoles in various regions of dog brain and cerebrospinal fluid, *J. Neurochem.* **15**:947–957.

Eccleston, D., Ashcroft, G. W., Crawford, T. B. B., Stanton, J. B., Wood, D., and McTurk, P. H., 1970, Effect of tryptophan administration on 5HIAA in cerebrospinal fluid in man, *J. Neurol. Neurosurg. Psychiatry* **33**:269–272.

Edvinson, L., Hakanson, R., Rönnberg, A.-L, and Sundler, R., 1973, Tryptophyl-polypeptides in rat brain, *J. Neurochem.* **20**:897–899.

Etienne, P., Young, S. N., and Sourkes, T.L., 1976, Inhibition by albumin of tryptophan uptake by rat brain, *Nature* **262**:144–145.

Fahn, S., Snider, S., Prosad, A. L. N., Lane, E., and Madadon, H., 1975, Normalization of brain serotonin by L-tryptophan in levodopa-treated rats, *Neurology* **25**:861–865.

Fernstrom, J. D., and Wurtman, R. J., 1971, Brain serotonin content: physiological dependence on plasma tryptophan levels, *Science* **173**:149–152.

Fernstrom, J. D., and Wurtman, R. J., 1972, Brain serotonin content: physiological regulation by plasma neutral amino acids, *Science* **178**:414–416.

Fernstrom, J. D., and Wurtman, R. J., 1974, Control of brain serotonin levels by the diet, *Adv. Biochem. Psychopharmacol.* **11**:133–142.

Flentge, F., Venema, K., and Korf, J., 1974, Automated assay of tryptophan at the nanogram level: determination of tryptophan in cerebrospinal fluid and of total and nonprotein bound tryptophan in serum, *Biochem. Med.* **11**:234–241.

Fonnum, F., Haavaldsen, R., and Tangen, O., 1964, Transamination of aromatic amino acids in rat brain, *J. Neurochem.* **11**:109–118.

Forn, J., Tagliamonte, A., Tagliamonte, P., and Gessa, G. L., 1972, Stimulation by dibutyryl cyclic AMP of serotonin synthesis and tryptophan transport in brain slices, *Nature (London) New Biol.* **237**:245–247.

Franklin, G. M., Dudzinski, D. S., and Cutler, R. W. P., 1975, Amino acid transport into the cerebrospinal fluid of the rat, *J. Neurochem.* **24**:367–372.

Frazer, A., Pandey, G. N., and Mendels, J., 1973, Metabolism of tryptophan in depressive disease, *Arch. Gen. Psychiatry* **29**:528–535.

Friedman, P. A., Kappelman, A. H., and Kaufman, S., 1972, Partial purification and characterization of tryptophan hydroxylase from rabbit hindbrain, *J. Biol. Chem.* **247**:4165–4173.

Frohman, C. E., Warner, K. A., Yoon, H. S., Arthur, R. E., and Gottlieb, J. S., 1969, The plasma factor and transport of indoleamino acids, *Biol. Psychiatry* **1**:377–385.

Fuxe, K., 1965, Evidence for the existence of monoamine neurons in the central nervous system. IV. Distribution of monoamine nerve terminals in the central nervous system, *Acta Physiol. Scand.* **64**: Suppl. 247.

Gál, E. M., 1972, 5-Hydroxytryptamine-*O*-sulphate: an alternate route of serotonin inactivation in brain, *Brain Res.* **44**:309–312.

Gál, E. M., 1974, Cerebral tryptophan-2,3-dioxygenase (pyrrolase) and its induction in rat brain, *J. Neurochem.* **22**:861–863.

Gál, E. M., 1975, Hydroxylation 4 tryptophan and its control in brain, *Pavlovian J.* **10**:145–160.

Gál, E. M., and Drewes, P. A., 1962, Metabolism of 5-hydroxytryptamine (serotonin). II. Effect of tryptophan deficiency in rats, *Proc. Soc. Exp. Biol. Med.* **110**:368–371.

Gál, E. M., and Roggeveen, A. E., 1973, Cerebral hydroxylases: stimulation by a new factor, *Science* **179**:809–811.

Gál, E. M., Armstrong, J. C., and Ginsberg, B., 1966. The nature of *in vitro* hydroxylation of L-tryptophan by brain tissue, *J. Neurochem.* **13**:643–654.

Gayford, J. J., Parker, A. L., Phillips, E. M., and Rowsell, A. R., 1973, Whole blood 5-hydroxtryptamine during treatment of endogenous depressive illness, *Br. J. Psychiatry* **122**:597–598.

Geddes, C., Martin, M. J., and Moir, A. T. B., 1973, The transport of L-tryptophan from cerebrospinal fluid in the dog, *J. Physiol.* **230**:595–611.

Gehlen, W., and Müller, J., 1974, Zur Therapie der Dopa-Psychosen mit L-Tryptophan, *Dtsch. Med. Wochenschr.* **99**:457–463.

Gessa, G. L., and Tagliamonte, A., 1974, Possible role of free serum tryptophan in the control of brain tryptophan level and serotonin synthesis, *Adv. Biochem. Psychopharmacol.* **11**:119–131.

Gessa, G. L., Biggio, G., Fadda, F., Corsini, G. U., and Tagliamonte, A., 1974, Effect of the oral administration of tryptophan-free amino acid mixtures on serum tryptophan, brain tryptophan and serotonin metabolism, *J. Neurochem.* **22**:869–870.

Gilboe, D. D., and Betz, A. L., 1971, Kinetics of amino acid transport in the isolated dog brain, Abstracts, 3rd Int. Meeting, International Society of Neurochemistry, Budapest (July, 1971), p. 94.

Grahame-Smith, D. G., 1971, Studies *in vivo* on the relationship between brain tryptophan, brain 5HT synthesis and hyperactivity in rats treated with a monoamine oxidase inhibitor and L-tryptophan, *J. Neurochem.* **18**:1053–1066.

Grahame-Smith, D. G., and Parfitt, A. G., 1970, Tryptophan transport across the synaptosomal membrane, *J. Neurochem.* **17**:1339–1353.

Green, A. R., and Curzon, G., 1968, Decrease of 5-hydroxytryptamine in the brain provoked by hydrocortisone and its prevention by allopurinol, *Nature (London)* **220**:1095–1097.

Green, A. R., and Curzon, G., 1970, The effect of tryptophan metabolites on brain 5-hydroxytryptamine metabolism, *Biochem. Pharmacol.* **19**:2061–2068.

Green, A. R., and Curzon, G., 1975, Effects of hydrocortisone and immobilization on tryptophan metabolism in brain and liver of rats of different ages, *Biochem. Pharmacol.* **24**:713–716.

Green, A. R., Koslow, S. H., and Costa, E., 1973, Identification and quantitation of a new indolealkylamine in rat hypothalamus, *Brain Res.* **51**:371–374.

Green, A. R., Sourkes, T. L., and Young, S. N., 1975*a*, Liver and brain tryptophan metabolism following hydrocortisone administration to rats and gerbils, *Br. J. Pharmacol.* **53**:287–292.

Green, A. R., Woods, H. F., Knott, P. J., and Curzon, G., 1975*b*, Factors influencing effect of hydrocortisone on rat brain tryptophan metabolism, *Nature (London)* **255**:170.

Greenwood, M. H., Friede, J., Bond, A. J., Curzon, G., and Lader, M. H., 1974, The acute effects of intravenous infusion of L-tryptophan in normal subjects, *Clin. Pharmacol. Ther.* **16**:455–464.

Greenwood, M. H., Lader, M. H., Kantameneni, B. D., and Curzon, G., 1975, The acute effects of oral (-)-tryptophan in human subjects, *Br. J. Clin. Pharmacol.* **2**:165–172.

Guchhait, R. B., Janson, C., and Price, W. H., 1975, Validity of plasma factor in schizophrenia as measured by tryptophan uptake, *Biol. Psychiatry* **10**:303–314.

Guroff, G., and Udenfriend, S., 1962, Studies on aromatic amino acid uptake by rat brain *in vivo, J. Biol. Chem.* **237**:803–806.

Hamon, M., and Glowinski, J., 1974, Regulation of serotonin synthesis, *Life Sci.* **15**:1533–1548.

Hartmann, E., 1974, L-Trytophan: a possible natural hypnotic substance, *J. Amer. Med. Assoc.* **230**:1680–1681.

Hartmann, E., Cravens, J., and List, S., 1974, Hypnotic effects of L-tryptophan, *Arch. Gen. Psychiatry* **31**:394–397.

Héry, F., Rouer, E., and Glowinski, J., 1972, Daily variations of serotonin metabolism in the rat brain, *Brain Res.* **43**:445–465.

Herrington, R. N., Bruce, A., Johnstone, E. C., and Lader, M. H., 1974, Comparative trial of L-tryptophan and E.C.T. in severe depressive illness, *Lancet* **2**:731–734.

Hess, S. M., and Doepfner, W., 1961, Behavioral effects and brain amine content in rats, *Arch. Intern. Pharmacodyn. Thér.* **134**:89–99.

Hillarp, N.-Å., Fuxe, K., and Dahlström, A., 1966, Central monoamine neurons, in: *Mechanisms of Release of Biogenic Amines* (U. S. Von Euler, S. Rosell, and B. Uvnäs, eds.), pp. 31–57, Pergamon Press, Oxford.

Hillier, J., Hillier, J. G., and Redgem, P. H., 1975, Liver tryptophan pyrrolase acivity and metabolism of brain 5HT in rat, *Nature (London)* **253**:566–567.

Himwich, H. E. (ed.), 1971, *Biochemistry, Schizophrenia and Affective Illnesses,* Williams and Wilkins, Baltimore.

Hirata, F., and Hayaishi, O., 1971, Possible participation of superoxide anion in the intestinal tryptophan-2,3-dioxygenase reaction, *J. Biol. Chem.* **246**:7825–7826.

Hirata, F., Hayaishi, O., Tokuyama, T., and Senoh, S., 1974, In vitro and in vivo formation of two new metabolites of melatonin, *J. Biol. Chem.* **249**:1311–1313.

Hirayama, C., 1971, Tryptophan metabolism in liver disease, *Clin. Chim. Acta* **32**:191–197.

Horwitt, M. K., Harvey, C. C., Rothwell, W. S., Cutler, J. L., and Haffron, D., 1956, Tryptophan–niacin relationships in man. Studies with diets deficient in riboflavin and niacin together with observations on the excretion of nitrogen and niacin metabolites, *J. Nutr.* **60**: Suppl. 1:1–43.

Ichiyama, A., Nakamura, S., Nishizuka, Y., and Hayaishi, O., 1970, Enzymatic studies on the biosynthesis of serotonin in mammalian brain, *J. Biol. Chem.* **245**:1699–1709.

Iwata, H., Okamoto, H., and Kuramoto, I., 1974, Effect of lithium on serum tryptophan and brain serotonin in rats, *Jpn. J. Pharmacol.* **24**:235–240.

Iwata, H., Okamoto, H., and Koh, S., 1975, Effects of various drugs on serum free and total tryptophan levels and brain tryptophan metabolism in rats, *Jpn. J. Pharmacol.* **25**:303–310.

Jacquez, J. A., 1961, Transport and exchange diffusion of L-tryptophan in Ehrlich cells, *Amer. J. Physiol.* **200**:1063–1068.

Jensen, K., Fruensgaard, K., Ahlfors, U.-G., Pihkanen, T. A., Tuomikoski, S., Ose, E., Dencker, S. J., Lindberg, D., and Nagy, A., 1975, Tryptophan/imipramine in depression, *Lancet* **2**:920.

Joanny, P., Barbosa, E., and Corriol, J., 1968, Accumulation active de quelques acides aminés dans les coupes de cerveau de Rat, *J. Physiol. (Paris), Supple.* **60**:265.

Joh, T. H., Shikimi, T., Pickel, V. M., and Reis, D. J., 1975, Brain tryptophan hydroxylase: purification of, production of antibodies to, and cellular ultrastructural localization in serotonergic neurons of rat midbrain, *Proc. Natl. Acad. Sci. U.S.A.* **72**:3575–3579.

Johnson, T. C., and Chou, L., 1973, Level and amino acid acceptor activity of mouse brain t-RNA during neural development, *J. Neurochem.* **20**:405–414.

Johnson, T. C., and Luttges, M. W., 1966, The effects of maturation on *in vitro* protein synthesis by mouse brain cells, *J. Neurochem.* **13**:545–552.

Jouvet, M., and Pujol, J.-F., 1974, Effects of central alterations of serotoninergic neurons upon the sleep-waking cycle, *Adv. Biochem. Psychopharmacol.* **11**:199–209.

Karobath, M., Diaz, J.-L., and Huttunen, M. O., 1971, The effect of L-dopa on the concentrations of tryptophan, tyrosine and serotonin in rat brain, *Eur. J. Pharmacol.* **14**:393–396.

Katzman, R., and Pappius, H. M., 1973, *Brain Electrolytes and Fluid Metabolism,* Williams and Wilkins, Baltimore.

Kiely, M. E., 1974, The transport of L-tryptophan into slices of rat cerebral cortex, Ph.D. thesis, McGill University, Montreal, p. 114.

Kiely, M. E., and Sourkes, T. L., 1972, Transport of L-tryptophan into slices of rat cerebral cortex, *J. Neurochem.* **19**:2863–2872.

Kim, J. H., and Miller, L. L., 1969, The functional significance of changes in activity of the enzymes, tryptophan pyrrolase and tyrosine transaminase, after induction in intact rats and in the isolated, perfused rat liver, *J. Biol. Chem.* **244**:1410–1416.

Kline, N. S., and Shah, B. K., 1973, Comparable therapeutic efficacy of tryptophan and imipramine: average therapeutic ratings versus "true" equivalence. An important difference, *Curr. Thera. Res. Clin. Exp.* **15**:484–487.

Knapp, S., and Mandell, A. J., 1973, Short- and long-term lithium administration: effects on the brain's serotonergic biosynthetic systems, *Science* **180**:645–647.

Knapp, S., and Mandell, A. J., 1975, Effects of lithium chloride on parameters of biosynthetic capacity for 5-hydroxytryptamine in rat brain, *J. Pharmacol. Exp. Ther.* **193**:812–823.

Knell, A. J., Pratt, O. E., Curzon, G., and Williams, R., 1972, Changing ideas in hepatic encephalopathy, in: *Eighth Symposium on Advanced Medicine* (G. Neale, ed.), pp. 156–170, Pitman, London.

Knell, A. J., Davidson, A. R., Williams, R., Kantamaneni, B. D., and Curzon, G., 1974, Dopamine and serotonin metabolism in hepatic encephalopathy, *Br. Med. J.* **1**:549–551.

Knott, P. J., and Curzon, G., 1972, Free tryptophan in plasma and brain tryptophan metabolism, *Nature (London)* **239**:452–453.

Knott, P. J., and Curzon, G., 1974, Effect of increased rat brain tryptophan on 5-hydroxytryptamine and 5-hydroxyindoleacetic acid in the hypothalamus and other brain regions, *J. Neurochem.* **22**:1065–1071.

Knott, P. J., and Curzon, G., 1975, Tryptophan and tyrosine disposition and brain tryptophan metabolism in acute carbon tetrachloride poisoning, *Biochem. Pharmacol.* **24**:963–966.

Knox, W. E., 1966, The regulation of tryptophan pyrrolase activity by tryptophan, *Adv. Enzyme Regul.* **4**:287–297.

Koslow, S. H., 1974, 5-Methoxytryptamine: a possible central nervous system transmitter, *Adv. Biochem. Psychopharmacol.* **11**:95–100.

Krnjevic, K., 1974, Chemical nature of synaptic transmission in vertebrates, *Physiol. Rev.* **54**:418–540.

Kuhar, M. J., Aghajanian, G. K., and Roth, R. H., 1972, Tryptophan hydroxylase activity and synaptosomal uptake of serotonin in discrete brain regions after midbrain raphe lesions: correlation with serotonin levels and histochemical fluorescence, *Brain Res.* **44**:165–176.

Lajtha, A., 1964, Protein metabolism of the nervous system, in: *International Review of Neurobiology,* Vol. 6 (C. C. Pfeiffer and J. R. Surythies, eds.), pp. 1–98, Academic Press, New York.

Lajtha, A., 1974, Amino acid transport in the brain *in vivo* and *in vitro,* in: *Aromatic Amino Acids in the Brain, Ciba Found. Symp. 22 (New Ser.),* (G. E. W. Wolstenholme and D. W. Fitzsimons, eds.), pp. 25–49, Elsevier, Amsterdam.

Lajtha, A., and Toth, J., 1962, The brain barrier system. III. The efflux of intracerebrally administered amino acids from the brain, *J. Neurochem.* **9**:199–212.

Lal, S., Aronoff, A., Garelis, E., Sourkes, T. L., Young, S. N., and de la Vega, C. E., 1974, Cerebrospinal fluid homovanillic acid, 5-hydroxyindoleacetic acid, lactic acid, and pH before and after probenecid in hepatic coma, *Clin. Neurol. Neurosurg.* **77**:142–154.

Lal, S., Young, S. N., and Sourkes, T. L., 1975, 5-Hydroxytryptamine and hepatic coma, *Lancet* **2**:979–980.

Lapin, I. P., and Oxenkrug, G. F., 1969, Intensification of the central serotoninergic process as a possible determinant of the thymoleptic effect, *Lancet* **1**:132–136.

Lees, G. J., and Weiner, N., 1975, An examination of the catalytic properties of several aminotransferases in brain and liver by means of an improved radiometric assay with dinitrophenylhydrazine, *J. Neurochem.* **25**:315–322.

Lehmann, J., 1966, Mental disturbances followed by stupor in a patient with carcinoidosis. Recovery with tryptophan treatment, *Acta Psychiatr. Scand.* **42**:153–161.

Lehmann, J., 1972, Mental and neuromuscular symptoms in tryptophan deficiency, *Acta Psychiatr. Scand.* **237**:1–28.

Lehmann, J., 1973, Tryptophan metabolism in levodopa-treated Parkinsonian patients, *Acta Med. Scand.* **194**:181–189.

Leklem, J. E., 1971, Quantitative aspects of tryptophan metabolism in humans and other species: a review, *Amer. J. Clin. Nutr.* **24**:659–672.

Levin, E., Nogueira, G., and Garcia Argiz, C. A., 1966a, Ventriculo-cisternal perfusion of amino acids in cat brain. I. Rates of disappearance from the perfusate, *J. Neurochem.* **13**:761–767.

Levin, E., Garcia Argiz, C. A., and Nogueira, G. J., 1966b, Ventriculo-cisternal perfusion of amino acids in cat brain. II. Incorporation of glutamic acid, glutamine and GABA into the brain parenchyma, *J. Neurochem.* **13**:979–988.

Levin, E., Sepulveda, F. V., and Yudilevich, D. L., 1974, Pial vessels transport of substances from cerebrospinal fluid to blood, *Nature (London)* **249**:266–268.

Lewander, T., and Sjöström, R., 1973, Increase in plasma concentration of free tryptophan caused by probenecid in humans, *Psychopharmacology* **33**:81–86.

Lineweaver, H., and Burk, D., 1934, Determination of enzyme dissociation constants, *J. Amer. Chem. Soc.* **56**:658–666.

Liu, C.-C., Chung, C.-H., and Lee, M.-L., 1973, Amino acid activation in mammalian brain: purification and characterization of tryptophan-activating enzyme from buffalo brain, *Biochem. J.* **135**:367–373.

Liu, Y. P., Ambani, L. M., and Van Woert, M. H., 1974. L-Dihydroxyphenylalanine: effect on levels of free amino acids in rat liver, *Proc. Soc. Exp. Biol. Med.* **147**:154–157.

Lorenzo, A. V., 1974, Amino acid transport mechanisms of the cerebrospinal fluid, *Fed. Proc. Fed. Amer. Soc. Exp. Biol.* **33**:2079–2085.

Lorenzo, A. V., and Cutler, R. W. P., 1969, Amino acid transport by choroid plexus *in vitro*, *J. Neurochem.* **16**:577–585.

Lovenberg, W., and Victor, S. J., 1974, Regulation of tryptophan and tyrosine hydroxylase, *Life Sci.* **14**:2337–2353.

McArthur, J. N., Dawkins, P. D., and Smith, M. J. H., 1971*a*, The displacement of L-tryptophan and dipeptides from bovine albumin *in vitro* and from human plasma *in vivo* by antirheumatic drugs, *J. Pharm. Pharmacol.* **23**:393–398.

McArthur, J. N., Dawkins, P. D., Smith, M. J. H., and Hamilton, E. B. D., 1971*b*, Mode of action of antirheumatic drugs, *Br. Med. J.* **2**:677–679.

McMenamy, R. H., 1964, The binding of indole analogues to defatted human serum albumin at different chloride concentrations, *J. Biol. Chem.* **239**:2835–2841.

McMenamy, R. H., and Oncley, J. L., 1958, The specific binding of L-tryptophan to serum albumin, *J. Biol. Chem.* **233**:1436–1447.

McMenamy, R. H., Lund, C. C., and Oncley, J. L., 1957, Unbound amino acid concentrations in human blood plasmas, *J. Clin. Invest.* **36**:1672–1679.

MacSweeney, D. A., 1975, Treatment of unipolar depression, *Lancet* **2**:510–511.

Madras, B. K., Cohen, E. L., Fernstrom, J. D., Larin, F., Munro, H. N., and Wurtman, R. J., 1973, Dietary carbohydrate increases brain tryptophan and decreases free plasma tryptophan, *Nature (London)* **244**:34–35.

Madras, B. K., Cohen, E. L., Messing, R., Munro, H. N., and Wurtman, R. J., 1974, Relevance of free tryptophan in serum to tissue tryptophan concentrations, *Metabolism* **23**:1107–1116.

Makino, K., Joh, Y., and Hasegauva, F., 1962, The detection of mausamine (5-hydroxykynuramine) in the brain of the mouse, *Biochem. Biophys. Res. Commun.* **6**:432–437.

Mandell, A. J., and Morgan, M., 1971, Indole(ethyl)amine *N*-methyltransferase in human brain, *Nature (London) New Biol.* **230**:85–87.

Mandell, A. J., Knapp, S., and Hsu, L. L., 1974, Some factors in the regulation of central serotonergic synapses, *Life Sci.* **14**:1–17.

Marsden, C. A., and Curzon, G., 1974, Effects of lesions and drugs on brain tryptamine, *J. Neurochem.* **23**:1171–1176.

Mattson, W. J., Iob, V., Sloan, M., Coon, W. W., Turcotte, J. G., and Child, C. G., 1970, Alterations of individual free amino acids in brain during acute hepatic coma, *Surg. Gynecol. Obstet.* **130**:263–266.

Mendels, J., Stinnett, J. L., Burns, D., and Frazer, A., 1975, Amine precursors and depression, *Arch. Gen. Psychiatry* **32**:22–30.

Miller, E. M., and Nieburg, H. A., 1974, L-Tryptophan in the treatment of levodopa-induced psychiatric disorders, *Dis. Nerv. Syst.* **35**:20–23.

Modigh, K., 1975, The relationship between the concentrations of tryptophan and 5-hydroxyindoleacetic acid in rat brain and cerebrospinal fluid, *J. Neurochem.* **25**:351–352.

Moir, A. T. B., 1975, Tryptophan concentration in the brain, in: *Aromatic Amino Acids in the Brain, Ciba Found. Symp. 22 (New Ser.)* (G. E. W. Wolstenholme and D. W. Fitzsimons, eds.), pp. 197–206, Elsevier, Amsterdam.

Moir, A. T. B., and Eccleston, D., 1968, The effects of precursor loading in the cerebral metabolism of 5-hydroxyindoles, *J. Neurochem.* **15**:1093–1108.

Morgan, W. W., Saldana, J. J., Yudo, C. A., and Morgan, J. F., 1975, Correlations between circadian changes in serum amino acids or brain tryptophan and the contents of serotonin and 5-hydroxyindoleacetic acid in regions of the rat brain, *Brain Res.* **84**:75–86.

Müller, W. E., and Wollert, U., 1975, Benzodiazepines: specific competitors for the binding of L-tryptophan to human serum albumin, *Naunyn-Schmiedebergs Arch. Pharmacol.* **288**:17–27.

Munro, H. N., 1968, Role of amino acid supply in regulating ribosome function, *Fed. Proc. Fed. Amer. Soc. Exp. Biol.* **27**:1231–1237.

Munro, H. N., Fernstrom, J. D., and Wurtman, R. J., 1975, Insulin, plasma aminoacid imbalance, and hepatic coma, *Lancet* **1**:722–724.

Murphy, D. L., Baker, M., Goodwin, F. K., Miller, H., Kotin, J., and Bunney, W. E., 1974, L-Tryptophan in affective disorders: indoleamine changes and differential clinical effects, *Psychopharmacol.* **34**:11–20.

Müting, D., 1962, Changes in the free amino acid composition of cerebrospinal fluid in liver disease, *Proc. Soc. Exp. Biol. Med.* **110**:620–622.

Oja, S. S., Lähdesmäki, P., and Vahvelainen, M.-L., 1974, Aromatic amino acid supply and brain protein synthesis, in: *Aromatic Amino Acids in the Brain, Ciba Found. Symp. 22 (New Ser.)* (G. E. W. Wolstenholme and D. W. Fitzsimons, eds.), pp. 283–297, Elsevier, Amsterdam.

Oldendorf, W. H., 1971, Brain uptake of radiolabelled amino acids, amines and hexoses after arterial injection, *Amer. J. Physiol.* **221**:1629–1639.

Orlowski, M., Sessa, G., and Green, J. P., 1974, γ-Glutamyl transpeptidase in brain capillaries: possible site of a blood-brain barrier for amino acids, *Science* **184**:66–68.

Paoletti, R., Sirtori, C., and Spano, P. F., 1975, Clinical relevance of drugs affecting tryptophan transport, *Annu. Rev. Pharmacol.* **15**:73–81.

Pardridge, W. M., and Oldendorf, W. H., 1975, Kinetic analysis of blood-brain barrier transport of amino acids, *Biochim. Biophys. Acta* **401**:128–136.

Paul, S. M., Hsu, L. L., and Mandell, A. J., 1975, Extrapineal *N*-acetyltransferase activity in rat brain, *Life Sci.* **15**:2135–2143.

Payne, I. R., Walsh, E. M., and Whittenburg, J. R., 1974, Relationship of dietary tryptophan and niacin to tryptophan metabolism in schizophrenics and nonschizophrenics, *Amer. J. Clin. Nutr.* **27**:565–571.

Perez-Cruet, J., Tagliamonte, A., Tagliamonte, P., and Gessa, G. L., 1972, Changes in brain serotonin metabolism associated with fasting and satiation in rats, *Life Sci.* **11**:31–39.

Perez-Cruet, J., Chase, T. N., and Murphy, D. L., 1974, Dietary regulation of brain tryptophan metabolism by plasma ratio of free tryptophan and neutral amino acids in humans, *Nature (London)* **248**:693–695.

Philips, S. R., Durden, D. A., and Boulton, A. A., 1974, Identification and distribution of tryptamine in the rat, *Can. J. Biochem.* **52**:447–451.

Pollin, W., Cardon, P. V., and Kety, S. S., 1961, Effects of amino acid feedings in schizophrenic patients treated with iproniazid, *Science* **133**:104–105.

Pond, W. G., Breuer, L. H., Loosli, J. K., and Warner, R. G., 1964, Effects of the α-hydroxy analogues of isoleucine, lysine, threonine and tryptophan and the α-keto analogue of tryptophan and the level of the corresponding amino acids on growth of rats, *J. Nutr.* **83**:85–93.

Prange, A. J., Wilson, I. C., Lynn, C. W., Alltop, L. B., and Stikeleather, R. A., 1974, L-Tryptophan in mania: contribution to a permissive hypothesis of affective disorders, *Arch. Gen. Psychiatry* **30**:56–62.

Roberge, A. G., Missala, K., and Sourkes, T. L., 1972, Alpha-methyltryptophan: effects on synthesis and degradation of serotonin in the brain, *Neuropharmacology* **11**:197–209.

Roberts, S., 1974, Effects of amino acid imbalance on amino acid utilization, protein synthesis and polyribosome function in cerebral cortex, in: *Aromatic Amino Acids in the Brain, Ciba Found. Symp. 22 (New Ser.)* (G. E. W. Wolstenholme and D. W. Fitzsimons, eds.), pp. 299–324, Elsevier, Amsterdam.

Rodgers, A. D., and Curzon, G., 1975, Melanin formation by human brain *in vitro, J. Neurochem.* **24**:1123–1129.

Rosenberg, L. E., and Scriver, C. R., 1969, Disorders of amino acid metabolism, in: *Duncan's Diseases of Metabolism*, Chapter 9, 6th Ed. (P. K. Bondy, ed.), W. 3. Saunders, Philadelphia.

Rubin, R. T., 1967, Adrenal cortical activity changes in manic-depressive illness. Influence on intermediary metabolism of tryptophan, *Arch. Gen. Psychiatry* **17**:671–679.

Rubin, R. T., and Mandell, A. J., 1966, Adrenal cortical activity in pathological emotional states, *Amer. J. Psychiatry* **123**:387–400.

Saavedra, J. M., and Axelrod, J., 1972*a*, Psychotomimetic *N*-methylated tryptamines: formation in brain *in vivo* and *in vitro, Science* **175**:1365–1366.

Saavedra, J. M., and Axelrod, J., 1972*b*, A specific and sensitive enzymatic assay for tryptamine in tissues, *J. Pharmacol. Exp. Ther.* **182**:363–369.

Saavedra, J. M., and Axelrod, J., 1973, Effect of drugs on the tryptamine content of rat tissues, *J. Pharmacol. Exp. Ther.* **185**:523–529.

Scapagnini, U., Preziosi, D., and DeSchaepdryver, A. F., 1969, Influence of restraint stress, corticosterone and betamethasone on brain amine levels, *Pharmacol. Res. Commun.* **1**:63–69.

Schubert, J., 1973, Effect of chronic lithium treatment on monoamine metabolism in rat brain, *Psychopharmacologia (Berlin)* **32**:301–311.

Seta, K., Sansur, M., and Lajtha, A., 1973, The rate of incorporation of amino acids into brain proteins during infusion in the rat, *Biochim. Biophys. Acta* **294**:472–480.

Shah, N. S., Stevens, S., and Himwich, H. E., 1968, Effect of chronic administration of cortisone on the tryptophan-induced changes in amine levels in the rat brain, *Arch. Intern. Pharmacodyn, Thér.* **171**:285–295.

Shaw, D. M., Johnson, A. L., Tidmarsh, S. F., MacSweeney, D. A., Hewland, H. R., and Woolcock, N. E., 1975, Multicompartmental analysis of amino acids. I. Preliminary data on concentrations, fluxes, and flow constants of tryptophan in affective illness, *Psychol. Med.* **5**:206–213.

Sidransky, H., Verney, E., and Sarma, D. S. R., 1971, Effect of tryptophan on polyribosomes and protein synthesis in liver, *Amer. J. Clin. Nutr.* **24**:779–784.

Siegel, F. L., Aoki, K., and Colwell, R. E., 1971, Polyribosome disaggregation and cell-free protein synthesis in preparations from cerebral cortex of hyperphenylalaninemic rats, *J. Neurochem.* **18**:537–547.

Snodgrass, S. R., and Horn, A. S., 1973, An assay procedure for tryptamine in brain and spinal cord using its [³H]dansyl derivative, *J. Neurochem.* **21**:687–696.

Snodgrass, S. R., and Iversen, L. L., 1974, Amino acid uptake into human brain tumors, *Brain Res.* **76**:95–107.

Sourkes, T. L., 1971, Alpha-methyltryptophan and its actions on tryptophan metabolism, *Fed. Proc. Fed. Amer. Soc. Exp. Biol.* **30**:897–903.

Sourkes, T. L., 1973, Enzymology and sites of action of monoamines in the central nervous system, *Adv. Neurol.* **2**:13–35.

Sourkes, T. L., Missala, K., and Oravec, M., 1970, Decrease of cerebral serotonin and 5-hydroxyindoleacetic acid caused by (-)-α-methyltryptophan, *J. Neurochem.* **17**:111–115.

Spano, P. F., Szyszka, K., Pozza, G., and Sirtori, C. R., 1974, Influence of clofibrate on serum tryptophan in man, *Res. Exp. Med.* **163**:265–269.

Spencer, R. P., and Samiy, A. H., 1960, Intestinal transport of L-tryptophan *in vitro:* inhibition by high concentrations, *Amer. J. Physiol.* **199**:1033–1036.

Stelmasiak, Z., and Curzon, G., 1974, Effect of electroconvulsive therapy on plasma unesteri-fied fatty acid and free tryptophan concentrations in man, J. Neurochem. 22:603–604.

Stoner, H. B., Cunningham, V. J., Elson, P. M., and Hunt, A., 1975, The effects of diet, lipolysis and limb ischaemia on the distribution of plasma tryptophan in the rat, Biochem. J. 146:659–666.

Tagliamonte, A., Biggio, G., and Gessa, G. L., 1971a, Possible role of "free" plasma tryptophan in controlling brain tryptophan concentrations, Riv. Farmacol. Ter. 11:251–255.

Tagliamonte, A., Tagliamonte, P., Perez-Cruet, J., Stern, S., and Gessa, G. L., 1971b, Effect of psychotropic drugs on tryptophan concentration in the rat brain, J. Pharmacol. Exp. Ther. 177:475–480.

Tagliamonte, A., Tagliamonte, P., DiChiara, G., Gessa, R., and Gessa, G. L., 1972a, Increase of brain tryptophan by electroconvulsive shock in rats, J. Neurochem. 19:1509–1512.

Tagliamonte, A., Tagliamonte, P., and Gessa, G. L., 1972b, Stimulation by dibutyryl cyclic AMP of serotonin synthesis and tryptophan transport in brain slices, Nature (London) New Biol. 237:245–247.

Tagliamonte, A., Biggio, G., Vargiu, L., and Gessa, G. L., 1973, Free tryptophan in serum controls brain tryptophan level and serotonin synthesis, Life Sci. 12:277–287.

Toda, N., 1975, Analysis of the effect of 5-hydroxykynurenamine, a serotonin metabolite, on isolated cerebral arteries, aortas and atria, J. Pharmacol. Exp. Ther. 193:385–392.

Tong, J. H., and Kaufman, S., 1975, Tryptophan hydroxylase: purification and some proper-ties of the enzyme from rabbit hindbrain, J. Biol. Chem. 250:4152–4158.

Tsuda, H., Noguchi, T., and Kido, R., 1972, 5-Hydroxytryptophan pyrrolase in rat brain, J. Neurochem. 19:887–890.

Turner, A. J., Illingworth, J. A., and Tipton, K. F., 1974, Stimulation of biogenic amine metabolism in the brain, Biochem. J. 144:353–360.

Vahvelainen, M. L., and Oja, S. S., 1972, Kinetics of influx of phenylalanine, tyrosine, tryptophan, histidine and leucine into slices of brain cortex from adult and 7-day-old rats, Brain Res. 40:477–488.

Vahvelainen, M.-L., and Oja, S. S., 1975, Kinetic analysis of phenylalanine-induced inhibition in the saturable influx of tyrosine, tryptophan, leucine and histidine into brain cortex slices from adult and 7-day-old rats, J. Neurochem. 24:885–892.

Warner, K. A., Frohman, C. E., and Gottlieb, J. S., 1972, The effect of the plasma factor on the uptake of amino acids by neural tissue, Biol. Psychiatry 5:173–180.

Weiss, B. F., Roch, L. E., Munro, H. N., and Wurtman, R. J., 1974, L-Dopa, polysomal aggregation and cerebral synthesis of protein, in: Aromatic Amino Acids in the Brain, Ciba Found. Symp. 22 (New Ser.) (G. E. W. Wolstenholme and D. W. Fitzsimons, eds.), pp. 325–334, Elsevier, Amsterdam.

Weiss, B. F., Liebschutz, J. L., Wurtman, R. J., and Munro, H. N., 1975, Participation of dopamine- and serotonin-receptors in the disaggregation of brain polysomes by L-dopa and L-5HTP, J. Neurochem. 24:1191–1195.

Wiley, W. R., 1970, Tryptophan transport in Neurospora crassa: a tryptophan binding protein released by cold osmotic shock, J. Bacteriol. 103:656–662.

Wu, P. H., and Boulton, A. A., 1973, Distribution and metabolism of tryptamine in rat brain, Can. J. Biochem. 51:1104–1112.

Wurtman, R. J., Shein, H. M., Axelrod, J., and Laren, F., 1969, Incorporation of ^{14}C-tryptophan into ^{14}C-protein by cultured rat pineals: stimulation by L-norepinephrine, Proc. Natl. Acad. Sci. U.S.A. 62:749–755.

Wyatt, R. J., Engelman, K., Kupfer, D. J., Fram, D. H., Sjoerdsma, A., and Snyder, F., 1970, Effects of L-tryptophan (a natural sedative) on human sleep, Lancet 2:842–846.

Yamamoto, S., and Hayaishi, O., 1967, Tryptophan pyrrolase of rabbit intestine: D- and L-tryptophan-cleaving enzyme or enzymes, J. Biol. Chem. 22:5260–5266.

Yanaihara, N., Hashimoto, T., Yanaihara, C., Tsuji, K., Kenmochi, Y., Ashizawa, F., Kaneko, T., Oka, H., Saito, S., Arimura, A., and Schally, A. V., 1973, Synthesis and biological evaluation of analogs of luteinizing hormone-releasing hormone (LH-RH) modified in position 2, 3, 4 or 5, *Biochem. Biophys. Res. Commun.* **52**:64–73.

Yarbro, M. T., and Anderson, J. A., 1966, L-Tryptophan metabolism in phenylketonuria, *J. Pediatr.* **68**:895–904.

Young, S. N., and Sourkes, T. L., 1974, Antidepressant action of tryptophan, *Lancet* **2**:897–898.

Young, S. N., and Sourkes, T. L., 1975, Tryptophan catabolism by tryptophan pyrrolase in rat liver: effect of tryptophan loads and changes in tryptophan pyrrolase activity, *J. Biol. Chem.* **250**:5009–5014.

Young, S. N., Lal, S., Martin, J. B., Ford, R. M., and Sourkes, T. L., 1973, 5-Hydroxyindole-acetic acid, homovanillic acid and tryptophan levels in CSF above and below a complete block of CSF flow, *Psychiatr. Neurol. Neurochir.* **76**:439–444.

Young, S. N., Garelis, E., Lal, S., Martin, J. B., Molina-Negro, P., Ethier, R., and Sourkes, T. L., 1974a, Tryptophan and 5-hydroxyindoleacetic acid in human cerebrospinal fluid, *J. Neurochem.* **22**:777–779.

Young, S. N., Oravec, M., and Sourkes, T. L., 1974b, The effect of theophylline on tryptophan pyrrolase in the hypophysectomized rat and some observations on the validity of tryptophan pyrrolase assays, *J. Biol. Chem.* **249**:3932–3936.

Young, S. N., Lal, S., Sourkes, T. L., Feldmuller, F., Aronoff, A., and Martin, J. B., 1975, Relationships between tryptophan in serum and CSF and 5-hydroxyindoleacetic acid in CSF of man: effect of cirrhosis of the liver and probenecid administration, *J. Neurol. Neurosurg. Psychiatry* **38**:322–330.

Young, S. N., Etienne, P., and Sourkes, T. L., 1976a, Relationship between rat brain and cisternal CSF tryptophan concentrations, *J. Neurol. Neurosurg. Psychiatry* **39**:239–243.

Young, S. N., Lal, S., Feldmuller, F., Sourkes, T. L., Ford, R. M., Kiely, M., and Martin, J. B., 1976b, Parallel variation of ventricular CSF tryptophan and free serum tryptophan in man, *J. Neurol. Neurosurg. Psychiatry* **39**:61–65.

Yuwiler, A., 1973, Conversion of D- and L-tryptophan to brain serotonin and 5-hydroxyindole-acetic acid and to blood serotonin, *J. Neurochem.* **20**:1099–1109.

Yuwiler, A., and Geller, E., 1973, Rat liver tryptophan oxygenase induced by neonatal corticoid administration and its effect on brain serotonin, *Enzyme* **15**:161–168.

Yuwiler, A., Geller, E., and Slater, G. G., 1965, On the mechanism of the brain serotonin depletion in experimental phenylketonuria, *J. Biol. Chem.* **240**:1170–1174.

Yuwiler, A., Wetterberg, L., and Geller, E., 1971: Relationship between alternate routes of tryptophan metabolism following administration of tryptophan peroxidase inducers or stressors, *J. Neurochem.* **18**:593–599.

CHAPTER 3

SUBSTANCE P AND SENSORY TRANSMITTER

MASANORI OTSUKA

Department of Pharmacology
Faculty of Medicine
Tokyo Medical and Dental University
Tokyo, Japan

1. INTRODUCTION

Two papers that appeared in the 1930s had an important influence on later studies of sensory transmitters and they have particular relevance in the present chapter. Euler and Gaddum (1931), while studying the distribution of acetylcholine in different organs, found that the extracts of equine brain and intestine contain an unidentified vasodilator and smooth-muscle-contracting agent. This agent was later referred to as substance P (Chang and Gaddum, 1933; Gaddum and Schild, 1934) and was shown to be of peptide nature (Euler, 1936). For many years, however, the chemical structure of substance P as well as its physiological function remained unknown (see below).

Dale (1935), on the other hand, in his Dixon Memorial lecture, suggested that the discovery and identification of the chemical transmitter released by peripheral endings of sensory neurons and mediating axon–reflex vasodilatation might indicate the nature of the transmission process at the central synapses formed by the same neurons. This statement was later formulated as the well-known *Dale's principle*, which states that the same chemical transmitter is released from all the synaptic terminals of a neuron (Eccles, 1957).

2. SEARCH FOR THE SENSORY TRANSMITTER

Following Dale's suggestion, much attention was focused on the vaso-dilator substances contained in sensory neurons. Kwiatkowski (1943) found appreciable amounts of histamine in sensory nerves and, on this basis, a possible role of histamine as a sensory transmitter was considered (Curtis *et al.*, 1961a). Electrophoretic application of histamine, however, produced not excitatory but inhibitory effects on the spinal neurons of the cat, which made histamine unlikely as the excitatory transmitter of spinal dorsal root fibers (Curtis *et al.*, 1961a; Phillis *et al.*, 1968).

Hellauer and Umrath (1947, 1948) used a unique approach in the search for a sensory transmitter. From the previous work of Loewi and Hellauer (1938), showing that ventral roots contain large amounts of acetyl-choline, whereas dorsal roots contain only negligible amounts, acetylcho-line seemed unlikely to be a sensory transmitter. Hellauer and Umrath (1948), therefore, tested the vasodilator activities of extracts from bovine dorsal and ventral roots on the denervated rabbit ear, on the assumption that this preparation should be sensitized to the sensory transmitter. They found that extracts from dorsal roots contained much larger amounts of noncholinergic vasodilator activity than extracts from ventral roots. Hel-lauer and Umrath (1948) considered the agent responsible for this activity to be the excitatory transmitter of spinal dorsal root fibers.

In 1953, Lembeck reported a similar discovery, namely, that bovine dorsal root contains a gut-contracting and vasodilator agent in much larger amounts than the ventral root. It now seems very likely that both Hellauer and Umrath (1947, 1948) and Lembeck (1953) were dealing with the same substance, but this point was controversial (Lembeck and Zetler, 1962). The properties of the gut-contracting agent in the dorsal root extracts were carefully examined by both Hellauer (1953) and Lembeck (1953), and the latter author assumed that the agent responsible was substance P. On this basis, Lembeck (1953) proposed the hypothesis that substance P might be a sensory transmitter of dorsal root fibers. This hypothesis, however, was not readily accepted, probably for the following reasons: First, the chemi-cal structure of the active agent in the extracts from dorsal roots was not clear and, second, the application of crude substance P onto neurons in the cat's central nervous system did not reveal any direct excitatory or inhibi-tory actions (Galindo *et al.*, 1967). Thus, almost 20 years after Lembeck

proposed his hypothesis, the general opinion about the physiological function of substance P was that it might probably act as a local vasodilator
rather than as a neurotransmitter (Eccles, 1964; Vogt, 1969; McLennan,
1970).

Holton and Holton (1952, 1954) found vasodilator activity in extracts
of both dorsal and ventral roots and suggested that the activity is due to
ATP and its breakdown products. Furthermore, ATP was shown to be
liberated on antidromic stimulation of the auricular nerve in rabbits (Holton, 1959a). Thus, ATP was proposed as a transmitter liberated at central
and peripheral endings of sensory neurons (Holton and Holton, 1954).
However, the study of the effects of ATP applied electrophoretically onto
spinal neurons did not reveal any excitatory effects (Curtis *et al.*, 1961a). In
the experiments of Galindo *et al.* (1967), ATP was found to have an
excitatory action on cuneate neurons, but this effect was attributed to the
Ca^{2+}-chelating action of ATP rather than to a genuine transmitter–receptor interaction (McLennan, 1970).

Recently, on the basis of neurochemical distribution studies in the
spinal cord of the cat, L-glutamate has been suggested as the transmitter
released from primary afferent fibers (Graham *et al.*, 1967). The excitatory
activity of L-glutamate in the central nervous system was first suggested by
Hayashi (1954). The electrophoretic application of glutamate onto feline
spinal neurons produced a fast and reversible excitatory response (Curtis *et
al.*, 1960). L-Glutamate is known to occur throughout nervous tissue, and
its content in the dorsal root, in particular, is higher than that in the ventral
root (Graham *et al.*, 1967; Duggan and Johnston, 1970), a fact that is
consistent with its putative role as sensory transmitter (Johnson and
Aprison, 1970; Johnson, 1972; Hammerschlag and Weinreich, 1972; Curtis
and Johnston, 1974; Krnjević, 1974). However, the dorsal/ventral root ratio
of the content of glutamate is only 1.3–2.3 (Johnson and Aprison, 1970),
which is much lower than the ventral/dorsal root ratio of 200 obtained for
acetylcholine (MacIntosh, 1941). Furthermore, L-glutamate is by no means
the most potent substance in causing depolarization of central neurons;
N-methyl-D-aspartic acid and kainic acid, e.g., are 70 and 300 times more
active, respectively, than L-glutamic acid in depolarizing spinal motoneurons
in the frog (Curtis *et al.*, 1961b; Otsuka, 1972).

3. THE MOTONEURON-DEPOLARIZING PEPTIDE IN DORSAL ROOT

Dorsal root fibers entering the spinal cord are generally believed to
form excitatory synapses with interneurons and motoneurons in the spinal

cord as well as with sensory relay neurons in the spinal cord and the brain stem (Eccles, 1957). When we embarked on an attempt to identify an excitatory transmitter of spinal dorsal root fibers, our first thought was to examine the motoneuron-depolarizing activity of dorsal root extracts, instead of focusing on the vasodilator activity as in previous studies. By analogy with other established transmitters, we expected the sensory transmitter substance to be much more concentrated in the dorsal root than in the ventral root and to display a high potency in depolarizing spinal motoneurons. Substance P seemed of obvious interest because it fulfilled the first qualification, although from the previous study of Galindo *et al.* (1967), it seemed less promising with regard to the second qualification. We thus wanted to retest the action of substance P on spinal motoneurons. For this purpose, the isolated spinal cord preparation from the frog seemed appropriate as a bioassay system because it can be easily maintained *in vitro,* where the drugs can be applied in controlled concentrations. It seemed, however, that one would not get a clear-cut result unless pure substance P was used in the experiment. This was a rather difficult task in 1970, when we started the present experiments, because the chemical structure of substance P was still unknown. However, what was already known was that certain undecapeptides, physalaemin and eledoisin, have structures similar to that of substance P (Erspamer and Anastasi, 1966). So studies were initiated by testing the action of physalaemin on motoneurons in the spinal cord of the frog. It was found that physalaemin exerted a powerful depolarizing action on the motoneurons, its potency being 500 times higher than L-glutamate on a molar basis (Konishi and Otsuka, 1971).

The finding of the powerful motoneuron-depolarizing activity of physalaemin further stimulated our interest in the possibility that a peptide similar to physalaemin might be present in the dorsal root and act as a sensory transmitter. So we collected bovine dorsal roots and extracted the tissue with acetone–HCl–water, fractionated the extract by column chromatography, and assayed each fraction for motoneuron depolarizing activity. Particular attention was directed to the agent described by Lembeck (1953) and Hellauer (1953). As a result of this survey, a fraction was found that displayed remarkable pharmacological activities, as shown in Figure 1(a). It elicited a contraction of atropinized guinea pig ileum, a fall of rat and rabbit blood pressure, and a depolarization of frog spinal motoneurons. All three activities were abolished after the fraction was incubated with chymotrypsin [Figure 1(b)], suggesting that the activities were due to a peptide. The latter was called *dorsal root peptide* (Otsuka *et al.,* 1972*a*).

The next step was to determine the chemical identity of the dorsal root peptide. Judging from its pharmacological properties, it resembled physalaemin as well as substance P although it was not known at this time whether substance P had motoneuron-depolarizing activity. At this stage in

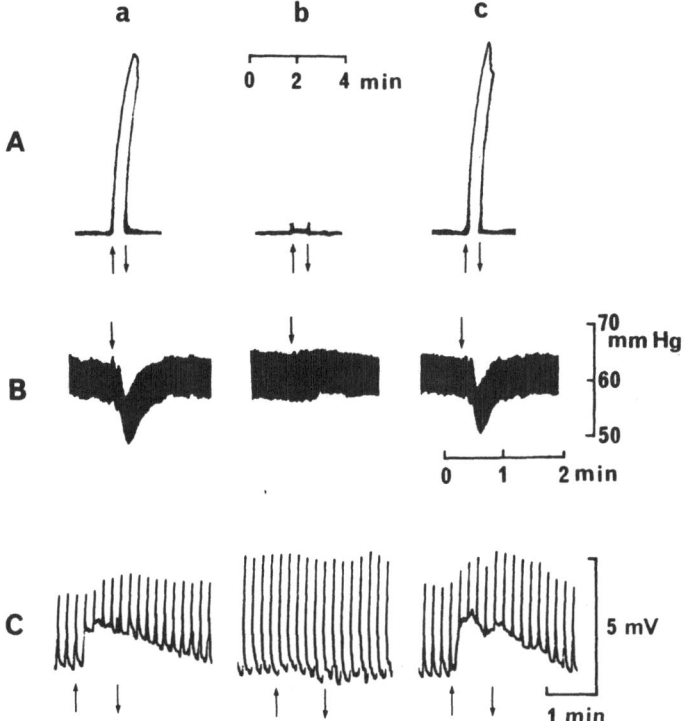

FIGURE 1. Comparison of the pharmacological activities of the dorsal root peptide and synthetic substance P. Dorsal root peptide was partially purified from the crude extract of bovine dorsal roots through Amberlite CG-50 and Sephadex LH-20 columns. (A) Contraction of guinea pig ileum treated with atropine, mepyramine, and tryptamine. (B) Blood pressure of the rabbit. (C) Effects on spinal motoneurons. Potentials were recorded from the 8th ventral root with an extracellular electrode in an isolated spinal cord of the frog soaked in 0.6 mM Ca Ringer's solution. Corresponding dorsal root was stimulated maximally at 0.1 Hz: depolarization upward. (a) Activities of the dorsal root peptide. Active fraction derived from 320 mg wet weight of dorsal root was added to 0.4-ml bath between arrows in (A); that from 200 mg tissue injected intravenously at arrow in (B); and that from 6 g tissue added to 0.3-ml bath between arrows in (C). (b) Activities of the dorsal root peptide after incubating with chymotrypsin. The amounts of the dorsal root extract were the same as in (a). (c) Activities of synthetic substance P; 16 pmol/0.4-ml bath in (A), 10 pmol injected in (B), and 390 pmol/0.3-ml bath in (C). From Takahashi et al. (1974).

our research work, synthetic substance P became available. Chang and Leeman (1970), almost 40 years after the discovery by Euler and Gaddum (1931), succeeded not only in isolating substance P from bovine hypothalamus and determining its structure as an undecapaptide (see Table 2), but also in synthesizing the peptide (Chang et al., 1971; Treagear et al., 1971).

A few hundred micrograms of synthetic substance P was generously given to us by Dr. Leeman, thereby enabling us to compare the properties of the dorsal root peptide with those of substance P as well as of several other synthetic peptides having more or less similar pharmacological activities. The chemical, pharmacological, and immunological properties and susceptibility to enzymatic degradation of the dorsal root peptide were found to be identical to those of the undecapeptide substance P (Table 1). Given such data, we concluded that the dorsal root peptide and substance P are identical (Otsuka *et al.*, 1972*b*; Takahashi *et al.*, 1974).

The action of synthetic substance P was examined on frog spinal cord, and as expected it had powerful motoneuron-depolarizing activity [Figure 1 (Cc)], being 200 times more active than L-glutamate (Otsuka *et al.*, 1972*b*; Konishi and Otsuka, 1974*a*). Other synthetic peptides with similar structures also displayed motoneuron-depolarizing activities, as shown in Table 2. It can be seen that a C-terminal sequence, -Phe-X-Gly-Leu-Met-NH$_2$ (where X = Ile, Tyr, or Phe), is important for the motoneuron-depolarizing activity (Otsuka *et al.*, 1972*a*; Konishi and Otsuka, 1974*a*).

The data in Table 3 show the comparisons of the amounts of substance P in bovine dorsal and ventral roots as estimated by bioassay on guinea pig ileum or by radioimmunoassay. In crude extracts, the contracting activity of dorsal root extract on the atropinized guinea pig ileum was 2.5 times higher than that of ventral root extract. This agrees with the results of Lembeck (1953) and Hellauer (1953). However, after partial purification by Sephadex LH-20 column chromatography, the dorsal/ventral root ratio estimated by bioassay increased to 27 (Otsuka *et al.*, 1972*b*; Takahashi *et al.*, 1974). Similar values were obtained with spinal roots from cats (Takahashi and Otsuka, 1975).

TABLE 1. Comparison of Chemical Properties of the Dorsal Root Peptide and Synthetic Substance P[a]

Property	Dorsal root peptide	Synthetic substance P
Molecular weight estimated by gel chromatography	1250 ± 150	1340
Electrophoretic mobilities		
Relative to L-glutamic acid at pH 2.0	1.17 ± 0.02	1.19 ± 0.01
Relative to L-lysine at pH 6.2	0.61 ± 0.01	0.59 ± 0.01
R_f in paper chromatography	0.62	0.62 ± 0.02
Enzymatic destruction		
Chymotrypsin	Destroyed	Destroyed
Trypsin	Not destroyed	Not destroyed
Carboxypeptidase A	Not destroyed	Not destroyed

[a]Each value represents mean $\pm$ S.E.M. From Takahashi *et al.* (1974).

TABLE 2. Structures of Substance P and Related Peptides, and Their Potencies in Depolarizing the Frog Spinal Motoneurons[a]

Peptide	Potency relative to L-glutamate
Arg-Pro-Lys-Pro-Gln-Gln-Phe-Phe-Gly-Leu-Met-NH₂ (substance P)	200
PCA-Ala-Asp-Pro-Asn-Lys-Phe-Tyr-Gly-Leu-Met-NH₂ (physalaemin)	1500
PCA-Pro-Ser-Lys-Asp-Ala-Phe-Ile-Gly-Leu-Met-NH₂ (eledoisin)	2000
Lys-Phe-Ile-Gly-Leu-Met-NH₂	770
Lys-Phe-Tyr-Gly-Leu-Met-NH₂	150
H₂N-(CH₂)₅-CO-Phe-Tyr-Gly-Leu-Met-NH₂	130
Lys-Phe-Ile-Gly-MeLeu-Met-NH₂	50
Lys-Phe-Ile-Gly-HyIc-Met-NH₂	30
Lys-Phe-Gly-Gly-Leu-Met-NH₂	2
Lys-Phe-Lac-Gly-Leu-Met-NH₂	2
Lys-MePhe-Ile-Gly-Leu-Met-NH₂	1
Lys-MePhe-Ile-Gly-MeLeu-Met-NH₂	1

[a] Determinations of the potencies were carried out in 0.4 mM-Ca Ringer's solution. Each value represents the mean of 2–5 determinations and is expressed on a molar basis. Abbreviations: PCA, pyrrolidone carboxylic acid; HyIc, α-hydroxy isocaproic acid; MeLeu, N-methyl leucine; Lac, lactic acid; MePhe, N-methyl phenylalanine. From Konishi and Otsuka (1974a).

4. DISTRIBUTION OF SUBSTANCE P IN THE SPINAL CORD AND DORSAL ROOT OF THE CAT

If substance P is the transmitter of primary afferent fibers, one would expect that the distribution of the peptide in the spinal cord would parallel that of primary afferent terminals. Therefore, we have examined the distribution of substance P in the lumbosacral spinal cord of the cat. In a preliminary study, the chemical properties of substance P in cat spinal cord were shown to be identical with those of the undacapeptide substance P of Chang and Leeman (1970). Substance P was highly concentrated in the dorsal horn, where most of the primary afferent fibers terminate and form synapses (Sprague and Ha, 1964). The highest content of substance P was found in the dorsolateral part of the dorsal horn, where the content was about 1000 times higher than in the ventral root (see the intact side of Figure 2).

In the experiment illustrated in Figure 2, the effect of surgical section of the dorsal roots on the distribution of substance P in the spinal cord was

TABLE 3.   Estimated Amounts of Substance P in the Extracts of Bovine Dorsal and Ventral Roots[a]

Purification step	Amount of substance P (pmol/g wet wt)		Dorsal/Ventral[b]	Method of determination
	Dorsal root	Ventral root		
Crude extract	160 ± 5.5	65 ± 8.5	2.5	Bioassay on ileum[c]
Crude extract	130 ± 8	< 12	> 11	Bioassay on ileum[d]
Crude extract	30	3.5	9	Radioimmunoassay
Amberlite CG-50 column	24	2.5	10	Radioimmunoassay
Sephadex LH-20 column	43 ± 2.1	1.6 ± 0.1	27	Bioassay on ileum[c]

[a] Values represent mean ± S.E.M.
[b] Ratio of amount of substance P in dorsal to that in ventral root extracts.
[c] Estimated from the total gut-contracting activity.
[d] Estimated from the gut-contracting activity that was inactivated by chymotrypsin. From Takahashi *et al.* (1974).

studied. The dorsal roots below L5 were unilaterally sectioned in anesthetized cats. About 10 days after the operation, the distribution of substance P in the L5–S1 spinal cord was examined. The level of substance P in the dorsal horn was markedly reduced on the sectioned side. In particular, the level in the dorsolateral part of the dorsal horn was lowered to one-tenth of that on the control side (Takahashi and Otsuka, 1974, 1975). These results suggest that substance P is located in the intraspinal axon terminals of

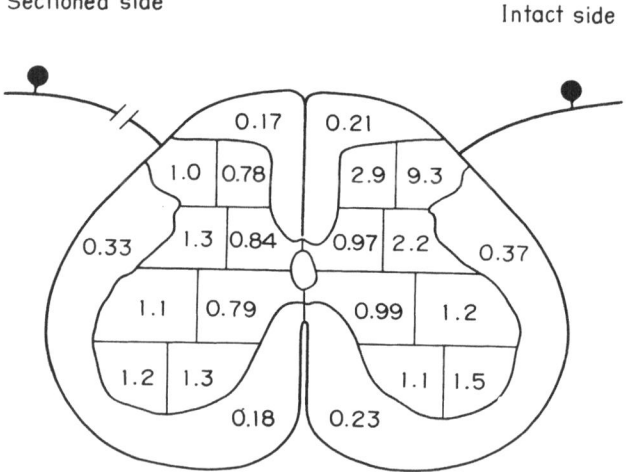

Sectioned side   Intact side

FIGURE 2. Distribution of substance P in the spinal cord (L5–S1) of a cat, 11 days after the unilateral section of the dorsal roots (below L5). Numbers in the figure indicate the content of substance P given in 10^{-10} mol per gram of wet weight. From Takahashi and Otsuka (1975).

primary afferent neurons, and this notion is consistent with the results of a recent immunohistochemical study by Nilsson *et al.* (1974), who showed that substance P-like immunoreactivity is located in nerve terminals of the substantia gelatinosa of rat spinal cord. Furthermore, previous subcellular fractionation studies have shown that substance P is concentrated in nerve-ending particles (synaptosomes) in brain and spinal cord homogenates (Inouye and Kataoka, 1962; Ryall, 1962; Cleugh *et al.*, 1964; Powell *et al.*, 1973).

Figure 3 shows the distribution of substance P in the cat dorsal root and the effect of ligation (Takahashi and Otsuka, 1975). About 10 days after the dorsal root was ligated and sectioned, a large amount of substance P had accumulated on the ganglion side of the ligature, whereas its level on the proximal side was markedly lowered [Figure 3(A)]. A similar phenomenon is known to occur in ligated adrenergic nerves (Dahlström, 1965). In contrast, no accumulation of L-glutamate was detected in the ligated dorsal root [Figure 3(B)]. Substance P was quite evenly distributed in the intact dorsal root [Figure 3(A)]. Holton (1959*b*) had already shown that substance P accumulated on the proximal side of a ligature in the auricular nerves of

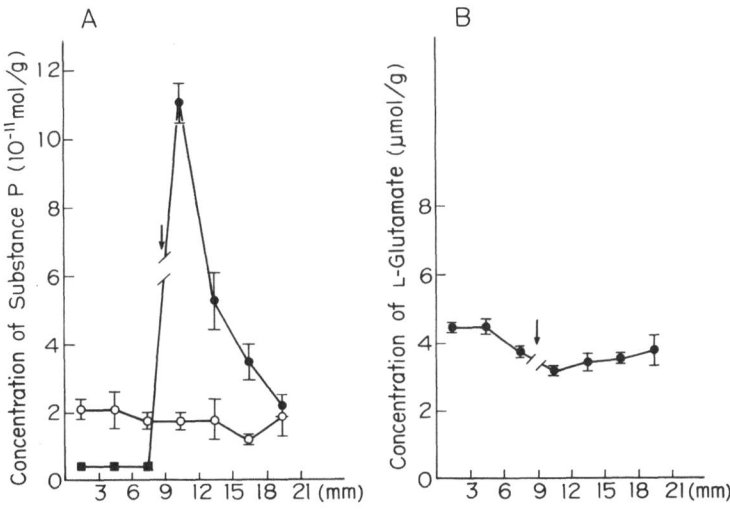

FIGURE 3. Distribution of substance P and L-glutamate in the ligated and intact dorsal roots 11–12 days after the operation. Ordinate: concentration of substance P in (A), and that of L-glutamate in (B). Abscissa: distance from the entry of the dorsal root into the cord. Site of ligation is indicated by arrows. Each point and vertical line represents mean ± S.E.M. obtained from three determinations derived from three cats in (A), and two determinations from two cats in (B). (●) substance P and L-glutamate in the ligated dorsal roots; (○) substance P in the intact roots; (■) threshold value of the assay for substance P (ligated roots). From Takahashi and Otsuka (1975).

the rabbit, whereas its level was reduced on the distal side. It thus appears that substance P synthesized in the cell bodies of sensory neurons is transported toward the central as well as peripheral nerve terminals, a notion that agrees well with the suggestion of Dale (1935).

5. ISOLATED SPINAL CORD OF NEWBORN RAT

Since we are considering the possible transmitter role of substance P in the mammalian CNS, it is naturally desirable to test the action of the peptide on mammalian central neurons. Here we would like to have a simple and stable bioassay system, where drugs can be applied in controlled concentrations so that a comparison of the potencies of applied drugs is possible. In this respect, *in vitro* experiments are preferable to *in vivo* ones. Although the mammalian spinal cord has long been a favorite preparation for electrophysiology and pharmacology, these experiments have been carried out exclusively *in vivo*. We have therefore developed a new preparation, the isolated spinal cord of newborn rat. This preparation survives well in the physiological solution probably for two reasons. First, it is of quite small size (diameter: <2 mm) which makes the penetration of oxygen to the neurons easy; second, it is known that the nervous tissue of newborn animals consumes less oxygen than the adult tissue (Greengard and McIlwain, 1955). The spinal cord of a 0–7 day old Wistar rat was isolated, hemisected sagittally, and placed in a bath perfused with oxygenated Krebs solution at 27°C. This preparation seems to be useful for the study of mammalian CNS for several reasons. Unlike the tissue-slice preparations which have so far been the only *in vitro* preparations of mammalian CNS for physiological experiments (Yamamoto and McIlwain, 1966), the isolated spinal cord of the newborn rat retains most of its neural connections intact. Monosynaptic and polysynaptic reflexes can be recorded from the ventral root, and the patterns of the reflexes are quite similar to those observed in cat spinal cord *in vivo*. Postsynaptic inhibition as well as the dorsal root potential, which represents presynaptic inhibition, can easily be recorded. These electrical activities can be recorded either extracellularly from the spinal roots or intracellularly from the motoneurons, using the conventional methods. The extracellular ion or drug concentrations, as well as the temperature of the external medium, can easily be changed. Finally, the preparation is quite stable and good reflex responses can be recorded for up to 10, and probably more, hours (Otsuka and Konishi, 1974).

The application of substance P in quite low concentrations in the bath induced a depolarization of spinal motoneurons and high-frequency spike discharges that could be recorded extracellularly from the ventral root or intracellularly from the motoneurons, and the potential recovered within a few minutes after the removal of the peptide [Figure 4(B)]. L-Glutamate produced a similar response, but a much higher concentration was needed [Figure 4(C)]. From a comparison of dose-response curves similar to the ones illustrated in Figure 4(A), it was estimated that substance P was 1000–9000 times more potent than L-glutamate in depolarizing spinal motoneurons in the rat. The time course of the depolarization was slightly slower with substance P than with L-glutamate, but the effect still appeared in a few seconds after the application of the peptide.

In order to test the possibility that substance P depolarized the motoneurons by some transsynaptic mechanism, the preparation was soaked in a modified Krebs solution containing low-Ca^{2+} (0.7 mM) and high-Mg^{2+} (7 mM), in which spinal reflexes were completely blocked. The response to substance P was only slightly smaller than that in normal Krebs solution, suggesting that the peptide acts directly on the motoneurons.

The effects of substance P on monosynaptic reflexes will be described elsewhere, but the results suggested that substance P activates certain

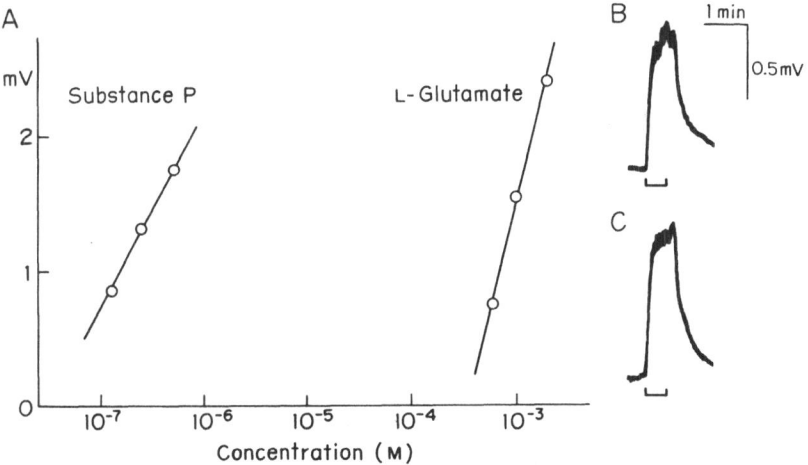

FIGURE 4. Comparison of the actions of substance P and L-glutamate on rat spinal motoneurons. Isolated spinal cord of the newborn rat was used. Extracellular recordings were made from the ventral root. (A) Dose–response curves of substance P and L-glutamate. Ordinate: potential change induced by the drugs and recorded from L4 ventral root. Abscissa: concentration of drugs on logarithmic scale. (B) and (C) Responses to substance P (2×10^{-7} M) and L-glutamate (7×10^{-4} M) recorded from L3 ventral root in another preparation. Drugs were applied during the periods marked under records. Upward deflection means positivity at the recording electrode. From Otsuka and Konishi (1976).

interneurons in the spinal cord (Konishi and Otsuka, 1974*b;* Otsuka and Konishi, 1976).

6. LIORESAL

No specific antagonist has been found either for the effects of substance P or for the excitatory transmitter of primary afferent neurons. The use of specific antagonists toward putative neurotransmitters has always provided a useful means for transmitter identification. If substance P is a transmitter of primary afferent neurons, an antagonist of substance P would be expected to block the spinal reflexes involving the primary afferent synapses. It has recently been shown that Lioresal (Ciba-Geigy, β-(4-chlorophenyl)-γ-aminobutyric acid; Figure 5) suppresses monosynaptic and polysynaptic

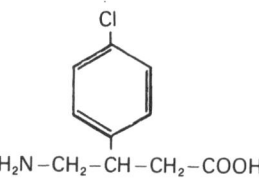

$$H_2N-CH_2-CH-CH_2-COOH$$

FIGURE 5. Lioresal, baclofen.

reflexes in the cat spinal cord (Pedersen *et al.,* 1970; Pierau and Zimmermann, 1973), and the dorsal root potential in the frog spinal cord (Davidoff and Sears, 1974). However, since Lioresal has been assumed to be a GABA agonist, it was not considered surprising that the drug suppressed spinal reflexes. In the isolated spinal cord preparation of the newborn rat, Lioresal did not produce any GABA-like action on the primary afferent fibers, nor did it affect GABA-induced depolarization of primary afferent fibers. Instead, a remarkable antagonistic action of Lioresal toward substance P was found (Figure 6). When Lioresal was applied in a quite low concentration to the isolated spinal cord of the rat, monosynaptic and polysynaptic reflexes, as well as the dorsal root potential, were greatly reduced or almost abolished within a few minutes. At this stage, the depolarizing response to substance P was almost completely abolished. Glutamate-induced depolarization was also antagonized by Lioresal, but to a much smaller extent. Glycine also induced a depolarizing response of motoneurons, and this response was not affected by Lioresal. Therefore the action of Lioresal toward substance P seems to some extent to be specific (Saito *et al.,* 1975; Otsuka and Konishi, 1976).

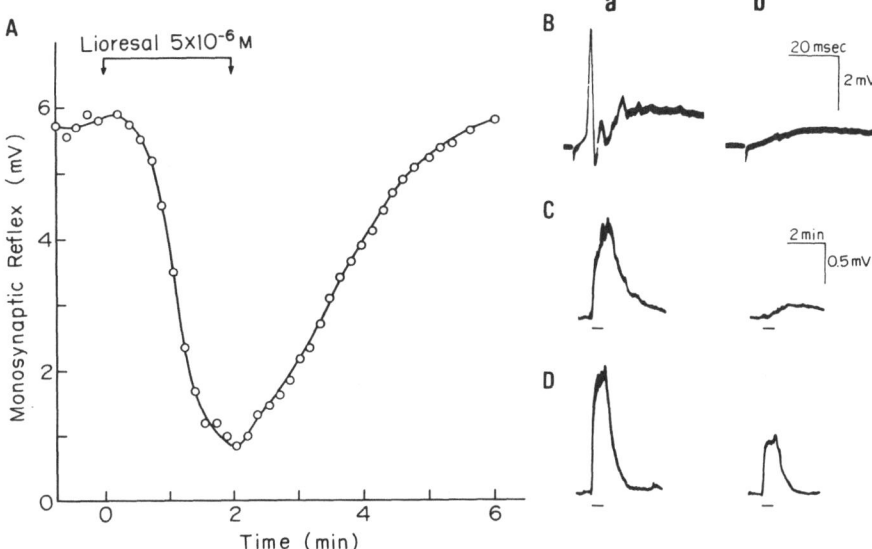

FIGURE 6. Effect of Lioresal on the ventral root responses in the isolated spinal cord of the newborn rat. (A) Effect of Lioresal on the amplitude of monosynaptic reflex. The drug was applied to the bath during the period marked in the graph. (B) to (D) Extracellular recordings from L3 ventral root in another preparation; (B) reflex activities induced by a single volley in L3 dorsal root, (C) responses to substance P (2×10^{-7} M); and (D) responses to L-glutamate (10^{-3} M). Horizontal bars under records mark the periods of drug applications. In each pair of records, (a) was obtained in normal Krebs solution, and (b) in the solution containing Lioresal (5×10^{-6} M). Scales for (C) apply also to (D). Positivity upward. From Otsuka and Konishi (1976).

7. EVIDENCE FOR SUBSTANCE P AND L-GLUTAMATE AS SENSORY TRANSMITTERS

At present, two substances may be regarded as candidates for the transmitters of the primary afferent neuron, i.e., substance P and L-glutamate. Table 4 summarizes the available evidence for these two substances. The distribution of substance P in the nervous tissues is quite selective. The content of substance P in the dorsal root is 9–27 times higher than that in the ventral root (Otsuka et al., 1972b; Takahashi et al., 1974; Takahashi and Otsuka, 1975). Furthermore, the content of substance P in the dorsolateral part of the dorsal horn of cat spinal cord is even 1000 times higher than that in the ventral root. Section of the dorsal roots results in a marked decrease in the content of substance P in the dorsal horn, particularly in its

TABLE 4. Comparison of Evidence for Substance P and L-Glutamate as Putative Transmitters of Primary Sensory Neurons

	Substance P	L-Glutamate
Presence		
Distribution	Selective	Ubiquitous
Dorsal/ventral root ratio		
of content	9–27	1–2
Effect of dorsal root		
section	Marked decrease	No effect
In nerve terminals	Concentrated	Not concentrated
Axonal flow	Yes	No
Action		
On spinal motoneurons	1000–9000	1
Antagonist	Lioresal	α-MG, GDEE[a]

[a] α-MG: α-methyl-DL-glutamate; GDEE: L-glutamic acid diethyl ester. For reference, see text. From Otsuka and Konishi (1976).

dorsal part (Takahashi and Otsuka, 1975). There is evidence that substance P is synthesized in the spinal ganglia and transported through the dorsal roots and stored in a high concentration in the nerve terminals of primary sensory neurons (Takahashi and Otsuka, 1974, 1975; Nilsson et al., 1974; Powell et al., 1973). The action of substance P in depolarizing rat spinal motoneurons is 1000–9000 times more potent than L-glutamate. Lioresal antagonizes the depolarizing action of substance P on the motoneurons and at the same time blocks the spinal reflexes involving primary afferent synapses (Konishi and Otsuka, 1974b; Saito et al., 1975; Otsuka and Konishi, 1976).

By contrast, L-glutamate is ubiquitous in nervous tissue. Although the content of glutamate in the dorsal root is higher than that in the ventral root, the dorsal/ventral root ratio is 1.3–2.3 (Johnson and Aprison, 1970). There is also no evidence that glutamate is transported by axonal flow (Takahashi and Otsuka, 1975). The content of glutamate in the nerve terminal fraction is not higher than those in other subcellular fractions (Mangan and Whittaker, 1966). Although many substances are known as antagonists of L-glutamate, their effects on mammalian central neurons are not very specific and, furthermore, their effects on spinal reflexes are not yet known (Curtis et al., 1972; Krnjević, 1974). Taken together, although no conclusive evidence is available for either substance, it seems safe to conclude that the evidence for substance P as a putative neurotransmitter of primary sensory neurons is much better than that for L-glutamate.

8. PEPTIDES AS NEUROTRANSMITTERS

It is well known that certain peptides are released from neurons as releasing factors or hormones. The term *peptidergic* was proposed by Bargman *et al.* (1967) for these neurons. Therefore, the idea that certain peptides function as neurotransmitters is not entirely novel. It is conceivable that peptides play both excitatory and inhibitory transmitter roles. In this connection, it was reported recently that thyrotropin releasing hormone exerts a depressant action on central neuronal activity (Renaud and Martin, 1975), and angiotensin, under certain experimental conditions, has a hyperpolarizing action on spinal motoneurons in frogs (Konishi and Otsuka, 1974*a*).

Whether all dorsal root fibers secrete only one sort of transmitter is only a matter of conjecture. Hökfelt *et al.* (1975) reported that only one-fourth of spinal ganglion cells displayed a positive immunohistochemical reaction for substance P. This may suggest some other transmitter(s) being involved. On the other hand, Lioresal blocked all spinal reflexes so far tested involving primary afferent synapses, which suggests that primary afferent transmission is mediated by a more or less uniform mechanism. The antagonistic action of Lioresal was also observed against a *C*-terminal hexapeptide analog of substance P, which suggests that the action of Lioresal may be related to the *C*-terminal amino acid sequence (Otsuka and Konishi, 1976).

Krnjević and his colleagues reported that the effect of electrophoretic application of substance P on cuneate and spinal neurons starts with a long delay and persists for 1–2 min after the cessation of injecting current and, on this basis, they postulated that substance P is unlikely to be an excitatory transmitter released by primary afferent terminals, but may be a modulator transmitting rather complex information (Krnjević and Morris, 1974; Henry *et al.*, 1975). According to the definition of Iversen (1974), a modulator is a substance that adjusts or regulates the synaptic transmission without interfering with the basic process of information transfer. In our experiments using the spinal cord of the rat, the time course of the excitatory action of substance P was under optimal conditions almost as fast as that of L-glutamate. Furthermore, substance P by itself displayed a quite potent and direct excitant action on the motoneurons, and the peptide did not modify the excitant action of L-glutamate (Otsuka and Konishi, 1976). Therefore, it seems unlikely that substance P is a modulator in the sense defined by Iversen (1974).

Although considerable evidence supports the hypothesis that the undecapeptide substance P is a sensory transmitter, a possibility that must be borne in mind is that substance P might be a precursor and be converted to the real transmitter after the removal of some amino acids. In this connection, among shorter C-terminal analogs of substance P, hexa- and hepta-peptides displayed much stronger activities in depolarizing rat spinal motoneurons than substance P (Otsuka and Konishi, 1976). In any case, after a long latency, the recent introduction of chemically defined substance P will hopefully open up a new era, not only in the field of peptide neuropharmacology but also in the field of synaptic transmission.

ACKNOWLEDGMENT

This review was written while on study leave at MRC Neurochemical Pharmacology Unit, Department of Pharmacology, Cambridge, England. It is a pleasure to thank Dr. L. L. Iversen of this Unit for his generous help and advice with the preparation of the manuscript.

9. REFERENCES

Bargmann, W., Lindner, E., and Andres, K. H., 1967, Über Synaspen an endokrinen Epithelzellen und die Definition sekretorischer Neurone, *Z. Zellforsch. Mikrosk. Anat.* **77**:282–298.

Chang, H. C., and Gaddum, J. H., 1933, Choline esters in tissue extracts, *J. Physiol. (London)* **79**:255–285.

Chang, M. M., and Leeman, S. E., 1970, Isolation of a sialogogic peptide from bovine hypothalamic tissue and its characterization as substance P, *J. Biol. Chem.* **245**:4784–4790.

Chang, M. M., Leeman, S. E. and Niall, H. D., 1971, Amino-acid sequence of substance P, *Nature (London) New Biol.* **232**:86–87.

Cleugh, J., Gaddum, J. H., Mitchell, A. A., Smith, M. W., and Whittaker, V. P., 1964, Substance P in brain extracts, *J. Physiol. (London)* **170**:69–85.

Curtis, D. R., and Johnston, G. A. R., 1974, Amino acid transmitters in the mammalian central nervous system, *Ergebn. Physiol. Biol. Chem. Exp. Pharmacol.* **69**:97–188.

Curtis, D. R., Phillis, J. W., and Watkins, J. C., 1960, The chemical excitation of spinal neurones by certain acidic amino acids, *J. Physiol. (London)* **150**:656–682.

Curtis, D. R., Phillis, J. W., and Watkins, J. C., 1961*a*, Cholinergic and non-cholinergic transmission in the mammalian spinal cord, *J. Physiol. (London)* **158**:296–323.

Curtis, D. R., Phillis, J. W., and Watkins, J. C., 1961*b* Actions of amino-acids on the isolated hemisected spinal cord of the toad, *Br. J. Pharmacol. Chemother.* **16**:262–283.

Curtis, D. R., Duggan, A. W., Felix, D., Johnston, G. A. R., Tebēcis, A. K., and Watkins, J. C., 1972, Excitation of mammalian central neurones by acidic amino acids, *Brain Res* **41**:283–301.

Dahlström, A., 1965, Observations on the accumulation of noradrenaline in the proximal and distal parts of peripheral adrenergic nerves after compression, *J. Anat. (London)* **99**:677–689.

Dale, H., 1935, Pharmacology and nerve-endings, *Proc. R. Soc. Med.* **28**:319–332.

Davidoff, R. A., and Sears, E. S., 1974, The effects of Lioresal on synaptic activity in the isolated spinal cord, *Neurology* **24**:957–963.

Duggan, A. W., and Johnston, G. A. R., 1970, Glutamate and related amino acids in cat spinal roots, dorsal root ganglia and peripheral nerves, *J. Neurochem* **17**:1205–1208.

Eccles, J. C., 1957, *The Physiology of Nerve Cells,* Johns Hopkins Press, Baltimore.

Eccles, J. C., 1964, *The Physiology of Synapses,* Springer-Verlag, Berlin.

Erspamer, V., and Anastasi, A., 1966, Polypeptides active on plain muscle in the amphibian skin, in: *Hypotensive Peptides* (E. G. Erdös, N. Back, F. Sicuteri, and A. F. Wilde, eds.), pp. 63–75, Springer-Verlag, New York.

Euler, U. S. von, 1936, Untersuchungen über Substanz P, die atropinfeste, darmerregende und gefässerweiternde Substanz aus Darm und Hirn, *Naunyn-Schmiedebergs Arch. Exp. Pathol. Pharmakol.* **181**:181–197.

Euler, U. S. von, and Gaddum, J. H. 1931, An unidentified depressor substance in certain tissue extracts, *J. Physiol. (London)* **72**:74–87.

Gaddum, J. H., and Schild, H., 1934, Depressor substances in extracts of intestine, *J. Physiol. (London)* **83**:1–14.

Galindo, A., Krnjević, K., and Schwartz, S., 1967, Micro-iontophoretic studies on neurones in the cuneate nucleus, *J. Physiol. (London)* **192**:359–377.

Graham, L. T., Jr., Shank, R. P., Werman, R., and Aprison, M. H., 1967, Distribution of some synaptic transmitter suspects in cat spinal cord: glutamic acid, aspartic acid, γ-aminobutyric acid, glycine, and glutamine. *J. Neurochem.* **14**:465–472.

Greengard, P., and McIlwain, H., 1955, Metabolic response to electrical pulses in mammalian cerebral tissues during development, in: *Biochemistry of the Developing Nervous System* (H. Waelsch, ed.), pp 251–260, Academic Press, New York.

Hammerschlag, R., and Weinreich, D., 1972, Glutamic acid and primary afferent transmission, in: "Studies of Neurotransmitters at the Synaptic Level," *Advances in Biochemical Psychopharmacology,* Vol. 6 (E. Costa, L. L. Iversen, and R. Paoletti, eds), pp. 165–180, Raven Press, New York.

Hayashi, T., 1954, Effects of sodium glutamate on the nervous system, *Keio J. Med.* **3**:183–192.

Hellauer, H., 1953, Zur Charakterisierung der Erregungssubstanz sensibler Nerven, *Naunyn-Schmiedebergs Arch. Exp. Pathol. Pharmakol.* **219**:234–241.

Hellauer, H., and Umrath, K., 1947, The transmitter substance of sensory nerve fibres, *J. Physiol. (London)* **106**:20–21P.

Hellauer, H. F., and Umrath, K., 1948, Über die Aktionssubstanz der sensiblen Nerven, *Pflügers Arch.* **249**:619–630.

Henry, J. L., Krnjević, K., and Morris, M. E., 1975, Substance P and spinal neurones, *Can. J. Physiol. Pharmacol.* **53**:423–432.

Hökfelt, T., Nilsson, G., and Pernow, B., 1975, Immunohistochemical localisation of substance P-like immunoreactivity in the nervous system, 6th Int. Congr. Pharmacol. Abstract (July 1975), p. 43.

Holton, F. A., and Holton, P., 1952, The vasodilator activity of spinal roots, *J. Physiol. (London)* **118**:310–327.

Holton, F. A., and Holton, P., 1954, The capillary dilator substances in dry powders of spinal roots; a possible role of adenosine triphosphate in chemical transmission from nerve endings, *J. Physiol. (London)* **126**:124–140.

Holton, P., 1959*a,* The liberation of adenosine triphosphate on antidromic stimulation of sensory nerves, *J. Physiol. (London)* **145**:494–504.

Holton, P., 1959*b,* Further observations on substance P in degenerating nerve, *J. Physiol. (London)* **149**:35P.

Inouye, A., and Kataoka, K., 1962, Sub-cellular distribution of the substance P in the nervous tissues, *Nature (London)* 193:585.

Iversen, L. L., 1974, Biochemical aspects of synaptic modulation, in: *The Neurosciences, Third Study Program* (F. O. Schmitt and F. G. Worden, eds.), pp. 905–915, The MIT Press, Cambridge, Massachusetts.

Johnson, J. L., 1972, Glutamic acid as a synaptic transmitter in the nervous system. A review, *Brain Res.* 37:1–19.

Johnson, J. L., and Aprison, M. H., 1970, The distribution of glutamic acid, a transmitter candidate, and other amino acids in the dorsal sensory neuron of the cat, *Brain Res.* 24:285–292.

Konishi, S., and Otsuka, M., 1971, Actions of certain polypeptides on frog spinal neurons, *Jpn. J. Pharmacol.* 21:685–687.

Konishi, S., and Otsuka, M., 1974a, The effects of substance P and other peptides on spinal neurons of the frog, *Brain Res.* 65:397–410.

Konishi, S., and Otsuka, M., 1974b, Excitatory action of hypothalamic substance P on spinal motoneurones of newborn rats, *Nature (London)* 252:734–735.

Krnjević, K., 1974, Chemical nature of synaptic transmission in vertebrates, *Physiol.Rev.* 54:418–540.

Krnjević, K., and Morris, M. E., 1974, An excitatory action of substance P on cuneate neurones, *Can. J. Physiol. Pharmacol.* 52:736–744.

Kwiatkowski, H., 1943, Histamine in nervous tissue, *J. Physiol. (London)* 102:32–41.

Lembeck, F., 1953, Zur Frage der zentralen Übertragung afferenter Impulse. III. Mitteilung, Das Vorkommen und die Bedeutung der Substanz P in den dorsalen Wurzeln des Rückenmarks, *Naunyn-Schmiedebergs Arch. Exp. Pathol. Pharmakol.* 219:197–213.

Lembeck, F., and Zetler, G., 1962, Substance P: a polypeptide of possible physiological significance especially within the nervous system, in: *International Review of Neurobiology*, Vol. 4 (C. C. Pfeiffer and J. R. Smythies, eds.), pp. 159–215, Academic Press, New York.

Loewi, O., and Hellauer, H., 1938, Über das Acetylcholin in peripheren Nerven, *Pflügers Arch.* 240:769–775.

MacIntosh, F. C., 1941, The distribution of acetylcholine in the peripheral and the central nervous system, *J. Physiol. (London)* 99:436–442.

Mangan, J. L., and Whittaker, V. P., 1966, The distribution of free amino acids in subcellular fractions of guinea-pig brain, *Biochem. J.* 98:128–137.

McLennan H., 1970, *Synaptic Transmission,* 2nd Ed., W. B. Saunders, Philadelphia.

Nilsson, G., Hökfelt, T., and Pernow, B., 1974, Distribution of substance P-like immunoreactivity in the rat central nervous system as revealed by immunohistochemistry, *Med. Biol.* 52:424–427.

Otsuka, M., 1972, γ-Aminobutyric acid in the nervous system, in: *The Structure and Function of Nervous Tissue* Vol. 4 (G. H. Bourne, ed.), pp. 249–289, Academic Press, New York.

Otsuka, M., and Konishi, S., 1974, Electrophysiology of mammalian spinal cord in vitro, *Nature (London)* 252:733–734.

Otsuka, M., and Konishi, S., 1976, Substance P and excitatory transmitter of primary sensory neurons, *Cold Spring Harbor Symp. Quant. Biol.* 40:135–143.

Otsuka, M., Konishi, S., and Takahashi, T., 1972a, The presence of a motoneuron-depolarizing peptide in bovine dorsal roots of spinal nerves, *Proc. Jpn. Acad.* 48:342–346.

Otsuka, M., Konishi, S., and Takahashi, T., 1972b, A further study of the motoneuron-depolarizing peptide extracted from dorsal roots of bovine spinal nerves, *Proc. Jpn. Acad.* 48:747–752.

Pedersen, E., Arlien-Soborg, P., Grynderup, V., and Henriksen, O., 1970, GABA derivative in spasticity (β-(4-chlorophenyl)-γ-aminobutyric acid, Ciba, 34.647-Ba), *Acta Neurol. Scand.* 46:257–266.

Phillis, J. W., Tebēcis, A. K., and York, D. H., 1968, Depression of spinal motoneurones by noradrenaline, 5-hydroxytryptamine and histamine, *Eur. J. Pharmacol.* **4**:471–475.

Pierau, F. K., and Zimmermann, P., 1973, Action of a GABA-derivative on postsynaptic potentials and membrane properties of cats' spinal motoneurones, *Brain Res.* **54**:376–380.

Powell, D., Leeman, S. E., Tregear, G. W., Niall, H. D., and Potts, J. T., Jr., 1973, Radioimmunoassay for substance P, *Nature (London) New Biol.* **241**:252–254.

Renaud, L. P., and Martin, J. B., 1975, Thyrotropin releasing hormone (TRH): depressant action on central neuronal activity, *Brain Res.* **86**:150–154.

Ryall, R. W., 1962, Sub-cellular distribution of pharmacologically active substances in guinea pig brain, *Nature (London)* **196**:680–681.

Saito, K., Konishi, S., and Otsuka, M., 1975 Antagonism between Lioresal and substance P in rat spinal cord, *Brain Res.* **97**:177–180.

Sprague, J. M., and Ha, H., 1964, The terminal fields of dorsal root fibers in the lumbosacral spinal cord of the cat, and the dendritic organization of the motor nuclei, in: *Organization of the Spinal Cord, Progress in Brain Research*, Vol. 11 (J. C. Eccles and J. P. Schadé, eds.), pp. 120–154, Elsevier, Amsterdam.

Takahashi, T., and Otsuka, M., 1974, Distribution of substance P in cat spinal cord and the alteration following unilateral dorsal root section, *Jpn. J. Pharmacol.* **24**(*Suppl.*):105.

Takahashi, T., and Otsuka, M., 1975, Regional distribution of substance P in the spinal cord and nerve roots of the cat and the effect of dorsal root section, *Brain Res.* **87**:1–11.

Takahashi, T., Konishi, S., Powell, D., Leeman, S. E., and Otsuka, M., 1974, Identification of the motoneuron-depolarizing peptide in bovine dorsal root as hypothalamic substance P, *Brain Res.* **73**:59–69.

Tregear, G. W., Niall, H. D., Potts, J. T., Jr., Leeman, S. E., and Chang, M. M., 1971, Synthesis of substance P, *Nature (London) New Biol.* **232**:87–89.

Vogt, M., 1969, Release from brain tissue of compounds with possible transmitter function: interaction of drugs with these substances, *Br. J. Pharmacol. Chemother.* **37**:325–337.

Yamamoto, C., and McIlwain, H., 1966, Electrical activities in thin sections from the mammalian brain maintained in chemically-defined media *in vitro, J. Neurochem.* **13**:1333–1343.

BIOCHEMISTRY OF SLEEP

MANFRED L. KARNOVSKY
AND PETER REICH

Departments of Biological Chemistry and Psychiatry
Harvard Medical School
Boston, Massachusetts

1. INTRODUCTION

The problem of defining sleep in biochemical terms is extraordinarily difficult, yet studies in this area are numerous and have a long history. Many of them focus on the general metabolism of the sleeping animal compared with the waking animal. They provide observations often really only peripheral to the central question: what biochemical manifestations are intrinsically causative of, or directly the result of, sleep?

The question, as it is framed above, is posed in terms that do not take into account the complexity and subtlety of the phenomenon. First, sleep is not a uniform state—but has a number of phases. Of these, slow-wave sleep (SWS) and paradoxical sleep, or rapid eye movement sleep (PS or REM), are the best known, although there are further subdivisions. Since these states are defined mainly by electrophysiological criteria, one would wish to have such criteria stated as an almost necessary refinement in any biochemical study of sleep. Further, the localization of biochemical observations to specific areas and cells of the brain—known to have special functions—would constitute a further *desideratum*. One would also hope to delimit the biochemical phenomena in terms of their subcellular locales. Finally, we will see that there are hints that the experiences of the animal just prior to a period of sleep may influence changes in the brain during sleep. There may thus be a need to control for the behavioral or emotional status of the subject.

Few, if any, studies in this field deal adequately with all the above possibilities. This causes us less concern than does the fact that many well-planned and executed studies often attempt to elucidate the biochemistry of sleep by examining the results of sleep *deprivation*. One might compare this approach with an examination of the role of a dietary component by studying animals maintained on a diet lacking that component. In the past, such nutritional studies led to important discoveries but also provided an enormous number of data on secondary or irrelevant phenomena. The will-o'-the-wisp nature of sleep has induced many investigators to use this approach rather than a more direct one. Such studies on *waking* animals avoid the many difficulties and pitfalls encountered in trying to focus on periods of natural sleep, which are often quite short in laboratory animals. Furthermore, in addition to being able to deprive animals of *total* sleep, one may selectively deprive animals of paradoxical sleep (Jouvet *et al.*, 1964; Cohen *et al.*, 1965; Morden *et al.*, 1967) by the "flowerpot," or pedestal, method, which will be discussed below. The hope then is to pinpoint the biochemical changes specific to that deprivation. However, as we will find repeatedly, it is extremely difficult to control for secondary or tertiary effects of the deprivation treatment.

Since so many recent biochemical studies have followed this approach, a brief word about the "flowerpot," or pedestal, method for PS deprivation might assist the reader to evaluate work described later. The animal is placed on a small platform surrounded by water. SWS may occur, during which loss of muscular tone does not take place. However, when the animal enters PS, muscular tone *is* lost, the animal falls into the water and is awakened (Jouvet *et al.*, 1964). One can well imagine that this could be a stressful situation, especially over a prolonged period of interference with PS (Heiner *et al.*, 1968). How should one control for variables other than deprivation of PS *per se*? Such controls are perhaps more imperative and more difficult in the case of biochemical studies than in electrophysiological studies. The latter determine phenomena, descriptive though they are in the main, that are closer to the functional activity of the CNS. The former, though ultimately more detailed and mechanistically revealing, are by virtue of their more basic nature likely to include observations irrelevant to PS deprivation *per se*. Such observations are not germane to the biochemical *basis* of PS. Though one may deduce the pattern of the head side of a coin by examining the tail—given a network of supportive information—we admit to a preference for direct experiments that study sleep, *per se*, compared with wakefulness. However, we recognize also that the constant refinement in devising control situations to screen out adventitious phenomena will make the deprivation-based approach ultimately an informative one.

With these general comments in mind, we state the objectives of this chapter as follows: We will attempt to describe the biochemical phenomena *in the brain* that are considered by various authors as relevant to the states of sleep contrasted with wakefulness. More general studies on systems other than the brain and studies that depend on drug-induced states will be omitted in the main.

Simply expressed, the questions to be discussed in this chapter are:

1. What are the changes in brain metabolism and general biochemistry that occur during sleep and that set this state apart from that of wakefulness? Can the definition of these changes shed light on the need of the animal for sleep and the physiological and psychological benefits that are clearly derived from sleep?
2. What is the role of the neurotransmitters in sleep? What are the cell–cell relationships in sleep that might differ from those in wakefulness?
3. What is it that brings on the sleeping state? In particular, are there humoral factors that accumulate during wakefulness and that ultimately reach such levels as to induce sleep? Are there substances that specifically lead to arousal? What is the nature of these humoral factors?

The first question, which concerns metabolic changes that have a potential specific benefit to the organism, will be dealt with rather fully in Sections 2 through 6. It has been the subject of detailed study for only a few years, even though some pioneer experiments were done nearly three decades ago. The second question has been systematically worked on, largely in laboratories such as that of Jouvet and his collaborators, and has been widely reviewed in the last decade or so. This topic will be touched on only lightly in Section 7 because of the extensive treatment it has received recently. The third topic (Section 8) has been sporadically attacked over many years and has recently achieved special prominence in the context of the numerous small molecules—peptides especially—that are elaborated by brain structures. It is also, of course, perhaps the most appealing and striking question in this field.

2. SOME PHYSIOLOGICAL ASPECTS OF SLEEP RELEVANT TO BIOCHEMICAL CHANGES

Although a detailed discussion of the physiology of sleep is not within the scope of this chapter, some physiological changes associated with the sleep–wakefulness cycle provide compelling, although indirect, evidence of changes in the patterns of biochemical processes during sleep (Kleitman, 1963).

Sleep is accompanied by a rise in alveolar CO_2 tension of 3–4 mm Hg, which reaches a peak after the first hour, while total ventilation falls with a concomitant small decrease in blood pH (Robin *et al.*, 1959). Birchfield *et al.* (1959) reported, in addition, a small decrease in arterial oxygen saturation from 95.1% (waking) to 93.7% (sleeping). The data indicate a decreased sensitivity of the respiratory center to CO_2. This can be confirmed by providing sleeping subjects with a respiratory CO_2 content of 4–6%. With this stimulus, only a moderate increase in ventilation is observed during sleep, while a marked increase occurs during wakefulness. Periodic breathing has also been observed during sleep. These basic changes of respiratory physiology are rapidly reversible upon awakening, indicating that their basis is neurological and not chemical *per se* (Robin *et al.*, 1959; Birchfield *et al.*, 1959; Reed and Kellogg, 1958, 1960). In addition, the basal metabolic rate falls during sleep by approximately 9% (Jana, 1965). The rate of oxygen consumption (whole body) varies with the stages of sleep, being highest in PS and lowest in SWS (Brebbia and Altschuler, 1965) (Figure 1). Other relevant changes in general physiology are observed in the diurnal temperature cycle and diurnal cycles in hormone secretions. The latter include the rise in adrenal cortical hormone secretion during the night and the association of the phases of sleep with the secretions of various pituitary hormones, growth hormone, luteinizing hormone, and antidiuretic hormone (Kripke, 1974).

More directly relevant to the biochemistry of the brain are the physiological changes that occur in the central nervous system itself. Twenty years ago Mangold *et al.* (1955) used a nitrous oxide technique to determine cerebral blood flow during sleep in 50 young men, ages 17–36. They were kept awake for 20 hr to allow sleep to occur with the experimental paraphernalia in place. In only six subjects were the experimental criteria, as laid down by the investigators, fulfilled. In these six instances, cerebral blood flow (CBF) increased during sleep from control values of 59 to 65 cc per 100 g per min ($P < 0.01$). When fatigued waking subjects were compared to rested controls, there were no differences in CBF. The rise in CBF in the sleeping subjects occurred despite a fall of mean arterial blood pressure from 94 to 90 mm Hg, indicating a reduction in cerebral vascular resistance during sleep. Cerebral AV oxygen differences and cerebral oxygen consumption showed no changes in sleep when compared with controls. The authors suggested that cerebral hyperemia in sleep cannot be explained by known increases in cerebral metabolism, although such changes may well be taking place. However, the findings rule out anoxia, ischemia, narcosis, or generalized depression in cerebral metabolism as explanations for sleep. The authors also found that onset of sleep was

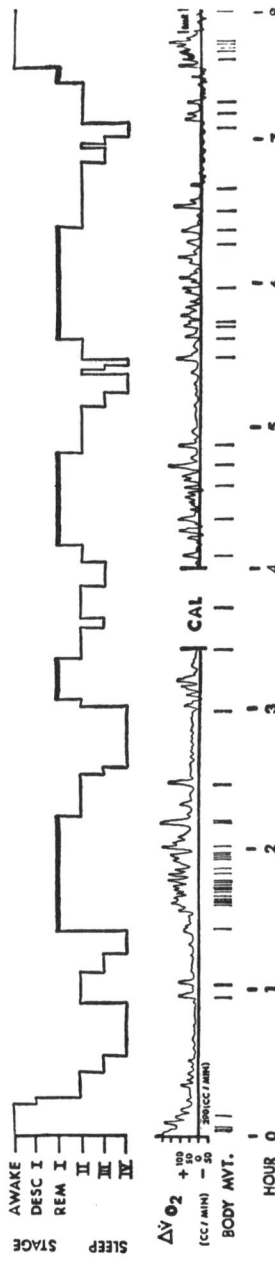

FIGURE 1. Changes in oxygen consumption ($\Delta V\ O_2$) during a typical night for one subject. Values for oxygen are in cubic centimeters per minute relative to a reference line that represents the basal oxygen consumption level of 290 cm^3/min (41.27 kcal m^{-2} hr^{-1}) for this individual. This slightly raised basal $V\ O2$ reflects the relatively elevated overall level of metabolism during the previous night. Stage of sleep and body movements are also represented. CAL marks the instrument calibration period, during which time the subject continued to sleep uninterrupted. From Brebbia and Altschuler (1965).

preceded by higher CO_2 tension and lower pH in both arterial and venous cerebral blood, suggesting some relationship between respiratory acidosis and the process of falling asleep.

Nine years later, Birzis and Tachibana (1964) estimated relative blood flow in small well-defined regions of brain through extrapolation from measurements of electrical impedance of tissue. With the onset of drowsiness, hypothalamic blood flow fell, but cortical and reticular formation blood flow rose and continued to rise progressively with the continuation of sleep. Thus, in light sleep (stage I), impedance increased in cortex and reticular formation. In SWS, impedance increased in all areas studied, and in PS, a picture similar to that of an hyperalert animal occurred with marked increase of hypothalamic amplitude and with reticular formation values either increased or decreased. This activity of the cerebral vascular pattern in PS seemed consistent with the active involvement of the ascending reticular system.

Reivich *et al.* (1968) made more refined observations on regional cerebral blood flow in cats during SWS and PS. Cats with chronic electrodes and with catheters in the femoral artery and vein were all deprived of sleep for 24–48 hr by walking on a treadmill that permitted little rest. This was done to encourage the occurrence of PS and SWS. Comparisons were made among waking controls (post sleep deprivation), SWS, and PS. Regional blood flow was measured by autoradiographic techniques. There were no differences in pCO_2, pO_2, pH, or hematocrit among the three groups. Figure 2 summarizes the findings of these authors. In PS, regional blood flow increased in all 25 regions studied, ranging from 62% increase in cerebellar white matter to 173% increase in the cochlear nuclei. During SWS, significant increases in blood flow occurred in 10 of the 25 regions studied. These were of smaller magnitude than those seen in PS. They varied from 26% increase in association cortex to 68% increase in superior olive. Most of the structures with more than 100% increase in blood flow during sleep were located in the brain stem and in the diencephalon, while those with the least change in blood flow were in the areas of the cortex and in the white matter. The authors support the possibility of increased cerebral metabolism as the most logical explanation for the increase in regional cerebral blood flow during sleep. The study emphasizes the importance of considering the various regions of the brain separately when searching for metabolic concomitants of sleep.

Further observations on cerebral blood flow during sleep and waking were made by Townsend *et al.* (1973). Blood flow in three regions of brain, temporal, precentral, and occipital, was studied in human subjects during SWS, PS, and W, using the clearance of inhaled ^{133}Xe as measured by extracranial monitoring. In all regions studied, there were significant variations among the three states. All brain regions showed decreases in blood

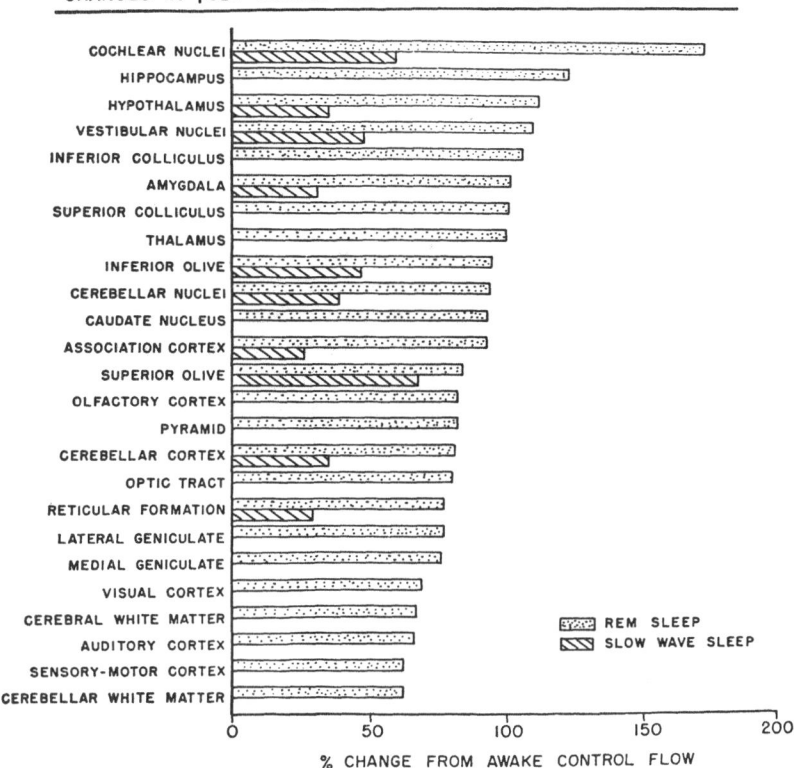

FIGURE 2. Percentage change from awake control flow. Those regions of the brain showing a significant change in flow during SWS and PS are plotted as a percentage change from the awake control flow (rCBF). From Reivich *et al.* (1968).

flow during SWS and increases during PS. The three regions differed in their pattern of change. Blood flow was greatest in the precentral region in all three states. Here the largest flow reduction during SWS occurred, but the smallest increase during PS. Conversely, the temporal region, where the lowest flow occurred during waking, showed the smallest decrease in SWS but the largest increase in PS. Again, the significance of regional differences should be emphasized. The authors point out that changes in cerebral metabolism are the most likely explanations for these changes in blood flow. As further evidence for such metabolic changes, they cite the alterations in cerebral unit activity that also occur during sleep.

There is increasing evidence that brain neuronal units exhibit characteristic patterns of spontaneous activity during SWS and PS (Huttenlocher, 1961; Evarts, 1964; Hobson *et al.*, 1974; McGinty *et al.*, 1974; Orem *et al.*,

1974). In general, unit activity decreases in SWS and increases in PS with respect to waking controls. There is considerable regional variation in these patterns and many brain areas have not been studied. It should also be noted that recording smaller neuronal units presents considerable technical difficulties.

Changes in brain temperature during sleep also provide indirect evidence of changes in cerebral metabolism. Kawamura and Sawyer (1965) recorded brain temperature in white rabbits, unrestrained and unanesthetized, in PS and SWS. Brain temperature dropped in most areas during SWS but during PS a marked rise of up to 0.4°C occurred. The longer periods of PS correlated with the higher elevations in brain temperature. There is an indication that brain temperature and cerebral blood flow are inversely related (Melzack and Casey, 1967).

Valatx et al. (1973) altered brain temperature by exposing rats to increasing ambient temperature ranging from 20 to 36°C. The animals were kept at the experimental ambience for 3-week periods. An increase in sleep time with rising ambient temperature was noted throughout, reading a maximum at the highest temperatures. The largest increase was in PS time. Thus, the PS/SWS ratio increased from 0.136 to 20°C to 0.194 at 34°C. Fluctuations of cerebral temperature corresponding to the S–W cycle were observed at 20°C. These were dampened at 34°C, while the mean cerebral temperature rose constantly to a high at 34°C.

Cooper and Hulme (1966, 1969) observed changes in intracranial pressure during sleep. They used patients who had intracranial hypertension because of disturbances in the circulation of CSF. They found that PS was associated with large increases of intracranial pressure, suggesting that a high rate of cerebral metabolism occurred during PS, producing cerebral vasodilatation and increases in intracerebral pressure. SWS showed lower changes in intracranial pressure. The effect of sleep on the electrolytes of the brain, and especially on specific structures, might be expected to be a topic of major concern to physiologists. However, there is little in the literature on this matter. Heiner et al. (1968) reported in a brief communication that after 10 days of PS deprivation (Jouvet et al., 1964), the brains of the experimental rats contained significantly less K^+ than those of control animals. Dry-weight, protein, and levels of Na^+, Ca^{2+}, or Mg^{2+} were not affected. The telencephalon was examined separately from the rest of the brain, but the findings pertained to both samples. Blood K^+ also dropped markedly, the bulk of the decrease being, of course, attributable to loss from red cells. Potassium excretion was not followed to rule out generalized K^+ loss. One cannot really ascribe any of these effects to deprivation of PS per se, partly because of the nonspecificity of the effect with respect to brain cells and partly because no stringently designed control measurements were made; the comparisons were simply to animals not subjec•:d to isolation on a pedestal or flowerpot. The degree of stress induced by the treatment is indicated by the authors in their description of the animals.

Some interesting studies on sleep, not restricted to physiological observations, might be mentioned here to indicate how different the various approaches are that have been taken. MacFayden *et al.* (1973) subjected ten young men to acute starvation while observing their sleep behavior. After four nights of starvation, they observed a significant rise in SWS with a fall in PS. Lucero (1970) found that the process of learning in a maze led rats to show an increase in PS time from 16% of total sleep to 24% of total sleep, the change being significant ($P < 0.001$). He postulated that PS is associated with brain chemical processes that are related to learning.

Thus, the evidence from various physiological and more general studies indicates that cerebral metabolism does change during sleep compared with quiet wakefulness. Some studies provide evidence that in SWS brain metabolic levels are greater than during W and more marked increases occur during PS. Other studies show a fall in brain metabolism during SWS with respect to waking controls. Despite this uncertainty, these studies do, however, support the assumption that brain metabolism *varies in specific brain regions* during sleep compared with wakefulness and give point to the experimental strategy of separating such regions and also of considering PS and SWS apart from each other in searching for metabolic changes during sleep.

3. CHANGES IN CEREBRAL ENZYMES DURING SLEEP

Comparison of the activities of cerebral enzymes during sleep and wakefulness should provide important and relevant information basic to comprehending metabolic shifts. On the other hand, the use of maneuvers such as sleep deprivation or of drugs to provide a window on the sleeping state could well introduce artifacts and false leads due to stress, changes in hormonal states, or even, in the case of drugs, direct effects on the enzymes *per se.*

Few studies on enzyme levels in brain during sleep have actually been undertaken. Those that are reported have importance (a) in comprehending the metabolism of amino acid precursors of transmitter substances implicated in sleep, (b) with respect to certain macromolecules (e.g., RNA) that have been considered as important in that state (see later), or (c) with reference to intermediary metabolism and metabolic energy.

(a) With respect to enzymes potentially linked to control of transmitter levels, one might note three investigations: *(i)* studies on tyrosine transaminase in sleep deprivation in mice (Francesconi and Mager, 1971), where stress effects make the assignment of a specific role of the enzyme in sleep dubious. *(ii)* Studies of tyrosine hydroxylase under conditions of PS deprivation in rats, where a marked and significant increase in activity was observed in lower brain stem and cerebral cortex. The increase in whole brain was less evident and was not significant for upper brain stem (Sinha *et al.*, 1973*a*). "Stress controls," i.e., measurements on animals subjected to the same general conditions as those deprived of PS, even though we question their absolute validity, would have made this study more valuable. Though the authors do comment on the contribution of stress, no data were given for animals that were placed on larger platforms and therefore did not fall into water during PS, or that were regularly placed in water to control for immersion (see later). *(iii)* A very recent paper on liver tyrosine transaminase in rats during sleep and after sleep deprivation has gone a substantial distance toward meeting such criticisms as those given, by employing adrenalectomized rats. Sleep (daytime) was accompanied by low levels of enzyme in the brain. Wakefulness was accompanied by higher levels (twofold). Sleep deprivation (SW and PS), achieved by gentle manipulation, caused the enzyme level to rise (daytime) and if the animals were then allowed to sleep, there was a precipitous fall. Though tryptophan pyrrolase did not show such a marked increase in normal rats that went from sleep (daytime) to wakefulness (night), sleep deprivation did cause at least a doubling of the enzyme activity. When the sleep-deprived animals were permitted to sleep, there was a steep drop in activity in this case as well (Deguchi *et al.*, 1975).*

In hypophysectomized and in adrenalectomized rats, there was indeed still an increase in the level of tyrosine transaminase from 11 A.M. to 10 P.M., i.e., as the animals went from sleep to wakefulness, but the effect was noticeably dampened compared to that in intact rats. Sleep deprivation accentuated the rise during the period mentioned, but no decline was observed when the animals were thereafter permitted to sleep. The activity of tryptophan pyrrolase, under the conditions of hypophysectomy or adrenalectomy, failed to change. Finally, after the manipulation of the sleep–wakefulness cycle, the authors noted no significant changes in the putative transmitters in the brain derived from each of these amino acids, i.e., norepinephrine or serotonin.

These interesting experiments, based on clear-cut patterns of sleep and wakefulness, leave one, however, puzzled by several matters. Though the situation with regard to tryptophan pyrrolase seems clear—i.e., it appears to be responsive to the stress of deprivation rather than intrinsically to the sleep–wakefulness cycle—that for tyrosine transaminase is ambiguous. The

---

*We are grateful to Dr. A. K. Sinha for providing a preprint of this article.

ambiguity stems from the diminution of the fluctuations in the adrenalectomized or hypophysectomized animals during the sleep–wakefulness cycle and, particularly, the failure to return to lower levels during sleep after sleep deprivation. Can one be sure that adrenalectomy or hypophysectomy totally eliminates the stress response? There are probably other sources of "antistress hormones" that affect the issue.

(b) When one considers enzymes concerned directly, or indirectly, with macro-molecules, the available information is minimal. Glushenko and Demin (1971) observed a slight activation of neutral proteinases in the cerebral hemispheres and midbrain of rats during natural sleep. The cerebellum was not affected. PS deprivation led to depression of neutral proteinase activity and increased activity of acid proteinases. Restoration of sleep following deprivation led to a normalization of proteolytic activity. It would be hard to relate these observations on hydrolases directly to biochemical changes intrinsic to the sleeping state, but the increases in acid hydrolases during PS deprivation make one wonder about the possible role of lysosomes, and this raises additional questions as to the nature of the cell types or brain nuclei involved.

(c) In considering enzymes involved in general intermediary metabolism, three groups of studies may be pertinent:

(*i*) Enzymes involved in nitrogen metabolism in brain responded to PS deprivation for 6 days as follows (Haulica *et al.*, 1970): glutamine synthetase—a small (15%) but significant rise; glutaminase—*ca.* 50% increase with a very high degree of probability; AMP-deaminase—greater than doubling, statistically highly significant; adenosine deaminase—no significant change. The last-mentioned two enzymes must also be considered in light of nucleic acid decreases during PS deprivation (see later). Unfortunately, the data just mentioned were not supported by control experiments. Enough has already been said to indicate that the study of sleep deprivation without "correction" for stress yields little or no real information about sleep *per se*.

(*ii*) A classic series of experiments on the variation of a key enzyme in the regulation of metabolic energy is that from Hydén's laboratory (Hydén and Lange, 1964, 1965*a,b*). These investigators studied succinoxidase during sleep and wakefulness. They employed rabbits as the experimental animal, and their techniques measured the enzyme activities on single cells taken from various areas of the brain. The animals were trained to sleep in individual boxes with the head protruding and were maintained in a quiet room with subdued light. They apparently slept readily and could be killed without being awakened, an hour and a half after being placed in the boxes. The control animals were kept active and awake by gentle handling for at least 1 hr, and a third group was simply left in their home cages without disturbance.

These methods are similar to those described by Sterner (1963) and would seem to us to represent a properly controlled system for the following reasons: first, the groups of animals are well chosen and natural sleep can be studied without any deprivation; second, extensive examination of the EEGs of at least the first two groups of animals had previously been carried out. It might perhaps have been better had EEGs also actually been followed in Hydén and Lange's experiments. Third, the examination of an enzyme by micromethods in single cells in various separate areas of the brain, differentiating between neurons and glia, is truly a refined approach that meets most of the *desiderata* mentioned in the Introduction.

In Table 1, which is from the paper of Hydén and Lange (1965 a), the important observations are as follows: In the nucleus reticularis gigantocellularis caudalis, succinoxidase in the nerve cells was increased threefold

TABLE 1. Succinoxidase Activity of Neurons and Glia Isolated from Rabbits[a]

	Nerve cells		Glia	
Group	Activity (10^{-4} μl O_2)	Number of analyses	Activity (10^{-4} μl O_2)	Number of analyses
	Nucleus reticularis giganto-cellularis			
1	3.41 ± 0.51	29	2.34 ± 0.18	25
2	1.30 ± 0.25	24	3.06 ± 0.24	28
3	2.74 ± 0.21	39	2.16 ± 0.18	33
	Nucleus reticularis pontis oralis			
1	6.38 ± 0.58	35	3.50 ± 0.30	19
2	4.01 ± 0.52	15	2.94 ± 0.21	17
3	5.41 ± 0.39	32	3.72 ± 0.58	12
	Nucleus trigeminus mesencephalicus			
1	2.68 ± 0.29	9	1.46 ± 0.23	8
2	3.08 ± 0.83	5	1.11 ± 0.12	5
3	3.21 ± 1.16	5	1.15 ± 0.15	13
	Nucleus hypoglossus			
1	1.03 ± 0.17	13		
2	0.83 ± 0.08	13		
3	0.68 ± 0.13	3		

[a] The results are expressed as 10^{-4} μl of oxygen per sample per hour. Group 1, sleep; group 2, wakefulness; group 3, cage controls. When wakefulness was compared with sleep, the difference between the enzyme values in nucleus reticularis giganto-cellularis proved to be significant both for nerve cells ($P < .001$) and glia ($P < .02$). Values for group 1 compared with those for group 3 were not significant. The difference between the values for the nerve cells in the nucleus reticularis pontis oralis of groups 1 and 2 was significant ($P < .01$). Data from Hydén and Lange (1965a).

during sleep compared with wakefulness, i.e., comparing groups 1 and 2. The third group, cage controls, was intermediate. The glial cells showed a significantly decreased level during sleep compared with wakefulness. There was no effect of sleep when the cage controls were compared with sleeping animals. It seemed that there was a rhythm of enzyme activity reflected by the ratio of enzymes in neurons and glia. The ratio was highest during sleep and lowest during wakefulness.

In the nucleus reticularis pontis oralis there was again an increase in activity of the enzyme in the nerve cells during sleep, but the effect was less than in the first-mentioned region. The glial data did not show significant changes here. It is of great interest that when the effect of arousal by weak sounds was studied there was a significant decrease in the enzyme titers in this region compared with sleep.

The authors also studied neurons and glia from an area of the brain stem not related to the reticular formation, i.e., the large neurons of the nucleus trigeminus mesencephalicus—a sensory region. No significant changes were found during sleep for either neurons or glia. Control neurons of a motor type, hypoglossal nerve cells, showed no significant differences that could be ascribed to sleep or wakefulness.

The authors concluded that the results indicate "that the caudal part of the reticular formation is a center of the brain stem which reflects in pronounced metabolic changes the biological clock behind sleep rhythm," and further, they believed that the neurons and glia form functional units that are energetically coupled and influence each other functionally (Hydén and Pigon, 1960). They felt that when functional demands increased, there was a change in the balance of the enzyme and, indeed, of protein content between neurons and glia. The thesis is attractive. However, it is difficult to comprehend why the "functional demands" that confronted the animals of group 2, i.e., the waking group, and, indeed, of group 3, the cage controls, should have actually lowered the enzyme activity in the nerve cells *per se*. We would prefer simply to accept the observations at this stage without too much rationalization.

In a companion study (Hamberger *et al.*, 1966), this laboratory noted that changes such as those described above for sleep and wakefulness did not occur in barbiturate-induced sleep; i.e., the levels of succinoxidase in the neurons were grossly depressed during barbiturate sleep compared with physiological sleep. In the glia, the level of the enzyme was depressed below that in wakefulness and not distinguishable from that during physiological sleep.

(iii) Glucose-6-phosphatase, another enzyme involved in intermediary metabolism, has recently been examined in the context of sleep and wakefulness. The experiments have a rather complex background. In 1967, Reich *et al.* (1967) studied the effect of sleep on the incorporation of

inorganic phosphate into components of the brains of 3-week-old rats. They observed a two- to threefold increase in the incorporation of $^{32}P_i$, administered by tail vein, into an acid-labile phosphate entity of the brain tissue that remained after removal of lipids and extraction with cold trichloracetic acid. Liver, tested as a control organ, did not show this effect. In the system used, the baby rats were kept awake for 90 min by gentle movement. If placed in a quiet cage, the animals would fall asleep spontaneously within 2–3 min and remained asleep for a period of ½ hr. The waking group was kept from sleeping by quiet manipulation for an equal period. This experimental design had been decided upon after careful examination of the daytime sleep–wakefulness cycle of rats of this age in which 30 min of sleep appeared to alternate with between 40 and 60 min of wakefulness. Thus, the rats were prepared, as described, by being kept awake through only one complete cycle, after which it was very easy to obtain a 30-min sleeping period. Because of the small size and the large number of experimental animals used, EEG recordings were not made; thus, the investigators were forced to rely on behavioral criteria for the sleeping or waking state. Further, the isotope was administered systemically rather than directly into the brain.

It was obviously necessary to conduct experiments to validate this phenomenon in adult rats with electrophysiological monitoring. Rats were consequently implanted with cortical and muscle electrodes and an external guide tube to one lateral ventricle (Goodrich *et al.*, 1969). The animals were allowed to recover for 3 weeks and were maintained on a 12-hr light–dark cycle. Two separate groups of animals were examined: those that were acclimated to the observation chamber and those that were not. In the acclimated series, the animals were simply placed in the observation chamber for 2 hr each day for 4 consecutive days prior to the experiment. In the experiment itself, the prepared animals were placed in the observation chamber in the morning and observed. After a while, regularly occurring sleep periods of 20–30 min duration began. In one such period, an infusion of radioactive inorganic phosphate was made into the ventricle, and the animal was allowed to sleep for 20 min. A matched animal (operated upon on the same day as the experimental animal) was maintained in the same manner and was kept awake by gentle auditory stimulation when necessary for the same period postinfusion. At the end of the period of sleep or wakefulness, a trap door was sprung and the animal was rapidly frozen by immersion in liquid nitrogen. The brains were later taken and analyzed. Data are given in Table 2 and are reproduced from Reich *et al.* (1973). It may be observed that in unacclimatized animals a situation comparable to that seen in the baby rats pertained; i.e., hot trichloracetic acid (TCA) did extract phosphate from a brain powder that had a significantly higher specific activity during sleep than during wakefulness. This situation was not observed in the acclimated animals and, furthermore,

TABLE 2. Specific Radioactivity of Phosphorus in Brain Fractions of Sleeping and Waking Animals[a]

Fraction	Not acclimatized (Group A)				Acclimatized (Group B)			
	Sleeping (c.p.m./μgP)	Waking (c.p.m./μgP)	$\dfrac{\text{Sleeping}}{\text{Waking}}$ (S/W)A [b]	P	Sleeping (c.p.m./μgP)	Waking (c.p.m./μgP)	$\dfrac{\text{Sleeping}}{\text{Waking}}$ (S/W)B [b]	P
Lipids	800 ± 369	652 ± 62	1.34 ± 0.33	NS	521 ± 145	829 ± 231	0.65 ± 0.05	<0.001
Cold TCA	33,756 ± 6772	30,944 ± 6187	1.39 ± 0.24	NS	28,349 ± 6277	27,902 ± 2696	0.98 ± 0.12	NS
Hot TCA	5714 ± 1069	4514 ± 938	1.79 ± 0.27	<0.02	6785 ± 1605	6184 ± 1413	1.08 ± 0.08	NS
Residue	8954 ± 1132	8574 ± 996	2.00 ± 0.49	<0.1	7040 ± 785	8716 ± 1581	1.02 ± 0.12	NS

[a] The results are given as the means ± S.E.M. in c.p.m./μg P for experiments with five pairs of animals that were not acclimatized and for six pairs that were acclimatized. Data from Reich et al. (1973).

[b] These values are the means ± S.E.M. of the ratios for each pair of animals in each group, i.e., twelve pairs in group A and six in group B. The probability, P, of the difference from 1.0 is given for the ratios; NS, not significant ($P > 0.05$).

here the lipid fraction exhibited a substantial diminution of labeling of phosphatides during sleep compared with wakefulness. One might add, parenthetically, that this is the only instance of an effect on brain lipids manifested in sleep that we have encountered in the literature.

This series of experiments, as is also true of those concerning carbohydrate metabolism reported later in this review, illustrates dramatically the effect of prior treatment on the sleep–wakefulness differences. However, there is no doubt that the unacclimated animals had normal sleep periods characterized by normal EEGs. It is the characterization of the acid-labile phosphate-containing entity that was affected in the unacclimated animals that is of further interest here.

When the delipidated, cold-TCA-extracted material was subjected to the action of proteases and the reaction mixture dialyzed, it was noted that small radioactive entities passed through the membrane and could be separated by paper chromatography. Ninhydrin reactivity and the presence of ^{32}P went hand in hand. The crude substance that contained acid-labile ^{32}P was also solubilized by deoxycholate or sodium dodecylsulfate, but not by nonionic detergents. This proved to be a key observation for ensuing work.

The preliminary data thus indicated that the entity that incorporated additional ^{32}P during sleep in unacclimated animals was a phosphoprotein. The final recognition of the substance involved has now been achieved (Anchors and Karnovsky, 1975) by the following stratagem: Rats that were permitted to sleep received inorganic ^{33}P$_i$; those that were maintained awake received inorganic phosphate ^{32}P$_i$. The specific isotope given was varied between animals in the two conditions to avoid the possibility of an isotope effect.

After the animals had been frozen, as described, the brains were removed and homogenized in chloroform methanol to remove lipids; the residue was extracted as before with cold trichloracetic acid. Equal amounts of the two remaining solid mixtures from sleeping or waking animals could now be solubilized in sodium dodecylsulfate and subjected to polyacrylamide gel electrophoresis. The ratio of the two isotopes, i.e., that of the sleeper to that of the waker, was now determined along the polyacrylamide gel, and a phosphoprotein entity of molecular weight approximately 28,000 was noted as the fraction in which the isotope of the sleeper reached a value severalfold that of the waker, with a high degree of probability. A study of the kinetics of hydrolysis of the phosphate and thin-layer chromatographic analysis of a base hydrolysate of the purified material indicated that the phosphate was attached to histidine. A second method of purification could be devised as follows: Delipidation of the brain was achieved by removal of myelin in sucrose gradient centrifugations. The sedimented material could then be extracted with sodium deoxycholate and subjected to gel filtration on Sephadex G-100 and chromatography on DEAE Sephadex. Such treatment yielded a material of the same amino acid composition as the relevant 28,000-dalton peak just mentioned and which, upon gel chromatography in SDS, yielded the same band. The molecular weight of the deoxycholate-extracted entity was, however, approximately four times that of the SDS-dissociated substance. Furthermore, the purified material obtained by the second method was enzymatically active. This observation was made in the following way: A rat brain was infused with inorganic $^{32}P_i$. After homogenization of the brain in buffer and centrifugation, the supernatant fluid was dialyzed so that small molecules could be obtained on the outside of the bag. When the purified protein obtained from brains of untreated rats was incubated very briefly with this mixture of labeled small molecular weight substrates and the protein was recovered and repurified, it was observed to have acquired radioactivity. A screening of potential enzymatic activities that could be associated with histidine-phosphate phosphoproteins revealed that the brain phosphoprotein of interest was glucose-6-phosphatase. Furthermore, incubation of this protein with ^{32}P-labeled glucose-6-phosphate transferred activity to the phosphoprotein (Feldman and Butler, 1969, 1972). Finally, it was noted that purification of the brain extract of an animal that had received labeled inorganic phosphate intraventricularly showed that the enzyme activity and radioactivity moved in parallel in chromatographic procedures.

The enzyme was present in neurons but not in glia and appeared to be associated with the membrane fraction. No real differences were observed between various areas of the brain though this aspect has certainly not been adequately pursued.

The authors of this work (Anchors and Karnovsky, 1975) considered the possible sleep-related roles of glucose-6-phosphatase, which is a multifunctional enzyme, i.e., catalyzes a large number of different reactions. For example, in addition to its ability to liberate inorganic orthophosphate from glucose-6-phosphate and inorganic pyrophosphate, this enzyme can act as a phosphotransferase. It can transfer phosphate to glucose from another molecule of glucose-6-phosphate or from inorganic pyrophosphate, carbamyl phosphate, phosphoenolpyruvate, and nucleoside triphosphates (Nordlie, 1974). The phosphorylated form of the enzyme is an intermediate in these reactions; i.e., the phosphate is transferred from the donor to a histidine on the enzyme and thence to an acceptor, such as glucose or water.

This enzyme may be involved in the transport of glucose during sleep and wakefulness, for which its membrane location makes it uniquely suitable (Anchors and Karnovsky, 1975). The stimulated labeling of the glucose-6-phosphatase during sleep was probably not due to an increased specific activity of the nucleotide pool. However, since the exact nature of the phosphate donor is not yet known, this matter is not resolved. The increased labeling of the phosphoprotein might be due to an increased flux of the substrate "through the enzyme" due to conditions other than a change in the activity of the enzyme *per se*. Finally, there might actually be a change in the activity of the enzyme, due, for example, to allosteric phenomena. It is certainly not easy to differentiate between the possibilities mentioned. However, some preliminary experiments (B. H. Lee and M. L. Karnovsky, unpublished data) have revealed that there is about a 20% increase in the total activity of glucose-6-phosphatase in the brains of sleeping animals compared with waking animals under conditions identical to those used in the labeling studies just summarized. Five pairs of animals were studied, and in all cases the enzyme in the brain of the sleeping animal was enhanced over that in the waking animal. A "control" enzyme, neutral phosphatase (using glycerol-2-phosphate as a substrate) was not affected by sleep.

Why a change in the transport of glucose might be relevant or important to the state of sleep is, at present, quite unknown. Van den Noort and Brine (1970) directly measured glucose in rat brain during sleep and wakefulness. In the case of unrestrained and unanesthetized animals, an increased glucose content of the brain was observed during sleep. This might reflect decreased utilization or an increased transport of glucose into the brain cells. The last-mentioned possibility would fit with the observations given concerning glucose-6-phosphatase, presumably behaving as a phosphorylating agent in the transport of glucose. That this is not too farfetched an idea is suggested by two papers on transport of sugars *in vitro* by synaptosomes. Diamond and Fishman (1973) found a high-affinity transport system for 2-deoxy-D-glucose (and thus for D-glucose) that was sodium-independent and insensitive to insulin. It was inhibitable by phloretin. They believe it to be a special property of neuronal cell membranes. Warfield and Segal (1974), also using rat synaptosomes, observed a transport system for D-galactose of a similar nature that was sensitive to phlorizin. Both authors observed lower affinity systems as well. Glucose-6-phosphatase is, of course, sensitive to inhibition by phloretin and its glycosides and is a relatively nonspecific enzyme.

That cerebral glucose metabolism and its control may be related to the state of sleep and wakefulness is also suggested by the observations of Panksepp *et al.* (1973), who noted that intraperitoneal injection of 2-deoxyglucose into cats caused a state of sleep, defined electroencephalographically as well as by behavioral criteria. The blood glucose then would be expected to rise dramatically after the injection, since this inhibitor blocks glycolysis and interferes with transport of D-glucose.

We do not know what the accumulation of 2-deoxyglucose or its phosphates was in the brain of the animals, nor does it seem to us opportune yet to offer explanations as to why an apparent *decrease* in glucose utilization by blockade with 2-deoxy-D-glucose should cause a sleeping or sleeplike state under conditions, at least in the rat, that would seem to indicate increased transport of glucose into the brain cells. Perhaps the accumulation of the phosphorylated sugar is the key matter.

A striking study that might have relevance in this context is that of Kennedy *et al.* (1975), who made an extremely interesting series of observations on the rates of glucose consumption of the central nervous system in terms of altered functional activity. These authors gave rats or monkeys a pulse of [^{14}C]deoxyglucose. They then examined cross sections of the central nervous system by autoradiography after altering the functional state of the animals. Sciatic nerve stimulation (rat) and seizures induced by penicillin (monkey) were followed by observing the content of the radioactive tracer in various areas of the central nervous system after sectioning the spinal cord and brain, respectively. Unilateral enucleation as a means of decreasing input was also studied. The stimulatory operations mentioned previously caused an increased content of 2-deoxyglucose metabolites, presumably 2-deoxy-D-glucose-6-phosphate, in the tissue in an asymmetric fashion. This approach may be used to map regions in CNS in which glucose utilization is altered in response to altered local functional activity.

One cannot help but be struck by these observations in considering the altered state of activity represented by sleep and the observations concerning glucose metabolism to be cited in Section 4. A quantitative approach has also been taken with this methodology (Sokoloff, 1975), though full details are not yet available.* Glucose utilization can be computed from [^{14}C]deoxyglucose-6-phosphate content after application of appropriate manipulations that take into account relative contents and turnover numbers of glucose and the inhibitor, as well as kinetic constants for transport and phosphorylation. Such an approach might indeed yield very significant data when applied to the question of altered glucose metabolism in the brain during sleep (Section 4).

4. CARBOHYDRATE METABOLISM AND METABOLIC ENERGY OF THE BRAIN DURING SLEEP

4.1. Glucose Catabolism and Labile Phosphates

A number of studies have now been performed concerning the metabolism of glucose during sleep by measuring levels of metabolites. The results obtained by the various workers differ somewhat. These differences may be ascribed to the methodology and conditions employed; i.e., they might be adventitious. One should not disregard the possibility, however, that

---

*For an extensive discussion, see Neurosciences Research Program Bulletin *14*, No. 4, 1976, "Neuroanatomical Functional Mapping by the Radioactive 2-Deoxy-D-Glucose Method" (F. Plum, A. Gjedde, and F. E. Samson, eds.), MIT Press.

those conditions might determine the metabolism of the brain during sleep in a real sense. As a guide, Table 3, modified from Reich *et al.* (1972), is presented.

Richter and Dawson (1948) and Cocks (1967), employing natural sleep in rats, found that brain lactate decreased during sleep. The former workers allowed sleep to be induced by strong sunlight, whereas the latter simply employed the natural sleep–wakefulness cycle and selected a period of sleep from that cycle. Van den Noort and Brine (1970) employed animals that had been exercising for long periods prior to sleeping. They observed a lactate diminution during the period of sleep that was somewhat smaller than that reported by the two groups mentioned first. The earlier observations and most of those of Van den Noort and Brine (1970) were based on animals that were not monitored electrophysiologically. The last mentioned investigators did, however, monitor *some* animals, and this approach was used by Shimizu *et al.* (1966) as well as by Reich *et al.* (1972). In these two sets of experiments, rather different results were observed from those previously obtained for lactate production; i.e., Shimizu and his colleagues (1966) observed no change during sleep, whereas Reich *et al.* (1972) noted that lactate production might increase or decrease depending on whether the animal had been acclimated to the observation chamber or not. No striking differences were observed with respect to the EEG between the acclimated and nonacclimated rats, except that the former animals appeared to sleep somewhat more and to have a somewhat higher content of PS in the period *prior to* the experiment *per se*.

In all cases except the experiments of Van den Noort and Brine (1970), the periods of observation on which measurements were based lasted ½ hr or less. In the experience of the present authors it is difficult to obtain a continuous period of sleep, or even one that could be labeled as a discrete period of sleep, that exceeds 30 min under normal conditions. Such a situation was, however, made possible in the experiments of Van den Noort and Brine (1970) by the prior very lengthy exercising of the rats, a stratagem that might also have less desirable sequelae in that the results might reflect the continuing effects of the period of exercise itself, as well as stress.

In general terms, the overall impact of all this work would indicate that in sleeping animals there is a significant, but not large, depression of glycolysis to lactate compared with the waking controls. In animals that were mildly apprehensive in terms of their lack of familiarity with the apparatus or observation chamber, there was a distinct increase of lactate production over the normal waking level. Reich *et al.* (1972) suggested that exercise to near exhaustion might reduce this "anxiety" effect and cited work of Hobson (1968) that correlated exercise with enhancement of SWS. Weyne *et al.* (1970) pointed out that anxious animals hyperventilate with reduction in blood pCO_2, cerebral vasoconstriction, and tissue anoxia. It is

TABLE 3. Summary of Metabolic Studies on Rat Brain during Sleep[a]

Source of data	Weight of rat (g)	Sleep			Method of killing	Control	Lactate[b]	Ratio and significance[c]	Pyruvate[b]	Ratio and significance[c]
		EEG monitor	Induction	Duration						
Richter and Dawson (1948)	30–40	No	Strong sunlight	30 min	Liq. air	Rest	S 1.36 (6) W 2.09 (10)	0.65 P < 0.01		
Shimizu et al. (1966)	30–45	Yes	Natural	20 min	Liq. N₂	Rest	S 2.78 (22) W 2.67 (7)	1.04 NS		
Cocks (1967)	Adult	No	Natural	30 min	Liq. N₂	Rest	S 2.25 (25) W 3.14 (30)	0.72 P < 0.005		
Van den Noort and Brine (1970)	150	No	4 hr exercise	60 min	Decap. into liq. N₂	4 hr exercise	S 4.51 (13) W 5.01 (13)	0.90 P < 0.05	0.21 (14) 0.25 (14)	0.84 NS
		Yes	4 hr exercise	60 min	Decap. into liq. N₂	4 hr exercise	S 4.08 (5) W 4.64 (6)	0.88 NS		
Reich et al. (1972)	250–300	Yes	Natural	20 min	Liq. N₂	Rest	A[d] S 4.02 (7) W 2.62 (7) B[d] S 2.28 (7) W 3.00 (7)	1.64 P < 0.05 0.79 P < 0.05	0.24 (7) 0.19 (7) 0.15 (7) 0.22 (7)	1.54 NS 0.67 P < 0.01

[a] Data from Reich et al. (1972). Abbreviations: Decap, decapitation; Liq. N₂, liquid nitrogen; S, sleeping; W, waking; NS, not significant ($P > 0.05$).
[b] The values represent the means (in μmol/g wet wt.) with the number of reported animals given in brackets.
[c] Ratios of S/W, with probabilities, P, of ratios differing from 1.0 indicated.
[d] (A) unacclimatized and (B) acclimatized groups of animals reported in the present study.

possible that the superposition of the sleeping state on this variable behavioral state might explain the differences between the results of the various experiments discussed.

In agreement with the observations made by Reich *et al.* (1972) concerning the necessity for close definition of the state of the animal with respect to its previous history, e.g., acclimation or nonacclimation, were some observations previously made by Van den Noort and Brine (1970). When their rats were restrained for recording purposes after recovery from brief ether anesthesia, the content of glucose during sleep compared with wakefulness decreased by some 30% from a level of 1.66 μmol/g brain. In rats that had *not* been anesthetized or restrained, brain glucose content increased by about 25% during sleep, from a level of 0.60 μmol/g brain. The overall differences in glucose levels in brain in this work, depending upon anesthesia and restraint, are noteworthy, but not explained.

It is difficult to compare the absolute levels of lactate formed in the various studies, but it would seem that the animals exercised by Van den Noort and Brine (1970) showed higher overall levels of lactate production during wakefulness and sleep than did any of the other rats studied in this context. The *ratio* of lactate to pyruvate did not differ during sleep and wakefulness for either set of animals, i.e., acclimated or nonacclimated in the experiments of Reich *et al.* (1972), indicating that the change in lactate production in unacclimated sleepers was due to an overall increase in glycolysis rather than a decrease in availability of oxygen.

With respect to high-energy phosphate compounds, the study of Van den Noort and Brine (1970) is an important one. These authors observed that in their groups of rats allowed to sleep after exercise, but without recording, the content of creatine phosphate and ATP in brain increased during sleep. This increase was rather small, perhaps less than 20%, but with a high degree of significance. On the other hand, ADP and AMP showed a diminution. These data are summarized in Table 4, taken from the work of Van den Noort and Brine (1970).

In the study of Reich *et al.* (1972) no significant changes were seen in creatine phosphate AMP, ADP, or ATP when the animals were sleeping compared with waking, whether the groups had been acclimated or not. These results are in contrast to the observations of Van den Noort and Brine (1970) with rats that had been exercised, or sleep-deprived, for at least 4 hr. Under the conditions without acclimation used by Reich *et al.* (1972) one might surmise that there had been an increased utilization of glucose during sleep. Further, in the observations of Van den Noort and Brine (1970), there was an increased amount of glucose present in the brain of the sleeping animals, the history of which most closely resembled that of the animals of Reich *et al.* (1972). All these facts suggest an increased accumulation of glucose by cells of the brain during natural sleep, i.e., an

TABLE 4. Comparison of Some Labile Components of Rat Brain from Sleep-
Deprived and Sleeping Rats[a]

Substance (μmol/g)	Sleep deprived	Asleep	P	Number of pairs
Creatine phosphate	1.50 ± 0.04	1.80 ± 0.05	<0.001	21
ATP	2.02 ± 0.05	2.23 ± 0.04	<0.005	21
ADP	0.76 ± 0.02	0.68 ± 0.02	<0.005	25
AMP	0.48 ± 0.02	0.38 ± 0.02	<0.005	25
Total adenine nu- cleotide	3.30 ± 0.08	3.28 ± 0.08		13
Glucose	0.53 ± 0.02	0.68 ± 0.04	<0.001	14
Lactate	5.01 ± 0.15	4.51 ± 0.19	<0.05	13
Glucose 6-phosphate	0.09 ± 0.01	0.08 ± 0.01		13
Fructose diphos- phate	0.43 ± 0.02	0.50 ± 0.01	<0.005	13
Triose phosphate	0.11 ± 0.01	0.13 ± 0.01		13
Phosphoenolpyru- vate	0.03 ± 0.01	0.02 ± 0.01		13
Pyruvate	0.25 ± 0.02	0.21 ± 0.02		14

[a] Data from Van den Noort and Brine (1970). Values are means ± S.E. Rats exercised for 4 hr with and without an hour of subsequent sleep without restraint or anesthesia.

increased transport of glucose. However, the mechanism and regulation of this phenomenon have not been explored. Information would be needed to determine whether the phenomenon (if it exists) is transient and what the balance of fluxes between various glucose pools might be. As has been discussed, this possibility may be tied up with the increased turnover of the phosphate moiety of a specific phosphoprotein—glucose-6-phosphatase— at least under conditions where the animals were unacclimated. Some thoughts concerning the possible role of this enzyme in glucose transport in the brain during sleep were expressed in Section 3.

4.2. Role of Brain Glycogen During Sleep

Little definitive work has been done on the role of glycogen in the brain during sleep *per se*. However, a specific investigation has been made of the effect of deprivation of paradoxical sleep on the levels of brain glycogen in different structures (Karadzik and Mrsulja, 1969; Mrsulja *et al.,* 1967). These workers utilized the so-called flowerpot method for paradoxical sleep deprivation, and the design of their experiments included a study of the effect of postdeprivational periods of recuperative sleep, as well as a control for stress. The findings were in general as follows: Glycogen of the brain decreased during deprivation of PS. This was true for the frontal cortex, the occipital cortex, and the nucleus caudatus, as well as in the hippocampus and the caudal brain stem. For the last three structures the effects were much greater than for the first pair mentioned.

In general, 3 hr of recuperative sleep was insufficient to restore the glycogen level, which continued to drop to a value of about 50% of normal. However, after 6–9 hr of recuperative sleep, the normal levels were attained. The effects were more obvious in bound glycogen than in the so-called free glycogen levels (Figure 3).

In this series of experiments, the authors used as a control an animal that had been allowed to be under the same conditions of humidity and crampedness as the experimental animals but was not on so small a platform that it would continually fall into the water during paradoxical sleep. These controls of animals on a larger platform showed no effect of the experimental system, i.e., were normal in the glycogen levels of the various structures compared with normally maintained cage controls. The authors therefore concluded that the diminutions of glycogen content that they saw were specific concomitants of deprivation of paradoxical sleep and that the restoration of these levels during the recuperative sleep, which is high in PS, was a specific effect.

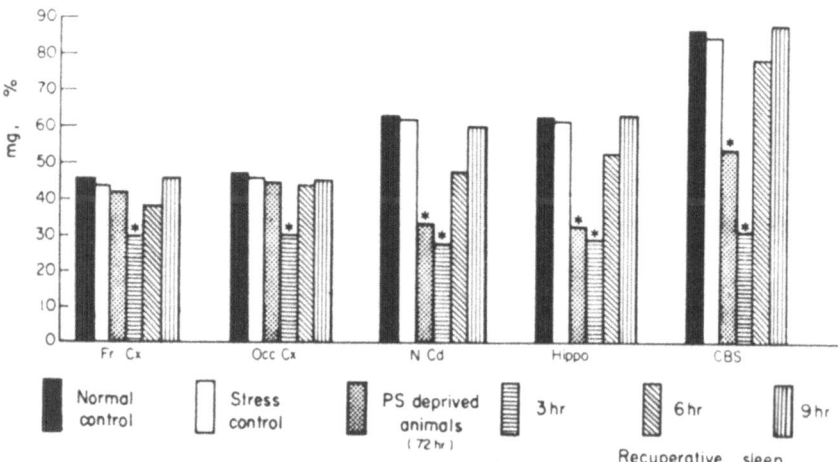

FIGURE 3. Influence of 72-hr paradoxical sleep deprivation and of 9 hr of recuperative sleep on the rat brain total glycogen levels. *Values significantly different ($P < 0.01$) from values of control animals. The structures labeled are (Fr Cx) frontal cortex, (Occ Cx) occipital cortex, (N Cd) nucleus caudatus, (Hippo) hippocampus, (CBS) caudal brain stem. Abscissa: glycogen in mg/100 mg. From Karadzic and Mrsulja (1969).

We wonder whether this is, indeed, adequately proved and note that the authors did adopt a conservative tone. However, additional controls are needed. For example, non-PS-deprived animals, which by some means are tipped into water at a frequency that resembles the situation in the actual "flowerpot" experiment, would present a control of stress that would be more comparable with what is really necessary. These experiments point up the grave difficulty confronting all investigators in this field concerning adequate control observations.

The same laboratory (Mrsulja and Rakic, 1970a) has extended this type of study to include observations on the influence of adrenergic and cholinergic blocking drugs on glycogen content of brain when rats were deprived of paradoxical sleep. In general, the same type of methods was used, and animals were pretreated with atropine, propranolol, or reserpine for 3 days prior to the experiment.

All three drugs blocked the glycogenolytic effects of PS deprivation in the subcortical structures of the brain of the rats. The authors also found that both atropine and propanolol blocked the glycogenolytic effect of physostigmine (Mrsulja et al., 1968).

The conclusion the authors reached was that probably cholinergic and adrenergic mechanisms both participate in the glycogenolytic effect of PS deprivation and that the first triggered the second. We would consider the conclusions very tenuous in such a complicated experimental situation so fraught with difficulty concerning adequate controls.

4.3. Effect of PS Deprivation on Glycolysis and Labile Phosphates in the Brain

In an extraordinarily interesting recent paper, Mendelson et al. (1974) have studied deprivation of PS and its effect on intermediary metabolism of the brain in rats (Table 5). These authors introduced an important "stress" control in addition to what one might call a "flowerpot control," which is simply the use of a larger flowerpot surface (12.5-cm pot). The latter controls for conditions of temperature, humidity, isolation, etc., identical to those of animals that were deprived of PS on the small (6.5-cm) pot. The important new control consisted of observations made on animals that were actually immersed in water for 1 hr each day (see Stern et al., 1971) ("swimming rats"), to control for the stressful effects of wetness, etc. This *stress control* has, of course, the disadvantage that the immersion occurs only for 1 hr of the day, and not at sporadic intervals when the PS-deprived animals may have fallen into the water. Such a schedule could perhaps be devised after careful observation of the number of immersions that the animals undergo during PS deprivation and the distribution of such

TABLE 5. Glycolysis, TCA Cycle Metabolites, and High-Energy Phosphates in Rat Brains[a]

	Baseline ($n = 13$)	6.5 cm Pot[b] ($n = 13$)	12.5 cm Pot[b] ($n = 8$)	Swimming rats ($n = 8$)	P[c]
Lactate	1.628 ± 0.047	2.097 ± 0.099*	1.842 ± 0.108	1.922 ± 0.115	<0.01
Pyruvate	0.090 ± 0.004	0.111 ± 0.003*	0.106 ± 0.004	0.104 ± 0.008	<0.05
L-Malate	0.282 ± 0.010	0.334 ± 0.010†	0.325 ± 0.009*	0.311 ± 0.010	<0.01
Glucose	1.391 ± 0.079	1.627 ± 0.116	1.642 ± 0.103	1.546 ± 0.128	NS
Glucose-6-phosphate	0.154 ± 0.004	0.160 ± 0.002	0.161 ± 0.006	0.169 ± 0.004	NS
Citrate	0.275 ± 0.007	0.300 ± 0.014	0.263 ± 0.016	0.266 ± 0.015	NS
	($n = 8$)	($n = 8$)			
Isocitrate	0.019 ± 0.001	0.020 ± 0.001	0.017 ± 0.001	0.018 ± 0.001	NS
	($n = 8$)	($n = 8$)			
α-Oxoglutarate	0.205 ± 0.005	0.204 ± 0.016	0.222 ± 0.008	0.201 ± 0.011	NS
	($n = 8$)	($n = 8$)			
Pyruvate/lactate	0.0568 ± 0.0044	0.0572 ± 0.0029	0.0582 ± 0.0021	0.0543 ± 0.0030	NS
ATP[d]	2.259 ± 0.042	2.308 ± 0.056	2.402 ± 0.050	2.460 ± 0.046	—
ADP[d]	0.570 ± 0.020	0.595 ± 0.011	0.595 ± 0.030	0.562 ± 0.028	—
AMP[d]	0.044 ± 0.003	0.049 ± 0.002	0.054 ± 0.009	0.055 ± 0.009	—
Creatine phosphate[d]	3.527 ± 0.099	3.433 ± 0.187	3.530 ± 0.117	3.602 ± 0.110	—

[a] Data from Mendelson et al. (1974). Concentrations in μmol/g wet wt. tissue ± S.E.

[b] Significance of difference between this individual value and baseline value ("Least Significant Difference" Test): * = P <0.05; † = P <0.01.

[c] Significance of changes among the four groups, by one-way analysis of variance (when significant, variance within groups is less than variance among groups).

[d] No significant changes among the four groups as determined by one-way analysis of variance (variance within any one group was not distinguishable from variance among the four groups).

during the 24-hr period. In other words, an even closer simulation of the control to the experimental conditions could be attempted.

A further refinement of a technical nature that was introduced in the studies just cited was the use of the "brain blower" (Veech *et al.*, 1973). This device very rapidly obtains the frozen brain sample and is a considerable advance over methods of dropping the animal into liquid nitrogen or Freon. It obviates, to a large degree, changes that occur during the period immediately before and after death. However, it must be remembered that this technique, which employs fairly cumbersome equipment, was applied here to *waking* animals, i.e., PS-deprived animals. Application to *sleeping* animals for the study of natural sleep would be very difficult and has not yet been made to our knowledge. Furthermore, it would not be an applicable technique if one wished to examine various areas and structures of the brain, since these are destroyed as such. Even with the last mentioned reservation, it is clear that application of this method to sleeping animals would be a very significant development.

Measurements of high-energy phosphates and of the redox state by assessing the levels of oxidized and reduced pyridine nucleotides showed no difference between the baseline animals (maintained simply in sawdust-lined containers) and those deprived of paradoxical sleep, controls on the larger flowerpot surface, or the new stress control (Mendelson *et al.*, 1974). However, when the metabolites of the glycolytic and TCA cycles were examined, significant differences for lactate, pyruvate, and malate were seen, as judged by one-way analysis of variance. When least-significance tests (Snedecor and Cochrane, 1967) were used, then the PS-deprived groups were significantly higher than the baseline groups for the three substrates, and the stress control groups were in between the low values of the baseline group and the high values of the PS-deprived group. They were not significantly different from either. Table 5 is reproduced from Mendelson *et al.* (1974).

The authors suggest that it is possible that changes in the three metabolites that were seen might be related both to the stress of the experimental condition and to the stress of PS sleep deprivation, perhaps in a cumulative manner. It is, of course, not currently possible to produce evidence for this, but once again, it is clear that the provision of adequate controls for comprehending the sleeping state is a frustrating matter.

One should note also the suggestion of the authors that the "signal-to-noise ratio" may have obviated the detection of a change in some fraction or pool of metabolites, or in some specific location of the brain, which might have been diluted out by the fact that the sample comprised most of the brain. The authors also recognize the influence of pretreatment of the animals on the outcome of the experiment—their results carry some overtones of the findings of Reich *et al.* (1972) regarding the influence of acclimation. In our opinion, this briefly reported but sophisticated study (Mendelson *et al.*, 1974) sets a standard for future work in this field.

5. NUCLEIC ACID METABOLISM

There has been evidence linking the levels of the nucleic acids, particularly RNA, in brain cells to the activity state of the cells. In a pioneer paper, Sokolova (1959) attempted to observe mitotic activity in the rat cornea during S and W. Sleep was not natural in these experiments but was induced with barbamil. After 1–3 hr of such sleep, no effect was seen. Prolonged sleep lasting 10–11 hr was accompanied by a rise in the mitotic activity. The nature of the sleep as well as the choice of the cornea for observation raise questions concerning the relevance of the data to the biochemical state of the CNS in natural sleep.

Older work was summarized by Einarson (1957), and a number of more recent papers present evidence that RNA levels and turnover are linked to sleep and wakefulness. For example, a Russian group, in particular, has shown that the content of neuronal and neuroglial cytoplasmic RNA in the supraoptic nucleus of the rat brain increases during natural sleep without effect on protein content (Voronka *et al.*, 1972; Demin and Rubinskaya, 1974). Evidence has also been given that RNA is depressed in glial cells of the red nucleus during sleep (Rubinskaya, 1973).

Deprivation of paradoxical sleep has been shown to produce a fall of the content of RNA in neurons but not of glial cells (Voronka *et al.* 1972; Demin and Rubinskaya, 1974). However, Rubinskaya (1973) reported a decrease in RNA of both types of cells. The depression was followed by a rise to greater-than-normal values, according to some workers. Nikaido (1961) also observed a decrease in RNA in rat brain following prolonged sleep deprivation—actually under really extreme conditions (cf. Turova, 1963).

Among the maneuvers that investigators have utilized to increase the possibilities of obtaining information on the putative linkage between sleep and RNA has been the administration of amphetamines to produce insomnia (Voronka *et al.*, 1972). This caused a decrease in the content of RNA of all cells. Sleep after deprivation, or amphetamine-induced insomnia, caused an increase in the content of RNA in the glial cells.

Earlier, Haulica *et al.* (1969) noted a decrease in DNA as well as RNA in rat brains after PS deprivation for 6–10 days—a heroic treatment. Among other stratagems used by investigators in studies of nucleic acids in brain were the induction of convulsions with metrazol (Chitre *et al.*, 1964) and forced motor activity (Tiplady *et al.*, 1974), as well as the previously mentioned use of a state of total exhaustion. The reader can hardly fail to be concerned about the difficulty of relating these observations to the process of natural sleep.

One of the most interesting papers in the field of nucleic acid metabolism during sleep is that of Vitale-Neugebauer *et al.* (1970). These investigators state a rationale for the study of macromolecules of CNS during sleep and use as their model the biosynthesis of RNA. Their concept embraces the possibility that sleep might not be a period of recovery in the brain that is narrowly limited to replenishment of a component or components, or the elimination of toxic substances, involving perturbations of intermediary metabolism. Biosynthesis, remodeling, or rearrangement of macromolecules or macromolecular units might instead constitute the advantageous recovery phenomenon.

Their general experimental design employed rabbits equipped with cortical EEG electrodes and EMG leads to the neck muscles. They did not see PS in these preparations. [^{14}C]orotic acid, the precursor of pyrimidines, was injected subarachnoidally. At the end of 1 hr, or almost 2 hr, the upper part of the cortex from which the EEG was derived was removed from the rabbit that had been stunned by a blow on the head and exsanguinated. (One might question the effects of such a procedure on the outcome of the experiment.) The sample, about 1.5 g, was then processed for the extraction of RNA. The RNA was separated into fractions by gradient centrifugation, and the profile of radioactivity in the various fractions was determined. The EEG recordings were correlated with the radioactive pattern observed in the fractions. There was a prevalence of rapidly labeled species of RNA in the heavy region of the gradient in the case of the synchronized cortex. In the case of the animals that were awake, there was a prevalence of labeled RNA in the more slowly sedimenting regions of the gradient. The results observed in this way were not to our mind entirely compelling, and the authors reported that they considered it indeed necessary to express the biochemical data in a more refined and quantitative way as a function of the degree of EEG synchronization. Consequently, they divided the gradient pattern into two regions, namely, 28–50 S and greater than 50 S. Since preribosomal RNA sediments at 45 S and 32 S are eventually converted into 28 S and 18 S ribosomal species, and DNA-like RNA is mainly to be found between 20 S and 80 S, the authors attempted to compare the preribosomal species with DNA-like RNA. They recognized that these fractions were not to be regarded as pure entities.

When the ratio of the radioactivity in the lighter fractions to that of the heavier fractions was correlated with a percentage of synchronization of the EEG in the various animals, it was clear that the ratio was greater at low percentages of synchronization than at high percentages; i.e., during sleep the ratio of activity in the lighter fractions to that in heavier fractions was decreased. The data at 105 min were much more clear-cut than those at 1 hr, and the authors attempted further refinement of their mode of expression by computing the moments of the distribution of counts in the gradients. Such calculations gave patterns that were susceptible to interpretation in terms of a model and allowed a firmer conclusion that the distribution of labeled RNA tends to prevail in the heavy region of the gradient when the electrical activity of the cortex is synchronized, and, conversely, when the EEG is of the activated type, i.e., during wakefulness, the labeling was prevalent in the intermediate and light regions.

The authors offer only suggestions as to the species of RNA involved in the effect. They believe that their data are in accord with the relative decrease of synthesis of ribosomal RNA during synchronized parts of the EEG or perhaps with a shift in the relative rates of synthesis of ribosomal and DNA-like RNA during sleep and wakefulness. On the other hand, the possibility of the involvement of a third RNA species whose variation was more important was suggested.

The indirectness of the approach to correlating sleep and a biochemical phenomenon—the biosynthesis of RNA-pyrimidine from orotic acid—through the use of derived data, leaves one somewhat uncertain. Yet, it should be pointed out that these authors adopted an appropriately stringent set of criteria for following sleep and for excluding animals the behavior of which might have indicated irrelevant phenomena after the injection of the orotic acid. In all events, it would seem that the weight of evidence does indicate that there are changes in the RNA, or more exactly, in particular species of RNA, in the brain during sleep. Measurement of the incorporation of a small precursor into the macromolecules determines the "sum" of many reactions, but the intriguing questions regarding rearrangements of the molecular or supramolecular level are not answered.

6. EFFECTS OF SLEEP ON CEREBRAL AMINO ACIDS AND PROTEINS

Biochemical studies of amino acids in the CNS during sleep and sleep deprivation offer an entrée into various metabolic processes that relate to sleep. These include aspects of the metabolism of proteins, biogenic amines, and possibly neurohumoral agents, as well as consideration of energy-related pathways. Investigations concerning amino acids in sleep have included the following facets: (1) measurements of the levels of free amino acids in brain; (2) dynamic aspects of amino acid metabolism, such as deamination, decarboxylation, and biosynthesis; and (3) incorporation of amino acids into protein.

6.1. Amino Acids

The levels, distribution, and uptake of amino acids in brain have been studied extensively in normal animals of various species (Battistin *et al.,* 1970) and in various physiological states (Battistin *et al.,* 1975). The *levels* of specific amino acids may relate to various processes, including cerebral permeability and transport processes (Battistin *et al.,* 1971), protein metabolism, and the activity of specific metabolic pathways. Amino acid levels vary in different anatomical regions of the brain, perhaps reflecting regional physiological function. Plasma amino acid levels vary according to a daily rhythm (Wurtman *et al.,* 1968). For example, in human beings, tryptophan, tyrosine, and phenylalanine have their lowest levels at 2:00 A.M., and reach their highest levels at 10:30 A.M.. Those amino acids that have the highest plasma concentration (alanine, glycine, and glutamic acid) have the least tendency to change. Levels of cerebral amino acids also vary on a circadian basis (Hery *et al.,* 1974); these variations may persist during sleep deprivation and thus may not reflect processes involving sleep *per se.* Against this background, studies of the levels of amino acid in brain during sleep and during sleep deprivation need to be evaluated cautiously.

Work on levels of cerebral free amino acids in sleep was first reported 10 years ago. Godin and Mandel (1965) studied 6-month-old rats under four conditions: (I) during normal waking; (II) after sleep of 20-min duration induced by strong illumination; (III) after 24 hr of total sleep deprivation; and (IV) after a 20-min sleep period following 24 hr of total sleep deprivation. Groups II, III, and IV all showed a 15% increase in levels of GABA when compared with group I. It is surprising that the level of this substance was the same whether the animals had slept or were sleep-deprived. Aspartic acid increased 29% after sleep, whether induced by light or by fatigue, but was unchanged with prolonged wakefulness alone. The content of glutamic acid, glycine, alanine, and lysine was essentially unchanged under all conditions and did not deviate from control values. Table 6 summarizes the mean values reported by Godin and Mandel (1965). The comments above are based on statistical evaluation of differences between the means given there.

In the same year, Jasper *et al.* (1965) reported a study of amino acids released from the surface of cerebral cortex in cats with various brain transections. In the animals with midbrain transections (cerveau isolé) manifesting EEG synchronization, the rate of release of GABA was three times greater than in those with cervical transection (encephale isolé) manifesting EEG "arousal," or in normal controls also with desynchronous EEG. In the case of glutamic acid, the pattern was reversed; i.e., the higher rate was associated with "arousal." The authors were able to demonstrate the same relationships in the two hemispheres of a split brain

TABLE 6. Effect of Sleep and Sleep Deprivation on Amino Acid Content of Rat Brain[a]

Group[b]	Content (μmol/g fresh weight)						
	GABA	Asp	Glu	Glu-NH$_2$	Gly	Ala	Lys
I	1.49	1.95	9.10	3.10	1.00	0.35	0.25
	±0.04	±0.24	±0.50	±0.45	±0.05	±0.05	±0.05
II	1.71[c]	2.51[c]	8.92	3.13	1.02	0.35	0.25
	±0.10	±0.17	±0.47	±0.33	±0.12	±0.03	±0.04
III	1.73[c]	2.10	9.11	3.06	0.98	0.33	0.28
	±0.07	±0.22	±0.40	±0.54	±0.08	±0.04	±0.03
IV	1.77[c]	2.52[c]	9.32	3.13	1.01	0.34	0.26
	±0.12	±0.25	±0.60	±0.45	±0.12	±0.08	±0.03

[a] Data adopted from Godin and Mandel (1965).
[b] Each group was made up of at least 6 animals. Group I: killed while awake after 1-hr exposure to bright light; II: killed after 20-min sleep, induced by bright light; III: sleep-deprived 24 hr, killed while awake; IV: sleep-deprived 24 hr, killed after 20-min sleep.
[c] Different from control (Group I) with $P < 0.001$.

preparation, with one hemisphere manifesting EEG synchrony and the other desynchrony. They designate the synchronous EEG pattern as "sleep" and the desynchronous as "waking"; whether the same changes in the release of amino acids occur in true physiological sleep is not established. We include these studies because of the wide use of such preparations in the construction of neurophysiological models (Moruzzi, 1972) and in the light of the wide interest in neurotransmitters in sleep (to be discussed below).

Soon after these studies appeared, there were several reports on the influence of the deprivation of paradoxical sleep on levels of amino acids in brain. Micic *et al.* (1967) deprived adult cats of paradoxical sleep for unspecified times by the flowerpot method. They looked specifically for changes in GABA, glutamic acid, and aspartic acid, because these substances appear to have functional significance in the nervous system. The brain was divided into six regions, frontal cortex, occipital cortex, caudate nucleus, colliculi, reticular formation, and thalamus to determine whether regional metabolism of the three amino acids varied in relationship to mechanisms regulating paradoxical sleep. In the case of GABA, the largest change occurred in the reticular formation (an increase of 39.4%) and the thalamus (increase of 26.2%). There was also a significant increase in frontal cortex (15%). However, the content of GABA decreased in the colliculi (29.1%) and caudate nucleus (21.9%) and also decreased slightly in occipital cortex but not with statistical significance. The content of aspartic acid increased significantly after deprivation of PS in thalamus (29%), frontal cortex (18.3%), and reticular formation (16.1%), but decreased in caudate nucleus (26%). With the exception of a slight increase in the thalamus (15%), none of the changes in glutamic acid was significant. The authors emphasize the regional specificity of the changes in amino acid levels, reflecting regional functions, and suggest that the fall in the level of GABA in the caudate nucleus following PSD might implicate that structure in PS. The nature of the "control" animal was not specified in this paper, but it was presumably a normal waking cat. These results should be considered in light of the comments previously made concerning controls.

Davis *et al.* (1969) reported a different profile of changes in content of cerebral amino acids following PS deprivation in the rat. Adult rats were deprived of PS for 72 hr by the flowerpot method. The authors analyzed whole brain rather than specific regions. They reported on levels of 25 amino acids in six experimental and four control animals. Of these, 15 showed significant increases in the experimental group, the most highly significant changes being in serine, glycine, alanine, isoleucine, and phenylalanine. These increases were two- to threefold. GABA did not show a significant increase after PS deprivation. Glutamine, glutamic acid, and aspartic acid all showed significant increases in the PS deprivation group. The authors comment that their use of whole brain may have masked significant regional fluctuations in GABA. They also point out that neither they nor Micic's group used a stress control group along with PS deprivation, noting that their experimental animals lost 30–40 g in weight and were clearly in a stressed state. Thus, they emphasized the difficulty in distinguishing between stress effects and those of sleep deprivation *per se.*

In other reports (Micic *et al.,* 1969; Karadzic *et al.,* 1971), the Yugoslavian investigators found changes in GABA, glutamic acid, and aspartic acid that were similar to those found in their previous study on PS deprivation in cats. In these studies they had three groups of eight cats: (1) normal control, (2) PS-deprived (flowerpot technique 72 hr), and (3) stress control, where the animals were kept on the pedestal as with group 2 (but the water was removed) for 8 hr a day to allow the animals to obtain PS. The authors explain that the main difference between the experimental and "stress control" groups is the amount of PS—stress was "presumed" to be equal. This experimental design appears less than adequate to us. The areas studied include frontal cortex, occipital cortex, hippocampus, caudate nucleus, thalamus, and mesencephalic reticular formation. Of 11 amino acids measured, 7 changed significantly in certain brain areas with PS deprivation. The content of GABA was decreased in the caudate nucleus but increased in frontal cortex and reticular formation. Glutamic acid, tyrosine, serine, and histidine showed no changes in content throughout. The following amino acids were increased in specific areas: aspartic in hippocampus, threonine in frontal cortex and thalamus, arginine in frontal cortex, glycine in caudate, hippocampus, and reticular formation, and lysine in frontal cortex and caudate. The investigators point out that PS deprivation seems to create conditions where GABA and its precursor, glutamic acid, are regulated independently. They also emphasized the possible involvement of GABA and glutamic acid in states of vigilance. As suggested by other studies, PS deprivation is a state of increased neural excitability.

An interesting comparison to the sleeping state is presented by the study of hibernating animals. Mandel *et al.* (1966) found that the level of GABA is elevated by 17% in the brains of hibernating mice, further supporting the possibility that it plays a role in sleeplike states. The content of aspartic acid decreased by 60% and that of glutamic acid, 24% whereas the level of glutamine increased by 29%. The authors interpret these data as consistent with diminished Krebs cycle activity during hibernation. They propose that accumulation of ammonia serves to awaken the animal periodically. However, it must be remembered that the intake of food is drastically reduced in this state.

Another comparison with studies on sleep is provided by a recent report on the level of amino acids as well as the uptake of amino acids in mouse brain and in specific regions of rabbit brain after metrazol-induced convulsions (Battistin *et al.*, 1975). In mice, metrazol (pentylenetetrazol) convulsions led to significant increases (approximately twofold) in the levels of alanine, isoleucine, and phenylalanine. GABA showed an average increase of twofold but the variability was high and the results were not significant. The influence of convulsions on the net uptake of three amino acids by mouse brain and by rabbit brain was also reported. These were histidine, methionine, and aspartic acid. In the mouse, convulsions did not change the uptake of histidine but greatly increased the uptake of methionine. Aspartic acid apparently did not cross from the blood to the brain in either control or experimental animals. In the rabbit, the uptake of histidine increased in the cortical areas while levels of methionine increased in the subcortical areas following convulsions. Again, aspartic acid was not taken up by brain under normal conditions but a small amount entered the subcortical areas after convulsions. These experiments demonstrated that convulsions altered the cerebral permeability and transport of amino acids and show that such changes vary with specific amino acids and with specific regions of the brain. These findings complicate the interpretation of studies of free amino acids in sleep and in sleep deprivation, especially since it is well known that sleep deprivation lowers the seizure threshold.

6.2. Amino Acid Metabolism

The possible significance of amino acids in sleep as precursors of neurotransmitters or as components for protein synthesis has given impetus to dynamic studies of the metabolism of amino acids in brain during sleep and during sleep deprivation. Mark *et al.* (1969) studied the influence of total sleep deprivation and of paradoxical sleep deprivation on the biosynthesis of amino acids from glucose. Total sleep deprivation was accomplished by forcing rats to walk for 48 hr on a running belt. If they left the belt, they received an electric shock, and thus they were trained to continue running. Some animals were incapable of adapting to the experimental conditions. Selective PS deprivation was carried out for 48 hr on another group of rats by the pedestal or flowerpot method. A control group was kept under unspecified physiological conditions and was used for comparison to both experimental groups. All animals were fed periodically. Randomly labeled glucose was injected subcutaneously, the experimental conditions were continued, and the animals were killed after 15, 30, 45, or 60 min. The highest percentage of radioactivity in control animals was found at all times in glutamate. The lowest percentage was found in GABA: 80% of the total radioactivity in the acid extract was found in glutamate, glutamine, aspartate, and GABA after 45–60 min. Neither total sleep deprivation nor PS deprivation modified the *distribution pattern* of ^{14}C among cerebral amino acids; nor did the two deprivation states lead to significant differences in the specific radioactivity of brain glucose in comparison to control rats. In totally sleep-deprived rats, the specific radioactivities of amino acids relative to that of glucose were significantly higher than in normal rats at all times studied (i.e., up to 60 min). With PS deprivation, specific activities of glutamine, aspartate, glutamate, and GABA relative to that of glucose were higher than in control rats at 15 min but fell to the control values at later times. The distributions of radioactivity within four separate brain regions were studied after total sleep (TS) deprivation. No significant differences among the regions emerged. However, specific activity of glutamine relative to glutamic acid increased in all brain regions with some suggestion of a greater increase in biosynthesis of glutamine in pons and medulla. The authors emphasized the importance of the increased specific activities of glutamine in totally sleep-deprived animals without a change in its concentration, suggesting an increased turnover of glutamine during PS deprivation. They acknowledge that stress controls were lacking and further point out that diffuse neuro-muscular activity can influence biosynthesis of glutamine in brain. Forced walking in TS deprivation and general stress in PS deprivation represent major physiological perturbations and, in themselves, may obscure the biochemical changes related to sleep deprivation *per se.* In general, the expectation that the incorporation of radioactivity would reflect subtle changes in the metabolism of cerebral amino acids during sleep deprivation was not fulfilled. Although this type of investigation is clearly exploratory, in that the changes sought were not rationalized *a priori,* they are clearly valuable in describing the dynamics of metabolism in sleep versus wakefulness or deprivation.

Haulica *et al.* (1970) have studied the influence of PS deprivation on cerebral ammonia metabolism, a relatively nonspecific aspect of nitrogen metabolism. In rats deprived of paradoxical sleep (flowerpot method) for 6 days and for 10 days, brain ammonia levels increased over threefold. In addition, incubation of a brain homogenate for 30 min resulted in greater ammonia formation from endogenous sources in the PS-deprived groups than in controls. The authors point out that cerebral ammonia levels are known to be lower in natural hibernation than in waking. Thus Godin *et al.* (1967) found that in natural hibernation in dormice, brain ammonia and urea both fall by approximately 50%. This may be due to the overall decrease in metabolic activity. However, brain ammonia has been shown to increase in nervous excitement and to decrease in inhibition. The role of ammonia or the role of metabolic processes producing ammonia may still be of significance in the sleep–wakefulness cycle. Haulica *et al.* (1970) speculate on whether ammonia may be an important humoral factor in producing the sensation of fatigue and again raise the issue of stress from PS deprivation as the primary variable.

Levental *et al.* (1972) looked more precisely at the effects of PS deprivation on the glutamine-ammonia-glutaminase-glutamine synthetase system in rat brain. Previous work in their laboratory had shown that free ammonia content increases in all brain regions in PS deprivation. They also found a decrease in the content of glutamine, especially in cerebellum and brain stem. These authors went on to study three brain regions (cerebrum, cerebellum, and brain stem) and three subcellular fractions (a soluble fraction, mitochondria, and microsomes). They included a stress control, animals that were allowed to sleep on a larger platform than the PS-deprived group. A significant increase in free ammonia in the soluble fractions of all three brain regions following PS deprivation was observed, with the highest increase in cerebellum. There was also a slight increase in the content of ammonia in the stress control when compared to nonstressed animals. When PS deprived animals were allowed 6 hr of recuperative sleep, the free ammonia level remained high in the soluble fraction of cerebellum but fell to normal in cerebrum and brain stem. After recuperative sleep in the stress control, free ammonia fell to normal levels throughout. In the mitochondrial fraction, free ammonia fell significantly in brain stem with PS deprivation and continued to decrease in brain stem during the recovery period. The same effect was noted to a lesser degree in the stress control animals. Free ammonia in microsomal fractions increased in both sleep-deprived and stress control animals but continued to increase in the PS-deprived animals during the recovery phase, while it fell to normal in the stress group. In cerebellum and brain stem, the ammonia decreased with both PS deprivation and stress control in microsomal fractions, but only in the PS deprivation group did it continue to decrease during recovery sleep. Both PS deprivation and nonspecific stress led to reduction in the amount of glutamine in all subcellular fractions in all three brain regions. Glutaminase activity was greatest in PS-deprived animals but was also elevated to a lesser degree in the stress controls. This increase was most marked in the mitochondrial fraction of cerebrum. The activity of glutamine synthetase also increased in both PS deprivation and in stress control animals in the microsomal fraction of cerebrum. The authors conclude that PS deprivation causes a significant imbalance in brain metabolism between the processes of formation and of removal of ammonia. They believe that they demonstrated significant differences between PS deprivation and stress controls. These differences are seen in the degree of rise of free ammonia and of enzyme activities during the experimental period and by the failure to return to normal levels during the recovery period after PS deprivation. Their report does not address the possibility that PS deprivation may simply represent a *more severe* stress than the stress-control situation and thus their findings should be considered in the context of a continuum of stress rather than of specific metabolic attributes of PS deprivation. The observations on the ammonia-glutamine system are of special interest in that it has been reported to be an important buffer control system in brain (Kelley and Kazemi, 1974).

Aspects of metabolism of amino acids in plasma and in other tissues outside the brain have been studied in relation to sleep and to sleep deprivation. For example, Francesconi and Mager (1971) investigated the effects of sleep deprivation on the periodicity of tyrosine metabolism in the mouse. In humans, Rodden *et al.* (1973) followed $^{14}CO_2$ elimination from labeled tryptophan and tyrosine measured in human sleep (S) and wakefulness (W). No significant difference was seen in tyrosine metabolism between S and W. With tryptophan, however, there were marked differences. In waking state the specific activity of CO_2 in the breath rose to a maximum in a mean of 42 min, with 1.3% of injected radioactivity expired in 60 min. In sleep, the specific activity of CO_2 reached its maximum in 125 min, with 0.6% of injected radioactivity expired in 60 min. The authors suggest that the most likely explanation is a difference in the cellular metabolism of tryptophan in sleep and wakefulness. The authors clearly indicate areas of uncertainty. For example, only a small and unknown proportion of expired CO_2 is accounted for by CNS, and most tryptophan is catabolized via kynurenine pathways.

A reader of these various analytical and metabolic studies of amino acids in sleep and sleep-related states is left disappointed, on the whole. The focus of investigation is often PS *deprivation,* a state defined by the experimental methods of Jouvet, described above as the flowerpot technique. This state is clearly associated with stress as demonstrated by many studies. When so-called stress controls were used, the effects of stress *per se* seemed intermediate between PS deprivation and normal control, suggesting that PS deprivation was simply a more extreme stress state. It would appear to us that valuable information on amino acid metabolism could be obtained during natural sleep and wakefulness (i.e., without deprivation, exercise, or such conditions as bright light). Further, in the brief periods of natural sleep available to the investigator, isotopic methods would offer the most promise.

6.3. Proteins

One theory of the function of sleep proposes that the restorative processes that take place during the sleeping state are related to protein synthesis (Oswald, 1969). It has also been proposed that SWS and PS differ with respect to protein synthesis, one state involving synthesis or restoration of protein and the other stage involving some form of patterning or processing of protein (Laborit, 1972). Although little hard information is at hand on such matters, as, for example, allosteric change in proteins during different functional states in the nervous system, these ideas are attractive.

At a less subtle level, incorporation of amino acids into brain proteins in sleep and related states provides one experimental approach. Shapot (1957), in a pioneer report, gave the results of an experiment in which he compared the uptake of [^{35}S]methionine into brain protein in the rat during sleep and during wakefulness. The rats were kept awake for 24 hr prior to the injection of labeled methionine into the subarachnoid space. The uptake of methionine into brain protein was twofold greater in animals allowed to sleep for 30 min after sleep deprivation than in normal controls and fourfold that in exhausted animals that were kept awake for the same period of time. Unfortunately, the experimental animals had been both treated with phenamine and teased to exhaustion.

More recently, Brodskii *et al.* (1974) measured incorporation of tritiated leucine into proteins during SWS and PS in the cat associative cortex. They devised a technique that allowed them to biopsy the brain with a special needle that extracted 50 mg of tissue without grossly affecting the physiological state of the animal. The pieces of tissue were homogenized and incubated for 45 min with [^{3}H]leucine. The results confirmed the high level of metabolism in PS and also showed a relative increase in protein metabolism in SWS when compared to waking controls. They postulated accelerated protein synthesis in PS with special aspects of nerve tissue repair occurring during SWS as well.

Bobillier *et al.* (1974) carried out a broad investigation of amino acid incorporation into brain proteins that involved as variables the age of the animals (rats), total sleep deprivation, PS deprivation, sleep after each of the deprivations mentioned, and normal waking controls. Measurements were made *in vivo* after subcutaneous (7-day-old animals) or a peritoneal (adult animals) injection of a mixture of tritiated amino acids. The authors also made observations *in vitro* with brain slices from the animals specified and also 24-day-old rats. These slices were incubated for 30 min with the labeled substrate after the manipulation of the sleep cycle previously mentioned. Finally, brains were separated into regions and some subcellular fractions were examined for protein labeling.

Experiments *in vitro* showed that sleep deprivation generally lowered the specific activity of proteins from the brain stem, and recovery sleep did cause a return to normal values. This was true of slices from both telencephalon and brain stem and for nuclear, mitochondrial, microsomal, and soluble proteins of the latter. (These fractions were monitored electron-microscopically, but were of unknown purity by biochemical criteria.) When protein specific activities were adjusted for TCA soluble activities, i.e., when the specific activity of the amino acid pool was included as a factor observed, differences were less marked. Table 7, taken from the paper of Bobillier *et al.* (1974), is given here as an example of the type of results these investigators achieved. It is perhaps not surprising that the data *in vivo* are less than revealing, probably because of the enormous dilution of substrate in such studies and the complex problems concerning the presentation of substrate to the brain after intraperitoneal injection. The studies *in vitro* are, of course, subject to other criticisms; e.g., one must question whether the brain slices adequately represent the state of the brain after several steps, such as giving the animals "a strong blow to the body" to kill them, decapitation, the slicing procedure *per se* and other operations. The authors' conclusions are conservative and properly focus on the great impact of stress in any studies that are based on deprivation.

Another approach to the relationship of protein synthesis to sleep has involved studies of the effects on sleep of agents known to inhibit protein synthesis in brain. Pegram *et al.* (1973) studied the effects of cycloheximide on the sleep of mice. At a dose known to inhibit cerebral protein synthesis by 90% within 30 min, there was significant decrease in PS and in waking time, with relative increase in SWS for the first 6 hr after injection. During the next 6 hr PS returned to control levels while SWS and waking changes persisted. No PS rebound was seen when the recordings were extended and more precise analysis showed that the average number of PS episodes was decreased but not the duration of individual PS periods. This suggested that the inhibitor of protein synthesis interfered with the mechanisms triggering PS rather than with metabolism during the PS periods themselves. Apparently there was no evidence that cycloheximide had nonspecific toxic effects.

In another study utilizing inhibitors of protein synthesis, Stern *et al.* (1972) studied the effects of cycloheximide and of puromycin on sleep in cats. The drugs were introduced by indwelling intraventricular cannulas to

TABLE 7. The *in Vitro* Incorporation (Duration 30 min) of L-[G-^{3}H]Amino Acid Mixture into the Brain Tissues and into the Brain Stem Subcellular Proteins of 24-Day-Old Rats[a]

Fractions examined	Specific activity of protein (d.p.m./mg protein)			Specific activity of TCA-soluble fraction (d.p.m./mg protein)			RSA: specific activity of protein / specific activity of TCA-soluble fraction		
	I	II	III	I	II	III	I	II	III
Telencephalon									
Brain stem									
Total	594 ± 38	438 ± 72	700 ± 43^c	5407 ± 231	4088 ± 616	5677 ± 336	0.110 ± 0.005	0.108 ± 0.005	0.126 ± 0.01
Homogenate	1374 ± 86	726 ± 56^b	1456 ± 140	12449 ± 523	8795 ± 696^b	14637 ± 737^{c,d}	0.111 ± 0.007	0.085 ± 0.006^d	0.098 ± 0.00
Nuclear	2080 ± 143	1429 ± 68	2504 ± 228						
Mitochondrial	876 ± 105	530 ± 53	896 ± 114						
Microsomal	1949 ± 179	1354 ± 104^b	2261 ± 208						
Soluble	1159 ± 82	690 ± 53^b	1519 ± 166						

[a] Data from Bobillier *et al.* (1974). The animals were divided into three groups: control animals (I), animals deprived of total sleep for 3 hr (II), and animals recuperating for 1.5 hr after an identical period of total sleep deprivation (III). Values are the means ± S.E. of 10 animals in each group. P values from Student's t test:
[b] <0.01 when compared with animals of group I.
[c] <0.01 when group III is compared with group II.
[d] <0.05 when compared with animals of group I.

prevent systemic effects. With cycloheximide, there was an initial decrease in W and an increase in SWS for the first day, followed by a marked increase in PS of 50–100% in all animals for 7–10 days, with a concomitant decrease in waking time and no change in SWS. Puromycin caused no change in PS during the 7–10 day period following the injection, but there was an acute decrease in both PS and W, with an increase in SWS during the first day. In further studies on mice instead of cats, these authors also found a decrease in PS and an increase in SWS on the day of injection of cycloheximide. Neither cycloheximide nor puromycin administered intra-peritoneally in the cat produced changes in sleep patterns. Furthermore, there were no differences in either norepinephrine or serotonin in 12 regions of the brain following intraventricular administration of cyclohex-imide in the experiments where clear changes in the sleep cycle were observed. They speculate that PS increases during times when protein synthesis patterns are returning to normal, and they link this change to the well-known observation that infant animals have a high percentage of PS during the time when their brains are maturing and when protein synthesis and protein patterning is occurring.

Another approach that has been pursued, especially in the Pavlov Institute in the U.S.S.R., had been to observe the *content* of proteins in individual neurons and glial cells in specific brain regions under various physiological conditions. Thus, Voronka (1971) studied the effects of phen-amine-induced prolonged insomnia and of the sleep that followed on the content of proteins in the neurons and their glial satellite cells of the brain supraoptic and red nuclei. The technique employed involved cytospectro-photometry and microspectrophotometry. Rats were kept awake 1, 2, and 4 days by amphetamines, with a resulting marked decrease of protein content in the neurons and neuroglia of both nuclei, the supraoptic in particular. The decrease was greatest at the end of the first day, with smaller decreases on subsequent days. The basic proteins varied in the same manner as the total protein content. With sleep, 15–20 min after 96 hr of insomnia, basic proteins increased to levels higher than normal in the glial cells. The volume of the cytoplasm of both neurons and glial cells was decreased with insomnia. Voronka *et al.* (1971) also described observations on the content of total protein and the quantity of basic proteins in neurons and neuroglia of the supraoptic and red nuclei of the rat brain in natural sleep and in deprivation of PS. Here, natural sleep was found to lead to an increase in total protein, especially in the basic proteins, in the neuroglia, and, to some extent, the neurons in the supraoptic nuclei. In the red nucleus, total protein dropped slightly with sleep. Deprivation of PS (24 hr) led to a marked drop of total protein in neurons and a smaller drop in neuroglia in the supraoptic nucleus. In the red nucleus, deprivation of PS also caused a rapid decline of total protein, although the drop was less in the neurons than it was in the neuroglial cells. The basic proteins decreased less than the total proteins in all observations of PS deprivation. In still another experiment along these lines, Voronka and Pevzner (1972) reported on the effect of barbiturate anesthesia on content of proteins in the neurons and neuroglia of supraoptic and red nuclei of the rat brain. Here, amytal at doses of 70 or 100 mg per kilogram produced a decrease in the content of total and basic proteins of neurons and neuroglia in both nuclei. This decrease was more pronounced in the lighter than in the deeper anesthesia. In the lighter anesthesia, the decrease was also somewhat greater in glial satellite cells, while in deep anesthesia the content of protein decreased equally in both neurons and neuroglia. It is really not possible to relate these observations to natural sleep. The drug used could well have affected specific enzymes, e.g., glucose-6-phosphatase, or flavine enzymes with important direct or indirect sequelae.

7. NEUROTRANSMITTERS

The evidence linking the neurotransmitters of the brain to the sleep–wakefulness cycle has been summarized thoroughly in recent reviews (Morgane and Stern, 1974; Jouvet, 1972; King, 1974). The pace of investigation in the area continues to be intense. Recent developments have tended to emphasize the complexities of the interrelationships among neurotransmitters and among the corresponding neuronal systems. It is becoming more difficult to sustain theories of specificity of neurotransmitters in the regulation of the S–W cycle. Instead, it has become necessary to view waking and the state of sleep as the resultants of complex neurophysiological interactions involving many regions of the brain. To add to the complexity, histofluorescent evidence suggests the presence of as yet undefined neurotransmitter substances in neuronal systems whose functions are unknown (Björklund et al., 1971). While some areas of the brain may exert a primary influence on the regulation of the S–W cycle, it seems likely that sleep affects neurotransmission throughout the brain. It follows that the metabolism of neurotransmitters would be affected by sleep and that alterations in metabolism of these substances would affect the structure of the S–W cycle.

Despite the trend away from specificity, there is still strong evidence implicating serotonin and the serotonin-containing neuronal systems in the control of sleep (Costa et al., 1974). Histofluorescence techniques (Falck et al., 1962; Björklund et al., 1971) have shown the serotonin-containing neurons to be concentrated in the raphe region of the medulla, pons, and mesencephalon. Lesions of the raphe nuclei lead to cortical desynchronization and behavioral arousal (Jouvet et al., 1966). The extent of wakefulness roughly correlates with the extent of the lesions in the raphe (Jouvet and Pujol, 1974). The effects of anatomical ablation of these pathways are replicated by metabolic disturbance of serotonin synthesis. Parachlorophenylalanine (PCPA) blocks serotonin synthesis at the tryptophan hydroxylase step, leading to a gradual decrease in cellular serotonin with corresponding decrease in sleep time, both SWS and PS. If the metabolic blockade is circumvented by the administration of 5-hydroxytryptophan (5-HTP), SWS and PS return within a few minutes (Koella *et al.*, 1968). Reserpine, known to deplete serotonin stores, suppresses SWS and PS (Matsumoto and Jouvet, 1964). In chickens, where there is no brain barrier to intravenous serotonin, the injection of serotonin induces a sleeplike state with EEG synchrony (Spooner and Winters, 1965). With rats, 5-HTP, a metabolic precursor of serotonin that is able to cross from blood to brain, also induces a sleeplike state somewhat resembling SWS (Macchitelli *et al.*, 1966). Oral administration of L-tryptophan, the amino acid precursor of 5-HTP and serotonin, decreases the latency to SWS in experimental animals and in man (Hartmann *et al.*, 1971).

There also are some contradictory findings. When PCPA is administered daily (Dement *et al.*, 1972), sleep returns after a period of insomnia despite the continued depletion of serotonin in brain. This sleep differs somewhat from the normal state in that ponto-geniculo-occipital (PGO) spikes intrude into W and SWS. This observation provides evidence for the role of serotonin in suppression of PGO spikes. SWS and PS return by the 6th or 7th day of chronic PCPA treatment (Henriksen *et al.*, 1974), raising some doubts about the role of serotonin in SWS.

Microelectrode studies of unit activity in the raphe nuclei show that unit activity in serotonergic neurons declines during SWS with respect to W, with a further decline during PS (McGinty *et al.*, 1974). When regional cerebral serotonin is measured in natural sleep and in wakefulness, serotonin is found to decrease during sleep (Sinha *et al.*, 1973). These puzzling results only serve to emphasize the difficulties encountered in attempting to relate neurochemical changes to the physiological events in sleep. Some of the biochemical assumptions regarding the metabolism of serotonin are also being questioned. For example, the raphe nuclei retain some capacity to hydroxylate tryptophan after treatment with PCPA (Deguchi *et al.*, 1973). In spite of the interest in the role of serotonin there have been few biochemical studies of cerebral serotonin during natural sleep. Thus, most of the information on the role of serotonin is indirect, coming from studies of sleep deprivation, ablation, stimulation, or enzymatic blockade.

When [^{14}C]tryptophan was given to human subjects during S and during W, the radioactivity that appeared in expired CO_2 was twofold greater in W than in S (see Section 6). A similar study of [^{14}C]tyrosine failed to show a metabolic difference between S and W (Rodden *et al.*, 1973). The level of 5-hydroxyindoleacetic acid (5-HIAA), a metabolic product of serotonin, rises in ventricular CSF about 5% in SWS and falls about 6% in PS with respect to waking levels in human subjects (Wyatt *et al.*, 1974). While this study provides more direct evidence of increased serotonin metabolism in SWS, the functional significance of this change is not clear.

Weiss *et al.* (1968) reported a decrease in serotonin and an increase in 5-HIAA levels in the brains of rats subjected to PS deprivation for 24 hr by the flowerpot method. Hery *et al.* (1970) observed an increase in the formation of [³H]serotonin from [³H]tryptophan, both *in vivo* and *in vitro*, in brains of rats when they were deprived of PS for 96 hr, again by the flowerpot method. The authors offered their findings as further confirmation of the presumed role of serotonin in mediating the onset of PS. Subsequently, two studies have demonstrated that stress alone produces effects on metabolism of serotonin similar to those seen with PS deprivation. Bliss *et al.* (1972) showed that a variety of stresses, acute and chronic, including foot shock, restraint, swimming, positioning on inverted pots, and noxious psychosocial conditions all increased levels comparably to those seen in PS deprivation by the flowerpot method. They also reported that the metabolism of serotonin did not vary with changes in physical activity. The elevations in levels of 5-HIAA secondary to stress persisted for several hours after the stressful conditions were terminated. They noted that social isolation has been shown to depress the serotonergic system and proposed a general role for serotonin in mechanisms mediating the state of arousal rather than a specific role in the mediation of sleep. Radulovacki (1973) compared the levels of 5-HIAA in CSF in cats subjected to PS deprivation to the levels in cats kept immobilized on pedestals that were large enough to allow the animals to obtain normal sleep. In both conditions, the levels of 5-HIAA rose significantly, indicating to the author that the change in the metabolism of serotonin observed with the flowerpot technique was a nonspecific stress effect.

The other neurotransmitters that have been implicated in the SW cycle are norepinephrine (NE), dopamine, and acetylcholine (Morgane and Stern, 1974). Each has been localized by histochemical techniques to specific brain areas and each has been studied by pharmacological and biochemical approaches. NE is associated with neuronal systems in the lateral portion of the brain stem and in the locus coeruleus complex of the pons (Jones *et al.*, 1973). NE is contrasted with serotonin in being an agent of arousal or wakefulness rather than of sleep induction (Torda, 1968). Lesions of the locus coeruleus and of other catecholamine-containing neuronal systems of the brain stem affect PS primarily (Jouvet, 1972). For example, bilateral lesions of the caudal part of the locus coeruleus suppress the descending motor inhibition associated with PS without affecting the ascending components of PS, such as activation of the EEG and rapid eye movements. More extensive lesions in these areas produce more general suppression of PS without changing the characteristics of SWS.

Pharmacological and biochemical studies implicating norepinephrine or dopamine in the maintenance of waking behavior are reviewed by Jouvet (1972). Much of the work was done in his laboratory. The evidence arises from such studies as observations on the effects of pharmacological alterations of catecholamine metabolism or of effects of direct deposition of crystals of neurotransmitter substances onto specific brain regions. As with serotonin, the evidence linking the catecholamines to the SW cycle is largely indirect (Jones, 1972; Jones *et al.*, 1973; Torda, 1969). Stress is also known to affect the metabolism of catecholamines, raising the question of the specificity of the relationships of the catecholamine-containing neuronal systems to the SW cycle (Bliss *et al.*, 1968).

Recent work indicates that interrelationships exist between serotonin-containing neurons and those associated with NE. Microelectrode studies show that both kinds of cells fire less in SWS and drop out almost entirely in PS. Stimulation of both types leads to arousal (McGinty *et al.*, 1974). Other studies in the last few years have shown neurophysiological interrelationships between these two classes of neuronal systems, indicating they should not be thought of as clearly differentiated and separated functionally. Lesions in the locus coeruleus in rats have produced increases in 5-HIAA concentration, and lesions in the raphe system have altered a metabolite of norepinephrine (Kostowski *et al.*, 1974). The administration of *a*-methyl paratyrosine, an inhibitor of catecholamine synthesis at the level of tyrosine hydroxylase, led to an *increase* in the concentration of 5-HIAA in several brain structures (Stein *et al.*, 1974). These and other recent studies (Blondaux *et al.*, 1973) indicate that the various neuronal systems associated with monoamine neurotransmitters should be considered, as intimated, in terms of their interrelationships rather than in terms of specific or exclusive functions.

Acetylcholine-containing neurons overlap the monoamine systems and are more widely distributed in the CNS. ACh seems to be involved in all aspects of sleep. Direct stimulation of numerous brain stem areas with micro amounts of crystals of cholinergic substances induces sleep (Hernández-Peón *et al.*, 1963). The release of ACh by brain tissue appears to be dependent on the SW cycle. In neurons of the caudate, ACh release was depressed in SWS but was increased in PS in the cat (Gadea-Ciria *et al.*, 1973). ACh release from the cortex has been shown to be decreased in SWS. Another study of microinjections of ACh into reticular formation demonstrated a state resembling PS (George *et al.*, 1964). In still another study sleeplike patterns were induced by ACh stimulation of the temporal cortex and striate structures (Hernández-Peón *et al.*, 1967). Circadian fluctuations of brain ACh have also been found (Saito, 1974). The rhythm correlated with the SW cycle and with motor activity. PS deprivation led to a 35% fall in ACh in telencephalon of the rat while other brain areas were not affected (Bowers *et al.*, 1966). Again, these studies of the neurotransmitter in sleep are largely indirect and it is too soon to bring them together into a composite picture.

Other neurotransmitters are active in brain and also appear to be involved in the SW cycle. GABA is said to increase following PS deprivation (Godin and Mandel, 1965). The function of GABA and, more specifically, its role in the SW cycle are not known. Melatonin appears to have sleep-inducing properties (Cramer *et al.*, 1974). Piperidine increases markedly in the brains of dormant mice (Stepita-Klauco *et al.*, 1974), and when piperidine is perfused into the brain stem of cats, it produces a decrease in sleep latency with an increase in PS duration (Drucker-Colin and Giacobini, 1975). However, there is some doubt as to whether the state induced by exogenous piperidine is actually sleep (Nixon *et al.*, 1976). Still other neurotransmitters appear to be present in the brain and may also be involved in the mediation of consciousness and in the SW cycle (Björklund, 1971).

We are left with a complicated and still somewhat confusing picture, suggesting the interaction of neuronal systems the functions of which are not yet well understood—a picture delineated by our observations on the neurotransmitters.

8. HUMORAL FACTORS AND THE CONTROL OF THE SLEEP–WAKEFULNESS CYCLE

The question of what it is that controls the cycle of sleep and wakefulness has been uppermost in the minds of investigators in this field for a long time. Indeed, as remarked in the Introduction, it constitutes the most provocative and mysterious question concerning sleep. Recently, attacks on the problem have been made utilizing techniques for separation of biological molecules that have been developed in such profusion in the last two decades. Although progress has been considerable, the tantalizing question of the exact nature of the substances that might accumulate in the brain during wakefulness, that might turn on the processes of sleep, or that might form during sleep until they are present in sufficient quantities to cause arousal have not yet been determined. It is not easy to make a clear distinction between humoral factors and transmitter substances potentially involved in sleep, as will be recognized from a reading of this section.

In large part, the slowness of progress is due to the probability that the active substances are present and active at minuscule concentration. Second, the modes of assay, involving, as they usually do, infusion into the brain and the quantification of sleep present many difficulties and opportunities for confusion through the introduction of artifacts.

In 1910 and subsequent years Legendre and Pieron (Legendre and Pieron, 1910, 1913; Pieron, 1913) described the effects of the transfusion of CSF from sleep-deprived dogs to the cerebrospinal system of animals not so deprived. The recipient dogs exhibited increased periods of sleep after infusion of the test material. The conditions of these experiments were extremely severe. The sleep deprivation of the donor dogs took place over some 10 days or more, and the animals were exercised at night. Furthermore, the withdrawal and administration of the CSF was carried out without anesthetics by a puncture of the atlanto-occipital membrane. The evaluation of the effect of the administered substance was by simple behavioral criteria.

These early workers did make basic observations on the nature of the substance(s) involved, with respect to heat lability and size as determined by dialysis; i.e., the substance was said to be heat-labile and nondialyzable.

A quarter of a century elapsed before the problem was reinvestigated by Schnedorf and Ivy (1939), who again transfused CSF from sleep-deprived dogs to normal recipients, but under considerably more reasonable conditions than had been possible in the earlier work. Though these investigators were able to confirm the Legendre-Pieron phenomenon, there was great variability in the extent and frequency of the response, and there was an associated hyperthermia. Consequently, they were not convinced of the relevance of the phenomenon to normal sleep. Furthermore, the necessity for the long sleep deprivation required to produce an active fluid also made them somewhat skeptical about the physiological significance of the phenomenon. The concept of neurohumoral agents that could be transfused from donor animals treated in a particular way to recipient animals who would then show specific physiological effects was further emphasized in the 1950s (e.g., Purpura, 1956).

However, the matter of a specific neurohumoral sleep factor was investigated again only after a further quarter century had elapsed from the time of the observations of Schnedorf and Ivy (1939). In 1964, Monnier and his collaborators began the publication of a series of communications that described the preparation of an active fluid in the sense that it induced sleep in rabbits. Donor rabbits were kept asleep by stimulation of the thalamic sleep area and a hemodialyzate was obtained. The state of the donors and of the recipient was determined by recording changes in the electrical activities of the brain, expressed as an increase in the amount of slow delta activity in the motor cortex. The controls included recording the effect of a fluid obtained in a preliminary dialyzate, i.e., obtained before stimulating the donor and also through use of a sham stimulation (i.e., with an electrode in place, but without current). The administration of the test substances was at first by the intravenous, and later by the intraventricular, route. (Monnier and Hösli, 1964; Monnier and Schoenenberger, 1973; Monnier *et al.*, 1972; Schoenenberger *et al.*, 1972*a,b.*)

The dialysis system employed was impermeable to ovalbumin, molecular weight 44,000, but permeable to myoglobin, molecular weight 17,800, and the investigators were able to initiate a series of studies to examine the physical–chemical properties and structural nature of their "factor delta." They observed that their active substance was somewhat sensitive to storage and to freezing and thawing; boiling at 100°C for 15 min completely abolished the sleep-inducing activity. In fact, this was also true at 70°C. Exposure to extreme acid or alkaline conditions indicated that factor delta was sensitive to pH.

After the crude dialyzate was desalted by gel chromatography on Sephadex G-10, it could be recovered separated from some nonactive peptide-containing fractions. Preparative thin-layer chromatography of the active fraction in acetone–water yielded six ninhydrin-positive entities, of which only one now contained the "hypnogenic" activity. This fraction again yielded six ninhydrin-positive substances when subjected to high-voltage electrophoresis on paper. Application of Sephadex G-15 gel-filtration to the active fraction from the thin-layer chromatography yielded three ninhydrin-positive peaks of which only one was active. By comparison with standard substances, a molecular weight greater than 355 and below 1500 was assigned. High-voltage electrophoresis of the complex peak from the Sephadex G-15 column still yielded four components, of which only one was biologically active. The best active fraction was shown to contain at least seven different amino acids upon amino acid analysis. These rather complicated patterns of ninhydrin-positive materials could be partly accounted for by contaminating amino acids derived from the paper. However, the evidence appears sound that a "hypnogenic" substance was being purified whose molecular weight was about 800–900. Estimates of the minimum effective dose vary as one employs figures from different publications of this group, but would appear to be as high as 10 nmol per gram of dry dialyzate. (Monnier *et al.*, 1975; Schoenenberger and Monnier, 1974; cf. Schoenenberger *et al.*, 1972*b*). The results of dansylation experiments indicated that the NH_2-terminal could be tryptophan or serine or, indeed, that a cyclic peptide was involved—a possibility not mentioned by the authors. Figure 4 is reproduced from Schoenenberger *et al.* (1972*b*). It illustrates the stimulation of the donor and the effects of the hemodialyzate and desalted preparation on the recipient.

Pappenheimer and his group in Boston (Pappenheimer *et al.*, 1967) took a different approach to the exploration of humoral sleep-promoting factor(s). They employed as the "donors" sleep-deprived goats from which CSF could be harvested atraumatically. This preparation took advantage of the peculiar anatomical nature of the cranium of goats and was based on

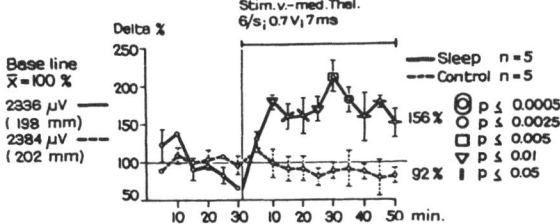

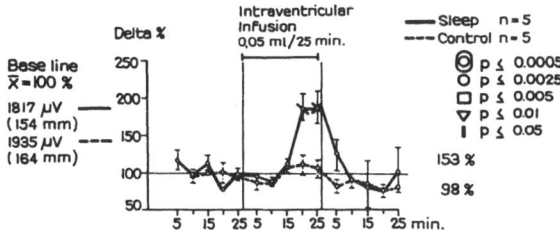

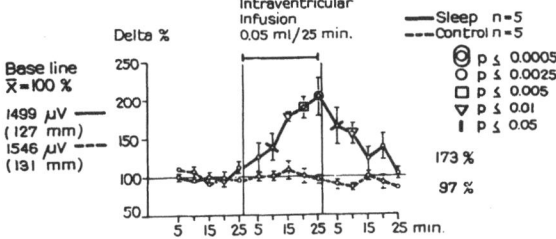

FIGURE 4. Assessment of sleep in donors and of the hypnogenic activity of sleep dialysate and its fraction in recipients. (a) Percentage of the increased delta amount in donor rabbits during electrical stimulation of the thalamic sleep area (—) compared with nonstimulated control animals (---). (b) Transmission of delta sleep by intraventricular infusion of sleep dialysate from donors to recipients (—); control equals infusion of dialysate from nonstimulated donors (---), (c) Increased delta amount in recipients (—) after infusion of the desalted fraction Pool 1 (Sephadex G-10). Controls (---). From Schoenenberger *et al.* 1972b).

work done several years previously in which the cerebral ventricular system of goats was perfused without anesthesia (Pappenheimer *et al.*, 1962). The assay system was based on depression of the nocturnal locomotor activity of rats and later on actual measurement of SWS in rats. The two measures were shown to correlate well (Fencl *et al.*, 1971) (Figure 5) with

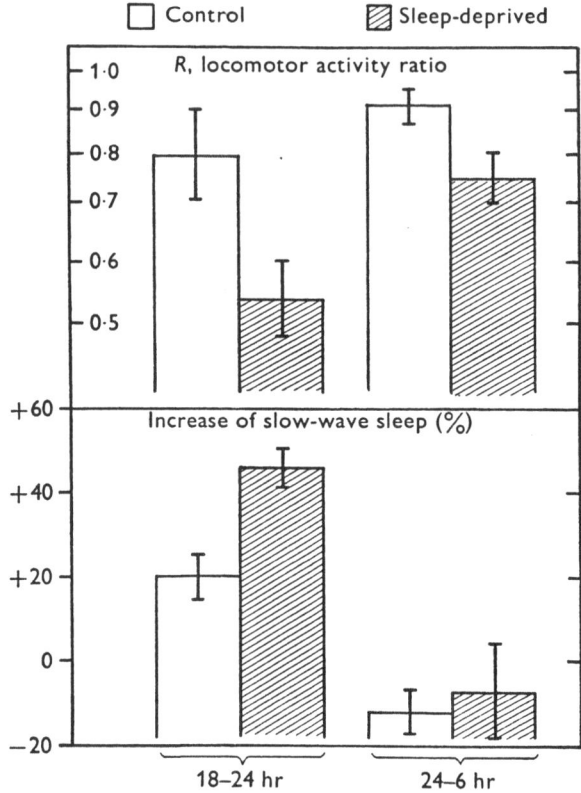

FIGURE 5. Effects on duration of slow-wave sleep and on locomotor activity of infusing 0.2 ml UM 05 ultrafiltrates of CSF from control and sleep-deprived goats. Data are from seven rats, each infused one or more times with UM 05-C and UM 05-S. From Fencl *et al.* (1971).

respect to the effectiveness of a substance of molecular weight less than 500 (nominal, by ultrafiltration) that was present in CSF taken from sleep-deprived goats (Fencl *et al.*, 1971). In a series of experiments with sleep-deprived rabbits, Ringle and Herndon (1969) were unable to find evidence for a sleep-promoting substance in the CSF when tested against rats (cf. Pappenheimer *et al.*, 1967). It is possible that technical aspects were responsible for the differences from the last work cited; e.g., the donor rabbits were electrocuted just prior to harvesting CSF. Species differences might also be of importance.

In a further communication (Pappenheimer *et al.*, 1975), the active substance in CSF of goats and humans (factor S) was reported to have been purified somewhat and was found to be directly extractable with acidified acetone from the brains of sleep-deprived goats and sheep. The active substance passed through a column of Sephadex G-10 just prior to the appearance of a labeled sucrose marker. In addition to the previous assays mentioned (Fencl *et al.*, 1971), SWS was determined in rabbits by EEG after intraventricular infusions (Figure 6).

The fluorescamine reaction, which quantitatively detects free amino groups (Udenfriend *et al.*, 1972), was found to be an extraordinarily useful adjunct to the methods previously used for following the fractionation of the crude active material, though rigorous correlation of this reaction with biological activity was not claimed. It was believed that amounts of peptide less than 500 pmol were active.

The evidence that the factor obtained by the Swiss and American groups is a peptide depends on the facts that (i) in the former case ninhydrin reactive fractions were active, (ii) the fluorescamine reaction has proved so helpful in the latter case, and (iii) factor S is inactivated by pronase (Pappenheimer *et al.*, 1975). A most interesting observation made by Pappenheimer *et al.* (1975) is that factor S not only increases the duration of slow-wave sleep, but also causes a dramatic increase in amplitude of EEG in SWS. The effect is most marked in the low-frequency range, being almost a doubling of control values. SWS after 28 hr of sleep deprivation in rabbits was characterized by similar high-voltage EEG patterns. This may,

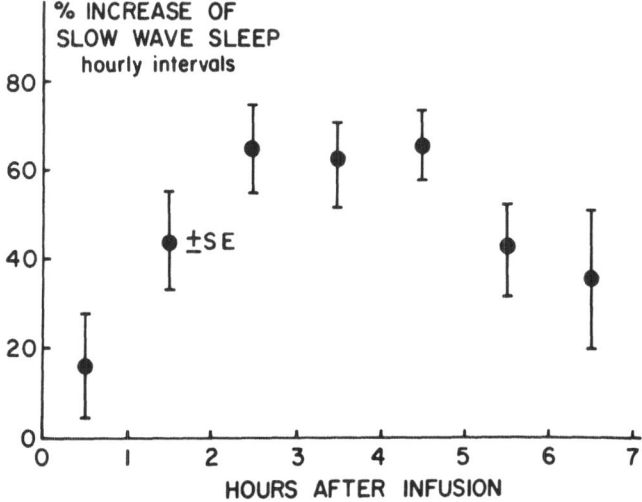

FIGURE 6. Hourly percentage excess SWS following intraventricular infusion of concentrated factor S from whole brains of sleep-deprived goats or sheep. Means ± S.E. of 13 assays on five rabbits. Approximate absolute values of excess duration may be calculated from the fact that normally the rabbits slept 35 ± 5% of any given hour. From Pappenheimer *et al.* (1975).

indeed, be some indication of "endogenous" accumulation of factor S during sleep deprivation (Pappenheimer *et al.*, 1975). In addition to substance S, the group in Boston has obtained good evidence for a peptide substance of about 2000 daltons that is apparently an "arousal" factor in that it has been shown to increase the nocturnal locomotor activity of rats (Pappenheimer *et al.*, 1974).

A third group, in Japan (Nagasaki *et al.*, 1974), has also prepared an active "hypnogenic" fraction—this time from extracts of the brain stem of sleep-deprived rats. The material was assayed by following nocturnal locomotor activity in normal rats after intraperitoneal injection of the preparation. Although there are some points of variance between the findings by the American and Japanese workers concerning amounts of material that would actually reach the brain, and the fact that the onset of the effect was much later in the case of the latter investigations than in the former, Pappenheimer *et al.* (1975) are of the belief that these two groups may be dealing with the same substance. On the other hand, the Swiss group appears to have produced a different factor. The "factor delta" is too large to be factor S—both on the basis of operational criteria and theoretical criteria with respect to the blood–brain barrier. Its activity is manifest *during* the infusion into rabbits (up to 1 hr) whereas the effects of factor S reach a peak 2 hr *after* the infusion and continue for hours. It is, of course, possible that differences in the conditions of the actual measurement are being seen, but at this point one has no basis for believing in the identity of factor S and factor delta. Apart from the species differences, it should be remarked that factor delta is obtained from sleeping and factor S, from waking sleep-deprived animals.

Among other substances that have been considered for a hypnogenic role is adenosine (Haulica *et al.*, 1973). Haulica and collaborators found that cerebral adenosine increased in rats deprived of paradoxical sleep. The accumulation depended on the length of PS deprivation—there was a 70% increase above normal after 72 hr. Further, when about 200 nmol of adenosine were administered intraventricularly to dogs, they exhibited, according to the authors, behavioral and EEG manifestations of sleep. The required dose appears very large in the context of a natural sleep-promoting substance (see above). Although some documentation of behavioral observations was given, EEG records were not provided. From the verbal description, the reader would infer that PS was the predominant pattern of sleep during the hour that the effects were said to last.

The authors mentioned the significant fact that adenosine is released from cerebral cortex slices during electrical stimulation (Pull and McIlwain, 1972*b*) and it might be involved in factor-delta- or factor-S-generated phenomena. However, it is too small a molecule for the former; on the basis of molecular weight, it could be considered as a candidate for the latter though other observations mentioned militate against this—especially the dose needed.

There are at least three other types of small molecules for which a sleep-initiating and/or prolonging role has been proposed. One is melatonin (Cramer *et al.*, 1974); a second type is a group of short-chain fatty acids (Samson *et al.*, 1956; Takagi and Matsuzaki, 1968). A third group comprises butyrolactone and 4-hydroxybutyrate (Hayashi, 1965; Hayashi *et al.*, 1967).

The studies cited have a predominantly "biochemical" goal; i.e., they aim to define the chemical nature of the "hypnogenic" substance(s). Among studies with a more "physiological" goal, in that the aim is to localize the structures of the brain involved in producing such substances and the conditions under which such production occurs, one might cite the observations of Hernández-Peón, Drucker-Colin, and their collaborators (e.g., Drucker-Colin *et al.*, 1970; Drucker-Colin, 1973). These investigators cross-perfused cats equipped with push–pull cannulae, i.e., from a donor treated in some specific way to an acceptor animal (cf. Kornmüller *et al.*, 1961). As mentioned, the study of "hypnogenic" factors comes close to the study of transmitter substances, and Drucker-Colin *et al.* (1970) have stated their general approach as follows (emphasis is ours):

> Of all criteria required for identifying a given substance as the transmitter of a synaptic pool, the most reliable would appear to be its extraction *as it is released* in the extracellular space during physiological activation. This can be accomplished by perfusing *localized areas of brain tissue* as it has been done at the cortical surface, in the lateral ventricles, and at subcortical regions. If we accept that sleep results from active inhibition of the vigilance midbrain reticular neurons located in the rostral part of the brain stem, it could follow that the presynaptic hypnogenic inhibitory terminals are located in the same anatomical region. If this reasoning is correct, it should be possible to remove from the brain the "hypnogenic inhibitory transmitter" by perfusing the midbrain reticular formation during sleep, and test the effect of the perfused substance on the same neurons, but on an awake subject.

It is clear that such an approach could fulfill some of the *desiderata* mentioned in the Introduction. Indeed, the results of perfusing from the mesencephalic reticular formation of a sleeping cat to a waking animal were encouraging in that the latency was decreased and duration increased in SWS in the recipient. The effects on PS were negative. Though this approach is beyond the limited scope of our review, one would believe that it could yield valuable information if it could be extended to several regions. One looks forward with great anticipation to the studies in physiology that would become possible when knowledge of the structure(s) of the sleep factor(s) is reasonably complete and pure material is available for experimentation in neurophysiology.

9. CONCLUSION

We have summarized many findings concerning biochemical changes in the brain that are linked to sleep. A number of the observations are contradictory, but in some cases there is reasonable agreement between laboratories. Despite conceptual and technical advances in neurophysiology, psychology, and biochemistry, investigators in this field may be said to be toiling more on the virgin plains than in orderly vineyards. Perhaps the matter of most importance is the realization that so many studies *have* been done, and many must be currently under way. This, coupled with increasing rigor of approach and imaginative design, leads one to the optimistic feeling that there is indeed a legitimate field to be explored. By "legitimate" we mean that real questions, susceptible of real answers, may be posed within the framework of contemporary science and technology. By contrast, up to a few years ago there was doubt as to the adequacy of background for an attack on the question of the biochemistry of sleep. Even today, one cannot but admit that some aspects of the field do have an Alice-in-Wonderland quality. The reader may recollect that at the Mad Hatter's tea party the Dormouse was constantly falling asleep, and was awakened by his friends by being sprinkled with tea. Finally, he was subjected to an attempt to immerse him in the teapot (cf. PS deprivation). For many years the world has recognized the verities in Alice's adventures. The reader should regard the analogy to the body of work described as complimentary to the investigators in the field of the biochemistry of sleep, despite numerous critical comments that were made in this review.

ACKNOWLEDGMENTS

We express our gratitude to our colleagues: Greg Koski, James M. Krueger, and John R. Pappenheimer. They were kind enough to review our manuscript and to make numerous suggestions and corrections.

The authors wish to acknowledge permission from the following for the republication of tables and figures. We are grateful to Drs. Hydén, Reich, Van den Noort, Mendelson, Godin, and Bobillier for permission to print Tables 1 through 5. We are also grateful to *Science,* the *Journal of Neurochemistry,* and the *American Journal of Physiology* for permission to republish these tables.

Drs. Brebbia, Reivich, Karadzic, Schoenenberger, Fencl, and Pappenheimer generously gave permission for the republication of Figures 1 through 6. The following journals authorized us to utilize these published figures: *Science, Journal of Neurochemistry, Pflügers Archives, Journal of Physiology,* and *Journal of Neurophysiology* for which we express our indebtedness.

This work was supported by United States Public Health Service Research Grant NS 07484-09.

10. REFERENCES

Anchors, J. M., and Karnovsky, M. L., 1976, Purification of cerebral glucose-6-phosphatase, an enzyme involved in sleep, *J. Biol. Chem.* **250**:6408–6416.

Battistin, L., and Lajtha, A., 1970, Regional distribution and movement of amino acids in the brain, *J. Neurol. Sci.* **10**:313–322.

Battistin, L., Grynbaum, A., and Lajtha, A., 1971, The uptake of various amino acids by the mouse brain *in vivo, Brain Res.* **29**:85–99.

Battistin, L., Varotto, M., and de Lorenzi, A., 1975, Amino acid uptake *in vivo* by the mouse brain and by various regions of the rabbit brain after drug-induced convulsions, *Brain Res.* **89**:215–224.

Birchfield, R. I., Sieker, H. O., and Heyman, A., 1959, Alterations in respiratory function during natural sleep, *J. Lab. Clin. Med.* **54**:216–222.

Birzis, L., and Tachibana, S., 1964, Local cerebral impedance and blood flow during sleep and arousal, *Exp. Neurol.* **9**:269–285.

Björklund, A., Falck, B., and Stenevi, U., 1971, Microspectrofluorimetric characterization of monoamines in the central nervous system: evidence for a new neuronal monamine-like compound, in *Progress in Brain Research [Histochemistry of Nervous Transmission]* (O. Eränkö, ed.), Vol. 34, pp. 63–73, Elsevier, Amsterdam.

Bliss, E. L., Ailion, J., and Zwanziger, J., 1968, Metabolism of norepinephrine, serotonin and dopamine in rat brain with stress, *J. Pharmacol. Exp. Ther.* **164**:122–134.

Bliss, E. L., Thatcher, W., and Ailion, J., 1972, Relationship of stress to brain serotonin and 5-hydroxyindoleacetic acid, *J. Psychiatr. Res.* **9**:71–80.

Blondaux, C., Juge, A., Sordet, F., Chouvet, G., Jouvet, M., and Pujol, J.-F., 1973, Modification du métabolisme de la sérotonine (5-HT) cérébrale induite chez le rat par administration de 6-hydroxydopamine, *Brain Res.* **50**:101–114.

Bobillier, P., Sakai, F., Sequin, S., and Jouvet, M., 1974, The effect of sleep deprivation upon *in vivo* and *in vitro* incorporation of tritiated amino acids into brain proteins in the rat at three different age levels, *J. Neurochem.* **22**:23–31.

Bowers, M. V., Jr., Hartman, E. L., and Freedman, D. X., 1966, Sleep deprivation and brain acetylcholine, *Science* **153**:1416–1417.

Brebbia, D. R., and Altschuler, K. Z., 1965, Oxygen consumption rate and electroencephalographic stage of sleep, *Science* **150**:1621–1623.

Brodskii, V. Ya., Gusatinskii, V. N., Kogan, A. B., and Nechaeva, N. V., 1974, Variations in the intensity of ^{3}H-leucine incorporation into proteins during slow-wave and paradoxical phases of natural sleep in the cat associative cortex, *Dokl. Akad. Nauk SSSR,* **215**:748–750.

Chitre, V. S., Chopra, S. P., and Talwar, G. P., 1964, Changes in the ribonucleic acid content of the brain during experimentally induced convulsions, *J. Neurochem.* **11**:439–448.

Cocks, J. A., 1967, Change in the concentration of lactic acid in the rat and hamster brain during natural sleep, *Nature (London)* **215**:1399–1400.

Cohen, H. B., and Dement, W. C., 1965, Sleep: changes in threshold to electroconvulsive shock in rats after deprivation of "paradoxical" phase, *Science* **150**:1318–1319.

Cooper, R., and Hulme, A., 1966, Intracranial pressure and related phenomena during sleep, *J. Neurol. Neurosurg. Psychiatry* **29**:564–570.

Cooper, R., and Hulme, A., 1969, Changes of the EEG intracranial pressure and other variables during sleep in patients with intracranial lesions, *Electroencephalogr. Clin. Neurophysiol.* **27**:12–22.

Costa, E., Gessa, G. L., and Sandler, M., (eds.), 1974, Serotonin—new vistas: Biochemistry and behavioral and clinical studies, *Advances in Biochemical Psychopharmacology,* Vol. II, Raven Press, New York.

Cramer, H., Rudolph, J., Consbruch, U., and Kendel, K., 1974, On the effects of melatonin on sleep and behavior in man, *Adv. Biochem. Psychopharmacol.* **11**:187–191.

Davis, J. M., Himwich, W. A., and Stout, M., 1969, Cerebral amino acids during deprivation of paradoxical sleep, *Biol. Psychiatry* **1**:387–390.

Deguchi, T., Sinha, A. K., and Barchas, J. D., 1973, Biosynthesis of serotonin in raphé nuclei of rat brain: effect of *p*-chlorophenylalanine, *J. Neurochem.* **20**:1329–1336.

Deguchi, T., Sinha, A. K., Dement, W. C., and Barchas, J. D., 1975, Enzyme activity in sleep and sleep deprivation, *Pharmacol. Biochem. Behav.* **3**:957–960.

Dement, W., Mitler, M., and Henriksen, S., 1972, Sleep changes during chronic administration of parachlorophenylalanine, *Rev. Can. Biol.* **31**(Suppl.):239–246.

Demin, N. N., and Rubinskaya, N. L., 1974, Neuronal and neurological protein and RNA content of supraoptic nucleus after deprivation of the paradoxical phase of sleep for 24 hours, *Dokl. Akad. Nauk USSR Ser. Biol.* **214**:940–942.

Diamond, I., and Fishman, R. A., 1973, High-affinity transport and phosphorylation of 2-deoxy-D-glucose in synaptosomes, *J. Neurochem.* **20**:1533–1542.

Drucker-Colin, R., 1973, Crossed perfusion of a sleep inducing brain tissue substance in conscious cats, *Brain Res.* **56**:123–124.

Drucker-Colin, R. R., and Giacobini, E., 1975, Sleep-inducing effect of piperidine, *Brain Res.* **88**:186–189.

Drucker-Colin, R., Rojas-Ramirez, J., Vera-Tueba, I., Monroy-Ayala, G., and Hernández-Peón, R., 1970, Effect of crossed-perfusion of the midbrain reticular formation upon sleep, *Brain Res.* **23**:269–273.

Einarson, L., 1957, Cytological aspects of nucleic acid metabolism, in: *Metabolism of the Nervous System* (D. Richter, ed.), pp. 403–421, Pergamon Press, New York.

Evarts, E. V., 1964, Temporal patterns of discharge of pyramidal tract neurons during sleep and waking in the monkey, *J. Neurophysiol.* **27**:152–172.

Falck, B., Hillarp, N. A., Thieme, G., and Torp, A., 1962, Fluorescence of catecholamines and related compounds condensed with formaldehyde, *J. Histochem. Cytochem.* **10**:348–354.

Feldman, F., and Butler, L. G., 1969, Detection and characterization of the phosphorylated form of microsomal glucose-6-phosphatase, *Biochem. Biophys. Res. Commun.* **36**:119–125.

Feldman, F., and Butler, L. G., 1972, Protein-bound phosphoryl histidine: a probable intermediate in the microsomal glucose-6-phosphatase/inorganic pyrophosphatase reaction, *Biochim. Biophys. Acta* **268**:698–670.

Fencl, V., Koski, G., and Pappenheimer, J. R., 1971, Factors in cerebrospinal fluid from goats that affect sleep and activity in rats, *J. Physiol.* **216**:565–589.

Francesconi, R. P., and Mager, M., 1971, Effects of sleep deprivation on the periodicity of tyrosine metabolism in mice, *Experientia* **27**:1273–1274.

Gadea-Ciria, M., Stadler, H., Lloyd, K. G., and Bartholini, G., 1973, Acetylcholine release within the cat striatum during the sleep-wakefulness cycle, *Nature (London)* **243**:518–519.

George, R., Haslett, W. L., and Jenden, D. J., 1964, A cholinergic mechanism in the brain stem reticular formation: induction of paradoxical sleep, *Int. J. Neuropharmacol.* **3**:541–552.

Glushenko, T. S., and Demin, N. N., 1971, Activity of proteolytic enzymes of various regions of the rat brain in natural sleep and when deprived of its paradoxical phase, *Dokl. Akad. Nauk SSSR Ser. Biol.* **197**:1222–1224.

Godin, Y., and Mandel, P., 1965, Distribution of free amino acids in the central nervous system of rats during sleep and in protracted wakefulness, *J. Neurochem.* **12**:455–460.

Godin, Y., Mark, J., Kayer, Ch., and Mandel, P., 1967, Ammonia and urea in the brain of garden dormice during hibernation, *J. Neurochem.* **14**:142–144.

Goodrich, C. A., Greehey, B., Miller, T. B., and Pappenheimer, J. R., 1969, Cerebral ventricular infusions in unrestrained rats, *J. Appl. Physiol.* **26**:137–140.

Hamberger, A., Hydén, H., and Lange, P. W., 1966, Enzyme changes in neurons and glia during barbiturate sleep, *Science* **151**:1394–1395.

Hartmann, E., Chung, R., and Chien, C-P., 1971, L-Tryptophan and sleep, *Psychopharmacologia* **19**:114–127.

Haulica, F., Ababei, L., Teodorescu, C., Costan, C., Rosca, V., Moisiu, M., and Haler, C., 1969, The effect of suppressing the paradoxical phase of sleep on the metabolism of cerebral ammonia, *Electroencephalogr. Clin. Neurophysiol.* **27**:637.

Haulica, I., Ababei, L., Teodorescu, C., Rosca, V., Haulica, A., Moisiu, M., and Haller, C., 1970, The influence of deprivation of paradoxical sleep on cerebral ammonia metabolism, *J. Neurochem.* **17**:823–826.

Haulica, I., Ababei, L., Branisteaunu, D., and Topoliceanu, F., 1973, Preliminary data on the possible hypnogenic role of adenosine, *J. Neurochem.* **21**:1019–1020.

Hayashi, T., 1965, Pavlov's sleep theory under a new light, *Keio J. Med.* **14**:135–144.

Hayashi, T., Hoshino, H., and Ootsuka, T., 1967, Chemoreceptors in brain to γ-hydroxybutyrate through cerebrospinal fluid in dogs, in: *Olfaction and Taste II* (T. Hayashi, ed.), pp. 599–608, Pergamon Press, Oxford.

Heiner, L., Godin, Y., Mark, J., and Mandel, P., 1968, Electrolyte content of brain and blood after deprivation of paradoxical sleep, *J. Neurochem.* **15**:150–151.

Henriksen, S., Dement, W., and Barchas, J., 1974, The role of serotonin in the regulation of a phasic event of rapid eye movement sleep: the ponto-geniculo-occipital wave, *Adv. Biochem. Psychopharmacol.* **11**:169–179.

Hernández-Peón, R., Chavez-Ibárra, G., Morgane, P. J., and Timo-laria, C., 1963, Limbic cholinergic pathways involved in sleep and emotional behavior, *Exp. Neurol.* **8**:93–111.

Hernández-Peón, R., O'Flaherty, J. J., and Mazzuchelli-O'Flaherty, A. L., 1967, Sleep and other behavioral effects induced by acetylcholine stimulation of basal temporal cortex and striate structures, *Brain Res.* **4**:243–267.

Hery, F., Pujol, J. F., Lopez, M., Macon, J., and Glowinski, J., 1970, Increased synthesis and utilization of serotonin in the central nervous system of the rat during paradoxical sleep deprivation, *Brain Res.* **21**:391–403.

Hery, R., Rouer, E., Kan, J. P., and Glowinski, J., 1974, The major role of the tryptophan-active transport in the diurnal variations of 5-hydroxytryptamine systems in the rat brain, *Adv. Biochem. Psychopharmacol.* **11**:163–167.

Hobson, J. A., 1968, Sleep after exercise, *Science* **162**:1503–1505.

Hobson, J. A., McCarley, R. W., Freedman, R., and Pivik, R. T., 1974, Time course of discharge rate changes by cat pontine brain stem neurons during sleep cycle, *J. Neurophysiol.* 37:1297–1309.

Huttenlocher, P. R., 1961, Evoked and spontaneous activity in single units of medial brain stem during natural sleeping and waking, *J. Neurophysiol.* 24:451–468.

Hydén, H., and Lange, P. W., 1964, Rhythmic enzyme changes in neurons and glia during sleep and wakefulness, *Life Sci.* 3:1215–1219.

Hydén, H., and Lange, P. W., 1965a, Rhythmic enzyme changes in neurons and glia during sleep, *Science* 149:654–656.

Hydén, H., and Lange, P. W., 1965b, Rhythmic enzyme changes in neurons and glia during sleep and wakefulness, in: *Sleep Mechanisms* (K. Akert, C. Bally, and J. P. Schade, eds.), pp. 92–95, Elsevier, Amsterdam.

Hydén, H., and Pigon, A., 1960, A cytophysiological study of the functional relationship between oligodendroglial cells and nerve cells of Deiter's nucleus, *J. Neurochem.* 6:57–72.

Jana, H., 1965, Energy metabolism in hypnotic trance and sleep, *J. Appl. Physiol.* 20:308–310.

Jasper, H. H., Khan, R. T., and Elliott, K. A. C., 1965, Amino acids released from the cerebral rat cortex in relation to its state of activation, *Science* 147:1448–1449.

Jones, B. E., 1972, The respective involvement of noradrenaline and its deaminated metabolites in waking and paradoxical sleep: a neuropharmacological model, *Brain Res.* 39:121–136.

Jones, B. E., Bobillier, P., Pin, C., and Jouvet, M., 1973, The effects of lesions of catecholamine-containing neurons upon monoamine content of the brain and EEG and behavioral waking in the cat, *Brain Res.* 58:157–177.

Jouvet, D., Vimont, P., Delorme, F., and Jouvet, M., 1964, Etude de la privation sélective de la phase paradoxale de sommeil chez le chat, *C. R. Soc. Biol. (Paris)* 158:756–759.

Jouvet, M., 1972, The role of monoamines and acetylcholine-containing neurons in the regulation of the sleep-waking cycle, *Ergeb. Physiol.* 64:166–307.

Jouvet, M., and Pujol, J. F., 1974, Effects of central alterations of serotoninergic neurons upon the sleep-waking cycle, *Adv. Biochem. Psychopharmacol.* 11:199–209.

Jouvet, M., Bobillier, P., Pujol, J. F., and Renault, J., 1966, Effets des lésions du système du raphé sur le sommeil et la sérotonine cerebrale, *C. R. Soc. Biol. (Paris)* 160:2343–2346.

Karadzic, V., and Mrsulja, B., 1969, Deprivation of paradoxical sleep and brain glycogen, *J. Neurochem.* 16:29–34.

Karadzic, V., Micic, D., and Rakic, L., 1971, Alterations of free amino acids concentrations in cat brain induced by rapid eye movement sleep deprivation, *Experientia* 27:509–511.

Kawamura, H., and Sawyer, C., 1965, Elevation in brain temperature during paradoxical sleep, *Science* 150:912–913.

Kelley, M. A., and Kazemi, H., 1974, Role of ammonia as a buffer in the central nervous system, *Respir. Physiol.* 22:345–359.

Kennedy, C., Des Rosiers, M. H., Jehle, J. W., Reivich, M., Sharpe, F., and Sokoloff, L., 1975, Mapping of functional neural pathways by autoradiographic survey of local metabolic rate with ^{14}C-deoxyglucose, *Science* 187:850–852.

King, C. D., 1974, 5-Hydroxytryptamine and sleep in the cat: a brief overview, *Adv. Biochem. Psychopharmacol.* 11:211–216.

Kleitman, N., 1963, *Sleep and Wakefulness*, University of Chicago Press, Chicago.

Koella, W. P., Feldstein, A., Cziman, J. S., 1968, The effect of parachlorophenylalanine on sleep of cats, *Electroencephalogr. Clin. Neurophysiol.* 25:481–490.

Kornmüller, A. E., Lux, H. D., Winkel, K., and Klee, M., 1961, Neurohumoral ausgelöste Schlafzustände an Tieren mit gekreuztem Kreislauf unter der Kontrolle von EEG-ableitungen, *Naturwissenschaften* 48:503–505.

Kostowski, W., Samanin, R., Bareggi, S. R., Viviana, M., Garattini, S., and Valzelli, L., 1974, Biochemical aspects of the interaction between midbrain raphé and locus coeruleus in the rat, *Brain Res.* **82**:178–182.

Kripke, D. F., 1974, Ultradian rhythms in sleep and wakefulness, in: *Advances in Sleep Research* (E. D. Weitzman, ed.), Vol. I. pp. 305–325, Spectrum Publications, Flushing, New York.

Laborit, H., 1972, Correlations between protein and serotonin synthesis during various activities of the central nervous system (slow and desynchronized sleep, learning and memory, sexual activity, morphine tolerance, aggressiveness, and pharmacological action of sodium gamma-hydroxybutyrate), *Res. Commun. Chem. Pathol. Pharmacol.* **3**:51–81.

Legendre, R., and Pieron, H., 1910, Des resultats histo-physiologiques de l'injection intra-occipito-atlantoidienne des liquid insomniques, *C. R. Soc. Biol. (Paris)* **68**:1108–1109.

Legendre, R., and Pieron, H., 1913, Recherches sur le besoin de sommeil couse cutif à une veille prolongée, *Z. Allg. Physiol.* **14**:235–262.

Levental, M., Rakic, L., and Rusic, N., 1972, Enzymes involved in the metabolism of glutamine in certain regions of the brain of paradoxical sleep-deprived rats, *Arch. Int. Physiol. Biochem.* **80**:861–870.

Lucero, M. A., 1970, Lengthening of REM sleep duration consecutive to learning in the rat, *Brain Res.* **20**:319–322.

Macchitelli, F. J., Fischetti, D., and Montararelli, N., 1966, Changes in behavior and electrocortical activity in the monkey following administration of 5-hydroxytryptophan, *Psychopharmacologia (Berlin)* **9**:447–456.

MacFayden, G. M., Oswald, I., and Lewis, S. A., 1973, Starvation and human slow-wave sleep, *J. Appl. Physiol.* **35**:391–394.

Mandel, P., Godin, Y., Mark, J., and Kayser, Ch., 1966, The distribution of free amino acids in the central nervous system of garden dormice during hibernation, *J. Neurochem.* **13**:533–537.

Mangold, R., Sokoloff, L., Conner, E., Kleinerman, J., Therman, P.-O. G., and Kety, S. S., 1955, Effects of sleep and lack of sleep on the cerebral circulation and metabolism of normal young men, *J. Clin. Invest.* **34**:1092–1100.

Mark, J., Godin, Y., and Mandel, P., 1969, Biosynthesis of aspartic, glutamic, γ-amino-butyric acids and glutamine in brain of rats deprived of total sleep or of paradoxical sleep, *J. Neurochem.* **16**:1263–1272.

Matsumoto, J., and Jouvet, M., 1964, Effets de réserpine DOPA et 5-HTP sur les deux états de sommeil, *C. R. Soc. Biol. (Paris)* **158**:2137–2141.

McElligott, J. G., and Melzack, R., 1967, Localized thermal changes evoked in the brain by visual and auditory stimulation, *Exp. Neurol.* **17**:293–312.

McGinty, D. J., Harper, R. M., and Fairbanks, M. K., 1974, Neuronal unit activity and the control of sleep states, in: *Advance in Sleep Research* (E. D. Weitzman, ed.), Vol. 1, pp. 173–216, Spectrum Publications, Flushing, New York.

Melzack, R., and Casey, K. L., 1967, Localized temperature changes evoked in the brain by somatic stimulation, *Exp. Neurol.* **17**:276–292.

Mendelson, W., Guthrie, R. D., Guynn, R., Harris, R. L., and Wyatt, R. J., 1974, Rapid eye movement (REM) sleep deprivation, stress and intermediary metabolism, *J. Neurochem.* **22**:1157–1159.

Micic, D., Karadzic, V., and Rakic, L., 1967, Changes of gamma-amino butyric acid, glutamic acid and aspartic acid in various brain structures in cats deprived of paradoxical sleep, *Nature (London)* **215**:169–170.

Micic, D., Karadzic, V., and Rakic, L., 1969, Behavior of free amino acid pool in various brain structures of cats deprived of paradoxical sleep, *Electroencephalogr. Clin. Neurophysiol.* **27**:554.

Monnier, M., Hatt, A. M., Cueni, L. B., and Schoenenberger, G. A., 1972, Humoral transmission of sleep VI. Purification and assessment of a hypnogenic fraction of "sleep dialysate" (factor delta), *Pflügers Arch.* **331**:257–265.

Monnier, M., Dudler, L., Gächter, R., and Schoenenberger, G. A., 1975, Humoral transmission of sleep IX Activity and concentration of the sleep peptide delta in cerebral and systemic blood fractions, *Pflügers Arch.* **360**:225–242.

Monnier, M., and Hösli, L., 1964, Dialysis of sleep and waking factors in blood of the rabbit, *Science* **146**:796–798.

Monnier, M., and Schoenenberger, G. A., 1973, Erzeugung, Isolierung, und Charakterisierung eines physiologischen Schlaffaktors "Delta," *Schweiz. Med. Wochenschr.* **103**:1733–1743.

Morden, B., Mitchell, G., and Dement, W., 1967, Selective REM sleep deprivation and compensation phenomena in the rat, *Brain Res.* **5**:339–349.

Morgane, P. J., and Stern, W. C., 1974, Chemical anatomy of brain circuits in relation to sleep and wakefulness, in: *Advances in Sleep Research* (E. D. Weitzman, ed.) Vol. I, pp. 1–131, Spectrum Publications, Flushing, New York.

Moruzzi, G., 1972, The sleep-waking cycle, *Ergeb. Physiol.* **64**:1–165.

Mrsulja, E. B., and Rakic, L. M., 1970a, The influence of adrenergic and cholinergic blocking drugs on the glycogen content of the brain in rats deprived of paradoxical sleep, *J. Neurochem.* **17**:455–456.

Mrsulja, B. B., and Rakic, L. M., 1970b, Changes in brain polyribosomes following an electroconvulsive seizure, *J. Neurochem.* **17**:457–460.

Mrsulja, B. B., Rakic, L. M., and Radulovacki, M., 1967, Influence of deprivation of paradoxical sleep on glycogen content in various brain structures of the cat, *Experientia* **23**:200–201.

Mrsulja, B. B., Terzic, M., and Varagic, V. M., 1968, The effect of physostigmine and neostigmine on the concentration of glycogen in various brain structures of the rat, *J. Neurochem.* **15**:1329–1333.

Nagasaki, H., Iriki, M., Inoué, S., and Uchizono, K., 1974, The presence of a sleep-promoting material in the brain of sleep-deprived rats, *Proc. Jpn. Acad.* **50**:241–246.

Nikaido, T., 1961, Chemical changes in brain tissue of rats caused by sleep deprivation, *Psychiatry Neurol. Jpn.* **63**:246–255.

Nixon, R., Geyer, S., and Karnovsky, M. L., 1976, Electrophysiological effects of intraventricular piperidine in the rat, *Sleep Res.* **4**:110.

Nordlie, R. C., 1974, Metabolic regulation by multifunctional glucose-6-phosphatase, in: *Current Topics in Cellular Regulation*, (B. L. Horecker and E. R. Stadtman, eds.), Vol. 8, pp. 33–117, Academic Press, New York.

Orem, J., Montplaisir, J., and Dement, W. C., 1974, Changes in the activity of respiratory neurons during sleep, *Brain Res.* **82**:309–315.

Oswald, I., 1969, Human brain protein, drugs and dreams, *Nature (London)* **223**:893–897.

Panksepp, J., Jalowiec, J. E., Zolovick, A. J., Stern, W. C., and Morgane, J. P., 1973, Inhibition of glycolytic metabolism and sleep-waking states in cats, *Pharmacol. Biochem. Behav.* **1**:117–119.

Pappenheimer, J. R., Heisey, S. R., Jordan, E. F., and Downer, J. de C., 1962, Perfusion of the cerebral ventricular system in unanesthetized goats, *Amer. J. Physiol.* **203**:763–774.

Pappenheimer, J. R., Miller, T. B., and Goodrich, C. A., 1967, Sleep-promoting effects of cerebrospinal fluid from sleep-deprived goats, *Proc. Natl. Acad. Sci. U.S.A.* **58**:513–517.

Pappenheimer, J. R., Fencl, V., Karnovsky, M. L., and Koski, G., 1974, Peptides in cerebrospinal fluid and their relation to sleep and activity, in: *Brain Dysfunction in Metabolic Disorders* (F. Plum, ed.), pp. 201–207, Raven Press, New York.

Pappenheimer, J. R., Koski, G., Fencl, V., Karnovsky, M. L., and Krueger, J., 1975, Extraction of sleep promoting factor S from cerebrospinal fluid and from brains of sleep-deprived animals, *J. Neurophysiol.* **38**:1299–1311.

Pegram, V., Hammond, D., and Bridgers, W., 1973, The effects of protein synthesis inhibition on sleep in mice, *Behav. Biol.* **9**:377–382.

Pieron, H., 1913, *Le Problème Physiologique du Sommeil*, Masson, Paris.

Pull, I., and McIlwain, H., 1972a, Metabolism of ^{14}C-adenine and derivatives by cerebral tissues, superfused and electrically stimulated, *Biochem. J.* **126**:965–973.

Pull, I., and McIlwain, M., 1972b, Adenine derivatives as neurohumoral agents in the brain. The quantities liberated on excitation of superfused cerebral tissues, *Biochem. J.* **130**:975–981.

Purpura, D., 1956, A neurohumoral mechanism of reticulo-cortical activation, *Amer. J. Physiol.* **186**:250–254.

Radulovacki, M., 1973, Comparison of effects of paradoxical sleep deprivation and immobilization stress on 5-hydroxyindoleacetic acid in cerebrospinal fluid, *Brain Res.* **60**:255–258.

Reed, D. J., and Kellogg, R. H., 1958, Changes in respiratory response to CO_2 during natural sleep at sea-level and at altitude, *J. Appl. Physiol.* **13**:325–330.

Reed, D. J., and Kellogg, R. H., 1960, Effect of sleep on CO_2 stimulation of breaking in acute and chronic hypoxia, *J. Appl. Physiol.* **15**:1135–1138.

Reich, P., Driver, J. K., and Karnovsky, M. L., 1967, Sleep: effects on incorporation of inorganic phosphate into brain fractions, *Science* **157**:336–338.

Reich, P., Geyer, S. J., and Karnovsky, M. L., 1972, Metabolism of brain during sleep and wakefulness, *J. Neurochem.* **19**:487–497.

Reich, P., Geyer, S. J., Steinbaum, L., Anchors, J. M., and Karnovsky, M. L., 1973, Incorporation of phosphate into rat brain during sleep and wakefulness, *J. Neurochem.* **20**:1195–1205.

Reivich, M., Issacs, G., Evarts, E., and Kety, S. S., 1968, The effect of slow wave sleep and REM sleep on regional cerebral blood flow in cats, *J. Neurochem.* **15**:301–306.

Richter, D., and Dawson, R. M. C., 1948, Brain metabolism in emotional excitement and in sleep, *Amer. J. Physiol.* **154**:73–79.

Ringle, D. A., and Herndon, B. L., 1969, Effects on rats of CSF from sleep-deprived rabbits, *Pflügers Archiv. Ges. Physiol.* **306**:320–328.

Robin, E. D., Whaley, R. D., Crump, C. H., and Travis, D. M., 1959, Alveolar gas tensions, pulmonary ventilation, and blood pH during physiologic sleep in normal subjects, *J. Clin. Invest.* **37**:981–989.

Rodden, A., Sinha, A. K., Dement, W. C., Barchas, J. D., Zarcone, V. P., MacLaury, M. R., and De Grazia, J. A., 1973, $^{14}CO_2$ elimination from [^{14}C]-tryptophan and [^{14}C]-tyrosine in human sleep and wakefulness, *Brain Res.* **59**:427–431.

Rubinskaya, N. L., 1973, RNA in neurons and their glial cell satellites in the rat red nucleus during natural sleep, deprivation of its paradoxical phase and phenamine insomnia, *Tsitologia* **15**:1471–1475.

Saito, Y., 1974, Circadian fluctuation of brain acetylcholine in cats, *Psychiatry Neurol. Jpn.* **76**:193–206.

Samson, F. E., Jr., Dahl, N., and Dahl, D., 1956, A study on the narcotic action of the short chain fatty acids, *J. Clin Invest.* **35**:1291–1298.

Schendorf, J. G., and Ivy, A. C., 1939, An examination of the hypnotoxin theory of sleep, *Amer. J. Physiol.* **125**:491–505.

Schoenenberger, G. A., and Monnier, M., 1974, Isolation, partial characterization and activity of a humoral "delta sleep" transmitting factor, in: *Brain and Sleep* (H. M. van Praag and H. Meinardi, eds.), pp. 39–69, De Erven Bohn B. Y., Amsterdam.

Schoenenberger, G. A., Cueni, L. B., Hatt, A. M., and Monnier, M., 1972a, Isolation and physical chemical characterization of a humoral sleep inducing substance in rabbits (factor "delta"), *Experientia* **28**:919–921.

Schoenenberger, G. A., Cueni, L. B., Monnier, M., and Hatt, A. M., 1972b, Humoral transmission of sleep VII. Isolation and physical–chemical characterization of the "sleep inducing factor delta", *Pflügers Arch.* **338**:1–17.

Shapot, V. S., 1957, Brain metabolism in relation to the functional state of the central nervous system, in: *Metabolism of the Nervous System* (D. Richter, ed.), pp. 257–262, Pergamon Press, New York.

Shimizu, H., Tabushi, K., Hishikawa, Y., Kakimoto, Y., and Kaneko, Z., 1966, Concentration of lactic acid in rat brain during natural sleep, *Nature (London)* **212**:936–937.

Sinha, A. K., Ciaranello, R. D., Dement, W. C., and Barchas, J. D., 1973a, Tyrosine hydroxylase activity in rat brain following "REM" sleep deprivation, *J. Neurochem.* **20**:1289–1290.

Sinha, A. K., Henriksen, S., Dement, W. C., and Barchas, J. D., 1973b, Cat brain amine content during sleep, *Amer. J. Physiol.* **224**:381–383.

Snedecor, G. W., and Cochran, W. G., 1967, *Statistical Methods*, Iowa State University Press, Ames, Iowa.

Sokoloff, L., 1975, Determination of local cerebral glucose consumption, in: *Blood Flow and Metabolism in the Brain* (A. M. Harper, W. B. Jennett, J. A. Miller, and J. O. Rowan, eds.), pp. 1–8, Churchill Livingstone, Edinburgh.

Sokolova, L. J., 1959, Mitotic acitivity in white rats during drug-induced sleep, *Byull Eksp. Biol. Med.* **48**:95–99.

Spooner, C. E., and Winters, W. D., 1965, Evidence for a direct action of monoamines on the chick central nervous system, *Experientia* **21**:256–258.

Stein, D., Jouvet, M., and Pujol, J. F., 1974, Effects of α-methyl-*p*-tyrosine upon cerebral amine metabolism and sleep states in the cat, *Brain Res.* **72**:360–365.

Stepita-Klauco, M., Dolexalova, H., and Fairweather, R., 1974, Piperidine increase in the brain of dormant mice, *Science* **183**:536–537.

Stern, W. C., Miller, F. P., Cox, R. H., and Maickel, R. P., 1971, Brain norepinephrine and serotonin levels following REM sleep deprivation in the rat, *Psychopharmacologia (Berlin)* **22**:50–55.

Stern, W. C., Morgane, P. J., Panksepp, J., Zolovick, A. J., and Jalowiec, J. E., 1972, Elevation of REM sleep following inhibition of protein synthesis, *Brain Res.* **47**:254–258.

Sterner, N., 1963, Testing of psychotropic agents by electroencephalography, *Farm. Revy (Sweden)* **62**:121–132.

Takagi, H., and Matsuzaki, M., 1968, Sleep state and its induction by sodium butyrate in acute "encephale isole" and "isolated midbrain-pons-medulla" preparation, *Jpn. J. Physiol.* **18**:380–390.

Tiplady, B., Glushchenko, T. S., and Pevzner, L. Z., 1974, Effect of forced motor activity on RNA content in neurons and neuroglial cells of the brain and spinal cord, *Dokl. Akad. Nauk USSR Ser. Biol.* **214**:973–976.

Torda, C., 1968, Effect of changes of brain norepinephrine content on sleep cycle in rat, *Brain Res.* **10**:200–207.

Torda, C., 1969, Biochemical and bioelectric processes related to sleep, paradoxical sleep and arousal, *Psychol. Rep. Monogr. Suppl.* 2-V24.

Townsend, R. E., Prinz, P. N., and Obrist, W. D., 1973, Human cerebral blood flow during sleep and waking, *J. Appl. Physiol.* **35**:620–625.

Turova, N. F., 1963, Ribonucleic acid in experimental exhaustion, *Tr. 1-90 (Pervogo) Mosk. Med. Inst.* **16**:225–233.

Udenfriend, S., Stein, S., Bohlen, P., and Dairman, W., 1972, A new fluorometric procedure for assay of amino acids, peptides and proteins in the picomole range, in: *Chemistry and Biology of Peptides* (J. Meienhofer, ed.), pp. 655–663, Ann Arbor Science Publishers, Ann Arbor, Michigan.

Valatx, J. L., Roussel, B., and Cure, M., 1973, Sommeil et temperature cerebrale du rat au cours de l'exposition chronique en ambiance chaude, *Brain Res.* **55**:107–122.

Van Den Noort, S., and Brine, K., 1970, Effect of sleep on brain labile phosphates and metabolic rate, *Amer. J. Physiol.* **218**:1434–1439.

Veech, R. L., Harris, R. L., Veloso, D., and Veech, E. H., 1973, Freeze-blowing: A new technique for the study of brain *in vivo, J. Neurochem.* **20**:183–188.

Vitale-Neugebauer, A., Giuditta, A., Vitale, B., and Giaquinto, S., 1970, Pattern of RNA synthesis in rabbit cortex during sleep, *J. Neurochem.* **17**:1263–1273.

Voronka, G. Sh., 1971, Effects of phenamine-induced prolonged insomnia and of following sleep on the content of proteins in the neurons and their glial-satellite cells of the brain supraoptic and red nuclei, *Fiziol. Zh. SSSR* **57**(7):962–968.

Voronka, G. Sh., and Pevzner, L. Z., 1972, Effect of barbamyl anesthesia of different degrees of the protein content in the neurons and neuroglia of supraoptical and red nuclei in rat brain, *Vopr. Med. Khim.* **18**(4):412–418.

Voronka, G. Sh., Demin, N. N., and Pevzner, L. Z., 1971, Total protein content and quantity of basic proteins in neurons and neuroglia of the supraoptic and red nuclei of the rat brain in natural sleep and deprivation of rapid-eye-movement sleep, *Dokl. Akad. Nauk SSSR Ser. Biol.* **198**(4):974–977.

Voronka, G. Sh., Demin, N. N., Rubinskaya, N. L., and Solov'eva, I. A., 1972, RNA content in neurons and their glial cell-satellites of the rat brain supraoptic nucleus during natural sleep, its REM-phase deprivation and amphetamine insomnia, *Ukr. Biokhim. Zh.* **44**:712–717.

Warfield, A. S., and Segal, S., 1974, D-Galactose transport by synaptosomes isolated from rat brain, *J. Neurochem.* **23**:1145–1151.

Weiss, E., Bordwell, B., Seeger, M., Lee, J., Dement, W., and Barchas, J. D., 1968, Changes in brain serotonin (5-HT) and 5-hydroxyindole 3-acetic acid (5-HIAA) in REM sleep deprived rats, *Psychophysiology* **5**:209.

Weyne, J., Demeester, G., and Leusen, I., 1970, Effects of carbon dioxide, bicarbonate and pH on lactate and pyruvate in the brain of rats, *Pflügers Archiv. Ges. Physiol.* **314**:292–311.

Wurtman, R. J., Rose, C. M., Chuan Chou, M. S., and Larin, F. F., 1968, Daily rhythms in the concentrations of various amino acids in human plasma, *N. Engl. J. Med.* **279**:171–175.

Wyatt, R. J., Neff, N. H., Vaughan, T., Franz, J., and Ommaya, A., 1974, Ventricular fluid 5-hydroxyindoleacetic acid concentrations during human sleep, *Adv. Biochem. Psychopharmacol.* **11**:193–197.

SENSORY RESPONSE IN BACTERIA

DANIEL E. KOSHLAND, JR.

Department of Biochemistry
University of California
Berkeley, California

1. INTRODUCTION

1.1. Description of the Bacterial Sensory System

Bacteria have a well-developed sensory system that allows them to swim toward attractants, which are usually nutrients, and away from repellents, which are usually indicators of toxic conditions. Thus, they respond to environmental conditions in ways that enhance their survival. In fact, it is quite easy to select for bacteria with improved motility by generating conditions that make much motility advantageous to the organism. Bacterial behavior, therefore, is a phenomenon that has many analogies to more complex behavior patterns but is contained in an extremely small living cell.

The bacterium we will discuss is shown in Figure 1. It is encased in a membrane that can be shown to be constructed from several distinct components. First, there is an outer membrane that is composed mainly of lipopolysaccharide and protein. Then there is a peptidoglycan layer that gives the cell wall its strength and, finally, an inner membrane, largely phospholipid, which contains many of the enzymes of oxidative phospho-

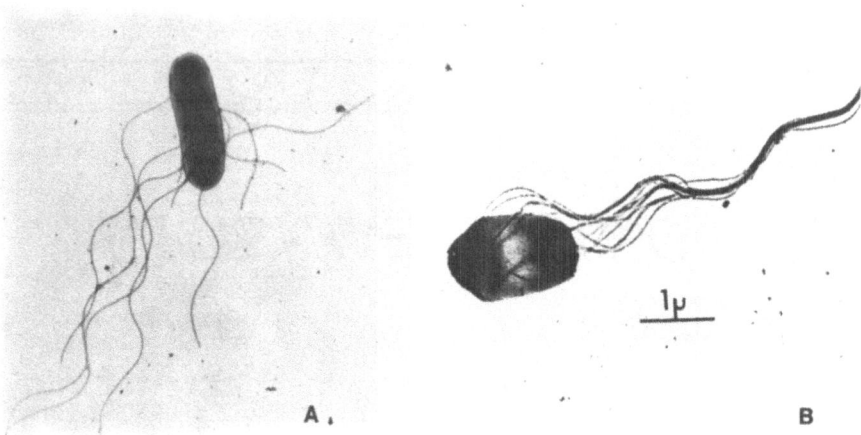

FIGURE 1. Electron micrographs of *Salmonella typhimurium* (*E. coli* appears essentially the same). (A) Bacterium with flagella spread out; (B) bacterium with flagella in a bundle.

rylation and other metabolic pathways. Inside the inner membrane is the cytoplasm, which does not contain a nucleus, at least in the conventional sense. The flagella in *Escherichia coli* and *Salmonella typhimurium* are called *peritrichous*, meaning that about six to eight of them are randomly spaced around the surface of the bacterium, as shown in Figure 1(A). They form a bundle behind the bacterium when it swims, as shown in Figure 1(B).

This is a relatively simple structure—organized, highly selected, but with few components relative to higher species. The sensory system that controls the behavior also appears relatively simple. The primary signal is activated by receptors in the membrane. Then the receptors transmit the signal by means of a "tumble regulator," the precise mechanism of which is unknown, but involves at least seven separate gene products. By regulating the frequency of tumbling, the net migration in a given direction is altered. The system, stated in this manner, seems rudimentary, and yet it is capable of detecting a gradient that varies by one part in 10^4 over the length of the bacterium.

1.2. Analogies to More Complex Systems

The formal analogy of the sensing system of bacteria to those of higher species is remarkable in many ways. As will be discussed, receptors that receive signals from the outside environment are located on the periphery of the bacteria in the same way that our eyes, ears, and nose are located on our surfaces. The receptors are specific for individual chemicals as are the sensors of higher systems and they respond to those environmental conditions that are important to the bacteria in the same way that we sense environmental changes important for our survival. The signal need not be metabolized, just as saccharin need not be metabolized to taste sweet.

The initial stimulus is analyzed and transmitted. Although the distances traversed between stimulus and motor response are small, it will be seen that bacteria analyze changes of concentrations, have a rudimentary "memory," and can evaluate a variety of signals in a "decision-making" way. The efficiency of this analyzing and transmitting system can be affected by chemicals, physical perturbants, and genetic alteration, just as in the case of higher species.

Finally, the signal has the ultimate purpose of altering the locomotor behavior in analogy to the motor response of mammalian systems. It appears to be an all-or-none event. It physically results from a reversal of flagella rotation, a switching event not unlike depolarization of a neuronal membrane, and is possibly even triggered by a bacterial membrane depolarization.

1.3. Advantages and Disadvantages of the Bacterial System

There are advantages and disadvantages to studies in bacterial sensing. One disadvantage is quite clearly that the system may be too simplified over that of higher species. The membrane structure is not precisely the same as mammalian structures; the sensory system involves a single cell in which the signal is transmitted at most from one end to the other of the cell, which is 3 μm in length contrasted with the communication between cells in higher species and the extremely long length of some neuronal cells. Quite clearly, there must be important differences.

On the other hand, past studies in biochemistry tend to stress that basic biochemical mechanisms repeat from *E. coli* to man. If so, the bacterial system has great advantages as a model sensory system because of its simplicity, the ease of isolating proteins, and its ease of genetic manipulation.

1.4. Scope of the Review

Chemotaxis in bacteria was discovered in the 1880s by the great biologists Engelmann (1881) and Pfeffer (1883, 1884, 1888). The phenomenon of chemotaxis is not only pervasive in bacteriological species [see the excellent reviews by Weibull (1960), Zeigler (1962), and Adler (1969)], but also occurs in all forms of life, from insects, which are guided by pheremones, to lymphocytes, which follow chemicals to their prey. We will make no attempts to cover the comparative biology of chemotaxis but will concentrate on bacterial chemotaxis because of its utility as a model sensory system. Adler (1969), who has contributed so much to the modern development of this system, has written an excellent review of the field, and more recently some more current aspects of the field have been summarized (Koshland, 1974; Adler, 1975). This article will therefore attempt to explain features of the system that will be relevant to higher sensory systems.

2. METHODS

If bacteria are examined in the microscope, a great deal can be learned. In fact, the early studies of Pfeffer and Engelmann utilized microscopic observation almost entirely. Watching *E. coli* or *S. typhimurium* for even a brief period leaves a clear impression of bacteria swimming in straight lines, tumbling, and heading in a new direction. Figure 2 shows a schematic representation of this observation in a two-dimensional plane showing the "runs" of roughly straight lines followed by abrupt changes in direction called "tumbles," "twiddles," or "turns." This movement can be recorded permanently by open-shutter photography with steady light (Vaituzis and Doetsch, 1969; Dryl, 1958) or stroboscopic light (Macnab and

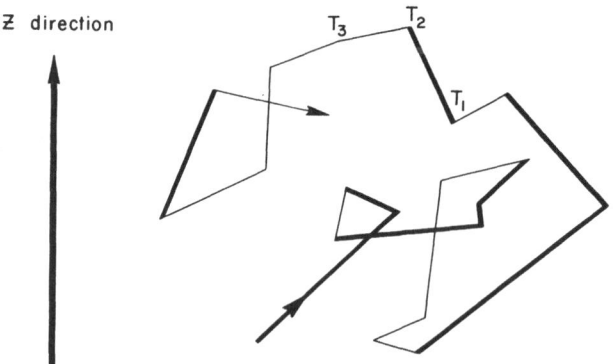

FIGURE 2. Idealized trajectory of a swimming bacterium. It consists of straight-line segments (runs) interrupted by discontinuous changes in direction (turns or tumbles).

Koshland, 1972). Such qualitative observation of bacteria has limitations and a number of more quantitative tools have been developed for other purposes.

2.1. The Capillary Assay

In the 1880s Pfeffer demonstrated chemotaxis by placing attractant in a capillary tube and observing the accumulation of bacteria near the mouth of the capillary and inside the capillary by microscopic observation [Figure 3(a)]. In the 1960s Adler (1969) made the method quantitative by counting the bacteria within the capillary by growing them on agar and counting the colonies. One form of Adler's method (Adler, 1973) involves a small chamber formed by laying a glass tube from a 5-cm length of melting-point capillary between a microscopic slide and a coverslip. The chamber is filled with about 0.2 ml of the bacterial suspension. After incubation of 1 hr, the capillary is removed and the exterior is rinsed with a thin stream of water from a wash bottle. The sealed end is then broken and the contents are delivered into a test tube containing tryptone broth at 0°C. Suitable dilutions are made and a sample is mixed with 2 ml soft tryptone agar at 45°C and then poured into a tryptone agar plate, which has a high plating efficiency. After incubation at 37°C, the colonies are counted. Averages of several runs are used to smooth out variability due to convection currents, temperature variations, etc. Variations are in the range of 27% between two values.

A typical concentration response curve is shown for L-serine as on an attractant to *E. coli* in Figure 3(B). Similar curves have been obtained for a large number of amino acids and sugars. In the serine case, the apparent threshold concentration is about 10^{-6} M, but the threshold value varies from one attractant to another.

The number of bacteria in the capillary will depend a little on the conditions of the experiment. If the experiment is allowed to go too long, the nutrient will diffuse out of the capillary tube completely. If it is carried out for too short a period, the attractant will not have time to diffuse throughout the solution and only a few bacteria will be attracted inside the capillary. Thus, a compromise between too short or too long periods of incubation is needed and is usually a period of about 1 hr. The decreased response of high concentrations in Figure 3(B) is not due to a repellent action but, rather, a function of the nature of the capillary assay. At the highest levels of a sugar or amino acid, the diffusion of the attractant will create a zone just outside the capillary mouth, which saturates the bacterial receptor. Therefore no further reason exists for swimming into the capillary and the number inside starts to decrease.

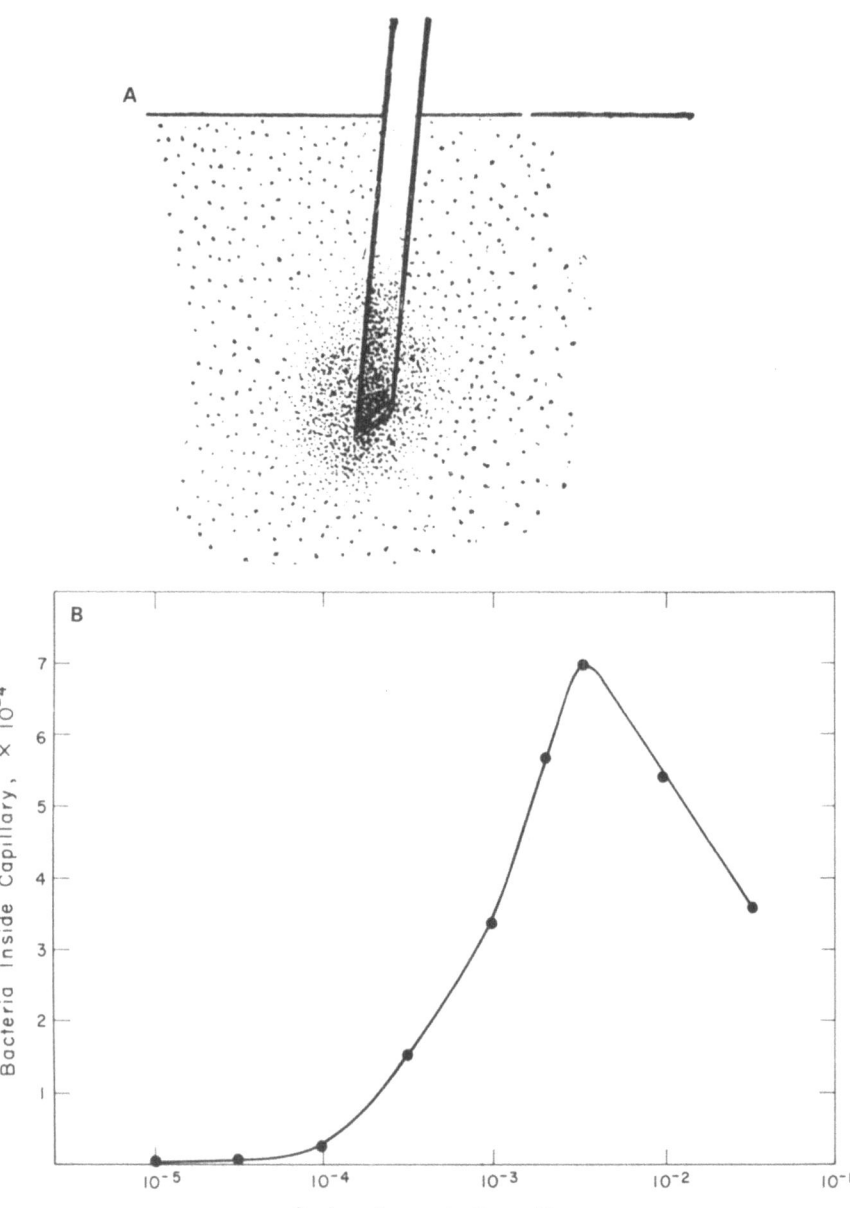

FIGURE 3. Capillary assay. (A) Original schematic drawing of Pfeffer, showing bacteria swimming into a capillary containing attractant. (B) Development of quantitative capillary assay by Adler (see text) allows plot of number of bacteria in capillary as a function of concentration of attractant inside capillary. Maximum is due to artifact of assay (see text) and does not mean excess concentrations repel.

Capillary curves can also be affected by metabolism of the attractant, since it is quite possible for the bacteria to digest a significant fraction of the attractant when it is as low as 10^{-6} M. This alters the concentration in the neighborhood of the mouth of the capillary and, hence, displaces the capillary curve. This situation does not occur with nonmetabolized compounds or mutants that are defective in metabolic pathways.

These complications make certain theoretical interpretation of the capillary assay results difficult, but, nevertheless, it has been an extraordinarily effective tool for rapid screening of compounds and the investigation of simple responses. Moreover, Adler and Epstein (1974) have shown it can be used for delineating quantitative relationships by careful analysis of sensitivity curves and dose-response relationships.

2.2. Trackers

The first tracker to be designed was that of Howard Berg (1971), which utilizes fiber optics, photomultiplier pairs, and an automatic feedback device to keep a bacterium in a fixed position in laboratory space. The optical fibers are arranged in the x, y, and z directions so that movement of the bacterium in any direction is recorded in the automatic machinery of the tracker, allowing it to follow the bacterium through three-dimensional space. As the bacterium moves on the stage of the microscope, its positions are recorded in a computer for subsequent analysis.

The tracking device can follow the movement of $E.$ $coli$ within the very small chamber quite accurately. The center of focus floats within the body length of the bacterium, which is 1 μm wide and approximately 2–5μm long. The optical systems do not allow the focus to stay on one fixed point within the bacterium and, hence, the center of focus tends to slide from one edge of the bacterium to another in the process of the tracking. To follow movements in a gradient (Berg and Brown, 1972), a capillary is introduced into the side of the vessel. Since the attractant or repellent is diffusing from the capillary into the bulk of the vessel, the gradient obviously varies with time and, therefore, calculation of the variation of the gradient over time must be allowed for in the analysis of the movement of the bacterium over time (Futrelle and Berg, 1972). Brown and Berg (1974) recently developed a method to vary the gradient of concentration in the tracker over time by enzymatic depletion of a chemical. This appreciably simplifies the mathematical analysis as compared with the complexities of diffusion from a capillary and makes the instrument more accurate.

The second type of tracker was developed by Lovely et al. (1974), operating on somewhat different principles. A stable gradient of attractant was produced in a relatively large reaction vessel (1 cm × 1 cm × 11 cm) by utilizing glycerol (a nonattractant) to stabilize the system against convection currents. An observer, looking through a long-range microscope, keeps the bacterium on a cross-hair and in focus by moving a pedal, which controls movement of the microscope stage in the z direction, and by operating a "joy stick," which controls movement in the x and y directions. The stage of the microscope is connected to a computer that records the three-dimensional movements of the bacterium.

The two instruments have complementary advantages and disadvantages. The tracker of Berg obtains more accurate data because the automatic recording device responds more rapidly to sudden changes in direction than a human observer. The instrument, however, has inertial characteristics that have limited it to a bacterium moving similarly to *E. coli,* and the accuracy of the data dropped rapidly when attempts were made to follow velocities such as those of *Salmonella,* which swim appreciably faster.

The tracker of Lovely *et al.* (1974) is less accurate in regard to the individual movements of bacteria but can follow *Salmonella* and has the advantage that bacteria can move in a stable gradient that can be watched over long periods of time. This type of tracker is particularly useful for examining "persistence time," i.e., the analysis of the movement of the bacteria in terms of the component in the direction of the gradient (see Macnab and Koshland, 1973). It should be noted that this instrument utilizes a one-dimensional gradient only, a simplifying advantage in the theoretical interpretation of data.

2.3. Temporal Gradient Methods

Our initial temporal gradient apparatus (Macnab and Koshland, 1972) was introduced to establish whether bacteria utilized a temporal sensing system (see later) but the approach has also become very useful for chemotaxis assays. The apparatus is described in Figure 4(A). Bacteria in one bottle, which may or may not contain attractant (or repellent), are rapidly mixed with a solution containing the same ingredients but a different concentration of attractant (or repellent). The two solutions are pushed rapidly through the mixing chamber and into an observation chamber on the stage of a microscope. When bacteria are observed, the attractant (or repellent) is uniformly (isotropically) distributed. Since the mixing time is rapid (on the order of 200 ms), the time in the mixing chamber is less than the memory time of the bacteria, and the bacteria receive a stimulus equivalent to swimming up or down a gradient rapidly [Figure 4(B)].

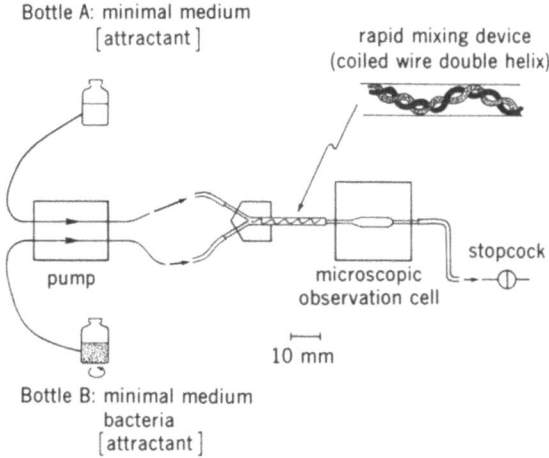

FIGURE 4A. The temporal gradient apparatus. Attractant concentrations or repellent concentrations can be placed in bottles A and B to yield increased gradients, decreased gradients, or zero gradients. The bacteria are suddenly mixed in the mixing chamber and projected into a microscopic observation cell. At the time of observation, they are in a uniform distribution of attractant or repellent following a mixing time of approximately 200 ms.

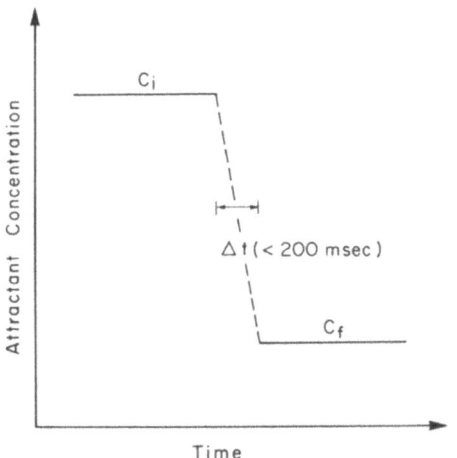

FIGURE 4B. Rationale of the temporal gradient apparatus. Bacteria in a uniform concentration of attractant C_i are suddenly mixed and find themselves in a uniform concentration of attractant C_f, which in this example is less than C_i. They have thus experienced a temporal decreased gradient equivalent to swimming down a spatial gradient of attractant. The observation chamber, however, has a uniform (isotropic) distribution of attractant.

The bacterial motility pattern is modified dramatically [Figure 5(A)]. Increased tumbling is observed as a result of increasing repellent or decreasing attractant concentration. Decreased tumbling frequency is observed as a result of increased attractant or decreased repellent. The altered swimming pattern then returns to normal motility over an interval of time.

By timing the return to the normal swimming patterns, a qualitative picture of the influence of the attractant or repellent is easily obtained [Figure 5(B)]. The instrument thus provides a very easy and rapid screening method for attractants and repellents. For rapid screening, the bacteria can also be placed on a microscope slide and a small amount of attractant or repellent can be added to the side of the cover glass with a micropipet. As the chemical diffuses into the liquid on the slide, the alteration in motility pattern is easily followed. The effect of methionine, dyes, uncouplers of oxidative phosphorylation, pH, etc., have been studied in this way (Tsang *et al.*, 1973; Taylor and Koshland, 1974; Taylor and Koshland, 1975). The time it takes a bacterium to return to normal is a function of the strength of the stimulus and the nature of the stimulant. Although not quantitative as in the procedure that will be described in the next section, these semiquantitative assays are frequently of great value in assessing mechanisms or testing unknown chemicals rapidly.

2.4.   Tumble Frequency Assay

The temporal gradient approach was made quantitative by measuring the tumble frequency (Spudich and Koshland, 1975) as a function of stimulus and time. In this method the responses of bacteria subjected to gradients are recorded photographically. By leaving a camera shutter open for 0.8 s, during which stroboscopic light at 5 flashes per second falls on the bacteria, photographs such as those shown in Figure 5(C) are obtained. The initial culture of a constantly tumbling mutant tumbles incessantly, so no smooth tracks are observed prior to addition of attractant. Right after addition ($t = 0.4$ min) all the bacteria are swimming smoothly and the stroboscopic light shows them at four separate positions appearing roughly in a straight-line track. Later ($t = 0.6$), the bacteria that tumble produce successive images near the same position and, hence, are shown as splotches of light. Sometimes these turn perpendicular to the plane of focus and disappear from observation. By counting the tracks in such a sequence of pictures, it is possible to quantitatively determine the number of bacteria that are swimming smoothly and the number that are tumbling in any interval of time. The raw data in Figure 5(C), therefore, can be plotted in Figure 6 to give an appropriate recovery curve, which serves as a quantitative assay. The analysis of such data is given in Section 3.3.

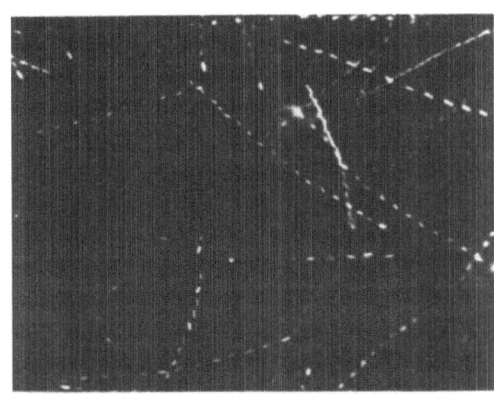

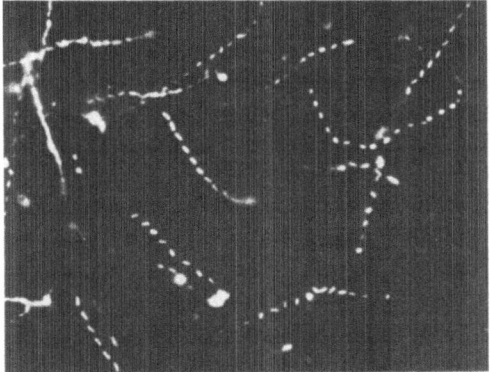

FIGURE 5A. Motility tracks of wild-type *S. typhimurium,* taken in the time interval 27 s after subjecting bacteria to a sudden (200 ms) change in attractant (serine) concentration in the temporal gradient apparatus. *(Top)* $\Delta C > 0$, $C_i = 0$, $C_f = 7.6 \times 10^{-4}$ M; smooth, linear trajectories. *(Center)* $\Delta C = 0$, $C_i = C_f = 0$ (control). Some changes in direction; bodies often show "wobble" as they travel. Bright spots indicate tumbling or nonmotile bacteria. This is the way wild-type bacteria behave when not stimulated. *(Bottom)* $\Delta C < 0$, $C_i = 10^{-3}$, $C_f = 2.4 \times 10^{-4}$ M; poor coordination; frequent tumbles and erratic changes in direction. (Photomicrographs were taken in dark field with a stroboscopic lamp operating at 5 pulses per second. Instantaneous velocity of bacteria in straight-line trajectories is of the order of 30 μm/s.)

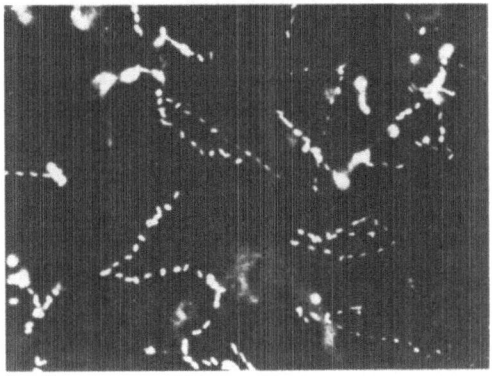

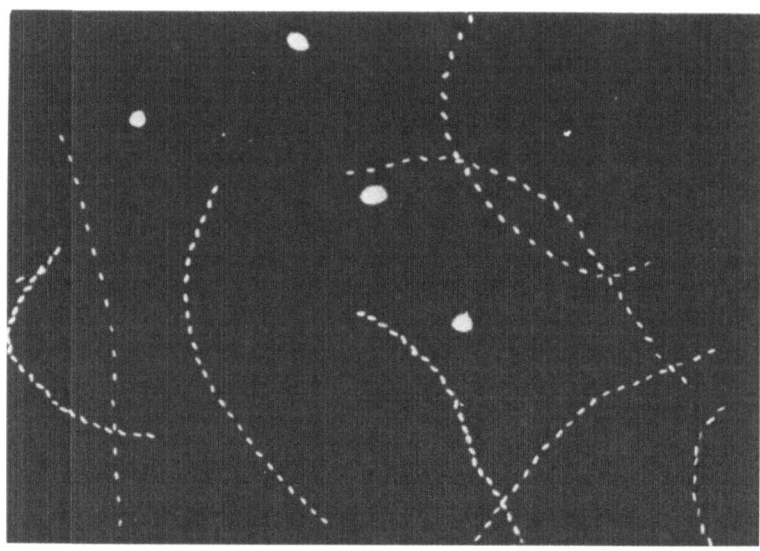

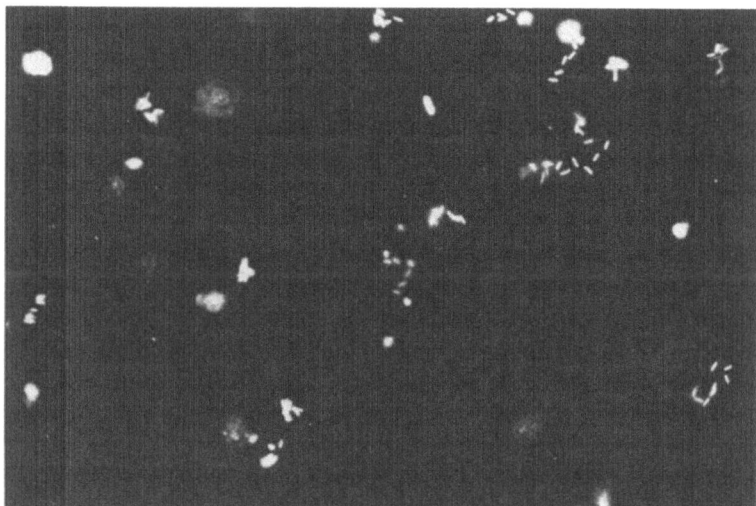

FIGURE 5B. Typical tracks of a smooth mutant (one that never tumbles) and a constantly tumbling mutant (one that never makes significant progress in a straight line.) Wild type looks like central panel of Figure 5A.

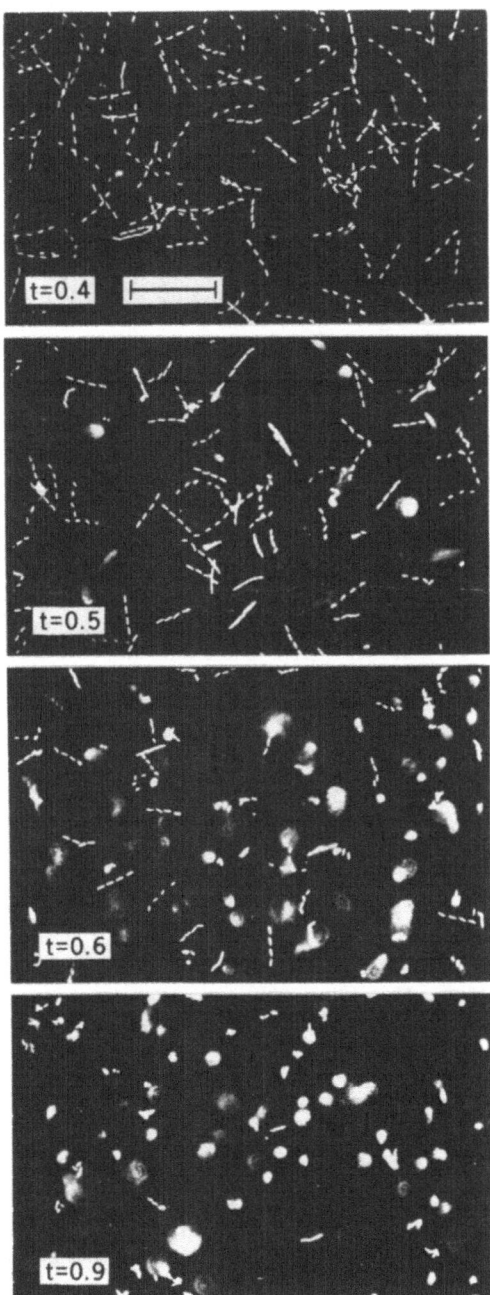

FIGURE 5C. Tumble frequency assay. The disappearance of tracks made by constantly tumbling mutant (Figure 5B) after an L-serine temporal gradient stimulus ($0 \rightarrow 0.02$ mM) is shown. A 0.9-ml aliquot of suspension of ST171 (a constantly tumbling mutant) was rapidly mixed with 0.1 ml of attractant to yield a final concentration of 0.02 mM L-serine. Photographs, at the indicated tenths of a minute after mixing ($t = 0$), were taken by 0.8-s open-shutter exposures to stroboscopic illumination. The length of the bar in the photograph at 0.4 min is 50 μm.

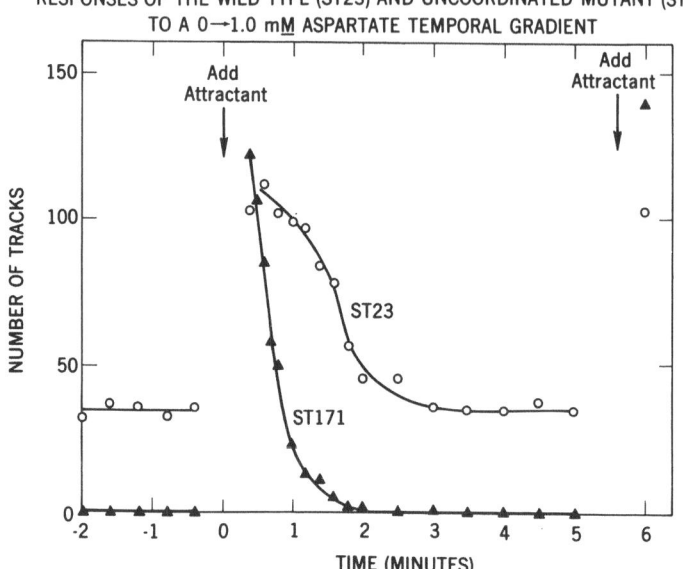

RESPONSES OF THE WILD TYPE (ST23) AND UNCOORDINATED MUTANT (ST171)
TO A 0→1.0 m$\underline{M}$ ASPARTATE TEMPORAL GRADIENT

FIGURE 6. Responses of wild type and a constantly tumbling mutant to the tumble frequency assay. Number of tracks measured in control prior to stimulus is zero for the constantly tumbling mutant (▲) and finite for the wild type (○). Plot of tracks as a function of time after stimulus leads to return to normal motility after a few minutes in both cases.

For routine use, it is convenient to use a constant tumbling mutant of the type shown in Figure 6, since the number of smooth tracks is zero initially and again after complete recovery. However, the method is applicable to wild-type populations as shown in Figure 6, but in that case the base is not zero.

2.5.   Rotation Assay

Silverman and Simon (1974*b*) recently showed that flagella rotate like propellors rather than propagate conformation waves. They did this by tethering bacterial cells to the walls of vessels by their flagella, and this caused the bacteria to rotate clockwise or counterclockwise. Tethered bacteria can therefore be subjected to additions of repellents or attractants and their changes in rotation can be observed under the microscope (Silverman and Simon, 1974*b*; Berg, 1974; Larsen *et al.,* 1974*b*). In order to observe such a bacterium in the case of *E. coli* and *S. typhimurium,* which have many flagella, it is necessary to grow the bacteria under nutritive conditions which suppress production of more than one flagellum. If this is not done, a second flagellum will prevent rotation. This has the difficulty of forcing study with bacteria under appreciably altered nutritive states, but the assay has been useful in studying the bacterial responses.

2.6. The Population Migration Assay

The population migration apparatus was devised to measure the movement of a population of bacteria in a defined gradient of attractant or repellent (Dahlquist *et al.*, 1972). It is shown schematically in Figure 7. An observation cell with an optical bottom and sides is filled with bacteria, usually uniformly distributed over the entire cell. A gradient of attractant or repellent is stabilized by means of glycerol concentrations, which maintain the gradient against thermal convection and mechanical vibrations. The gradient of attractant or repellent can be altered in various ways by appropriate mixing devices, and some of the more useful gradients are linear, step gradients and an exponential gradient. A laser beam is directed through the bottom of the vessel and a recording device reads the optical density of the scattered light at various positions in the observation vessel. From the intensity of the scattered light, the concentration of bacteria at any position can then be determined.

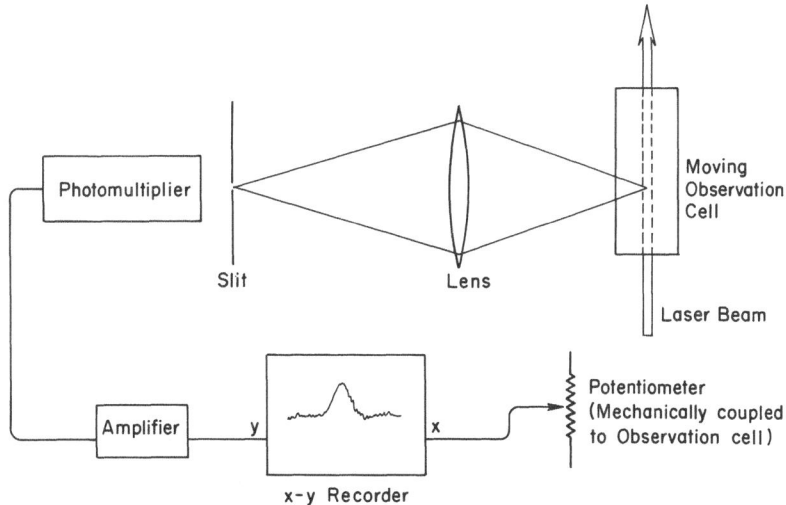

FIGURE 7. Schematic representation of the apparatus used to measure bacterial migration by light scattering. A laser light beam is scattered by the bacteria in an observation cell. The scattered light is recorded as a function of position in the observation cell on an *x-y* recorder.

A typical scan is shown in Figure 8. The gradient of attractant is exponential. As a result, bacteria in the gradient swim at uniform rates up the gradient until they accumulate in the plateau region. The area representing accumulation is proportional to the bacteria that have migrated and thus the change of this area as a function of time can give the migration velocity of the population (Figure 9). Gradients of various steepnesses, mixtures of

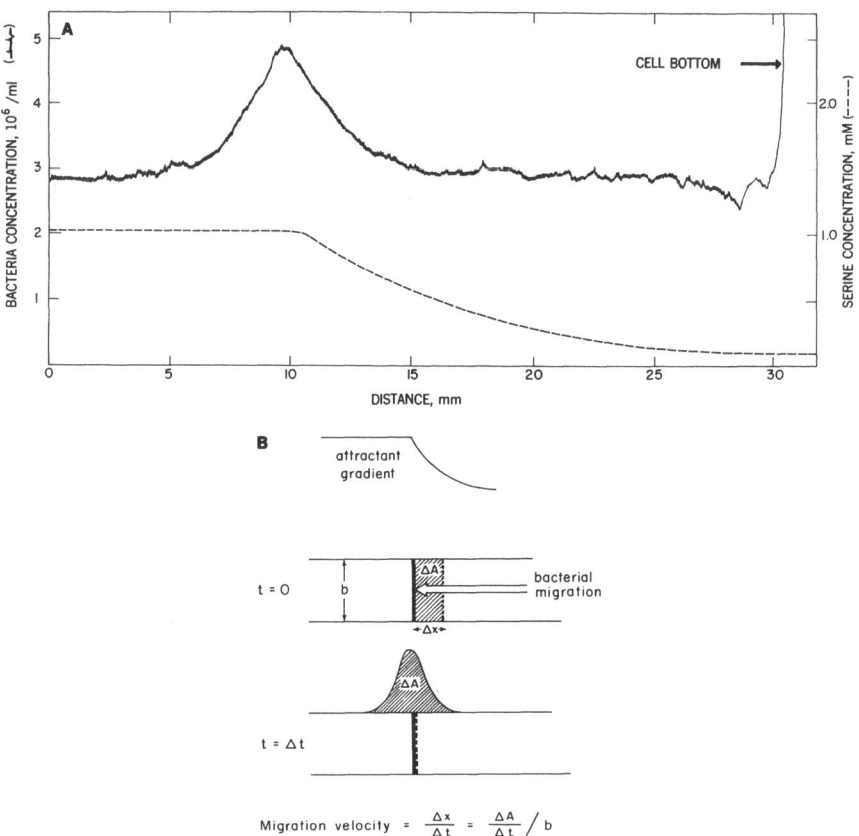

FIGURE 8. Response of *S. typhimurium* to an exponential gradient of L-serine. The plateau concentration of serine employed was 10^{-3} M and the decay distance of the gradient was 6.4 mm. (A) The dashed line represents the initial serine concentration. The trace represents the bacteria distribution after 15 min. The initial trace was a straight line over the whole observation cell. Bacteria have swum at uniform rates in the exponential gradient and have accumulated in the plateau region. The bacteria at the bottom of the cell are depleted but recording of this is obscured by light scattered from the bottom of the cell. (B) Schematic illustration of manner in which data of (A) can be used to calculate migration velocity.

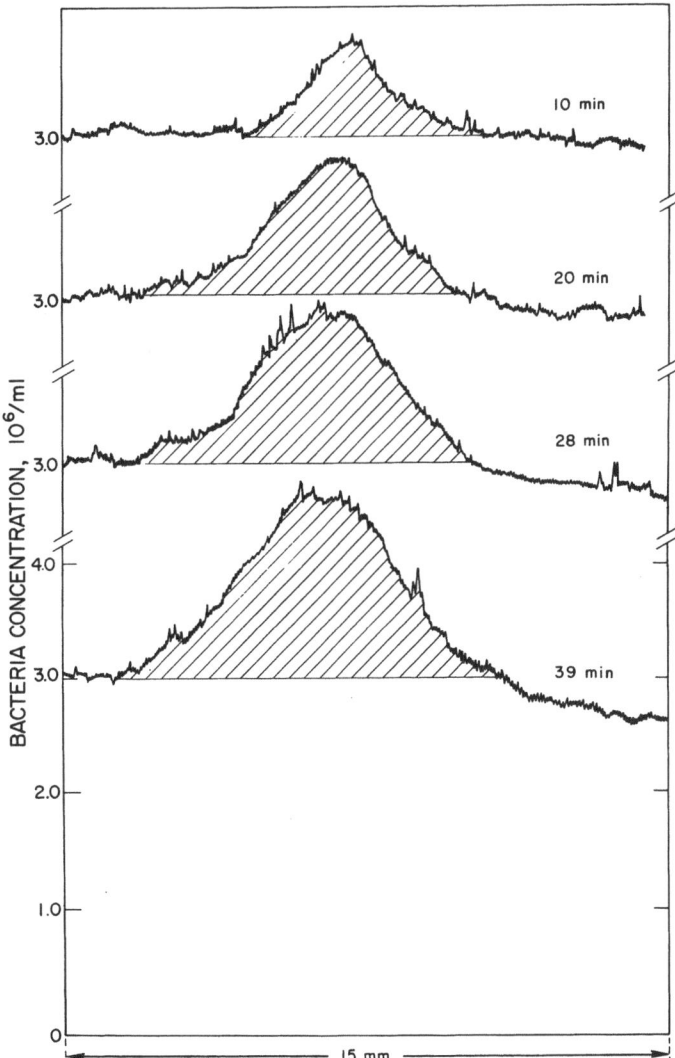

FIGURE 9A. Development of a response peak to $10^{-3}M$ L-serine plateau and exponential decay distance of 16.4 mm (cf. Figure 9B). The hatched area represents the number of bacteria that have passed a unit plane located perpendicular to the gradient direction at the right extreme edge of the peak. In each case, the scale is constant and the vertical displacement was used so that all the traces could be shown conveniently.

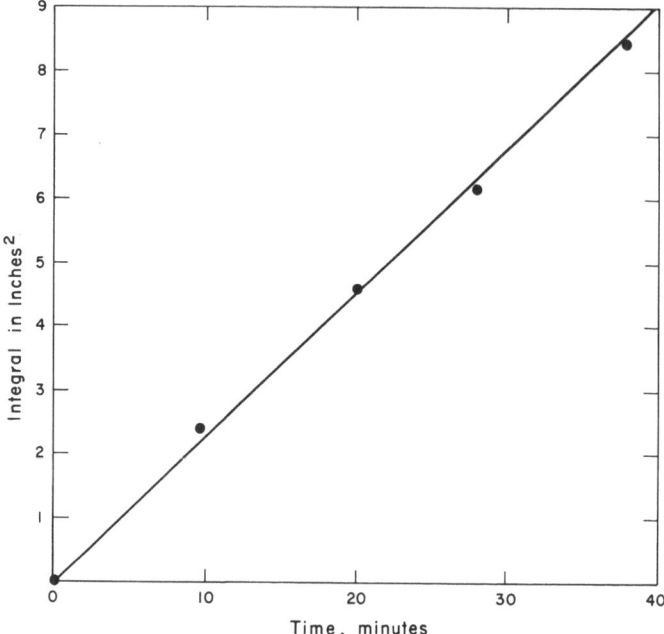

FIGURE 9B. Plot of the areas shown in Figure 9A as a function of time. The conversion factor for square inches to microns is 3.0×10^{-3} in.^{2} μm^{-4}.

attractants and repellents, or repellents plus attractants can be used (see Figure 15).

The population migration apparatus is appreciably more cumbersome to set up than the capillary assay and, hence, should not be used for routine screening of attractants or repellents. On the other hand, it provides a stable gradient, which can be defined accurately and maintained for long intervals, and measures bacterial movement quantitatively. Hence the movement of the bacteria in such a gradient is not only quantitatively determinable, but can be analyzed theoretically in ways that are not available in the more empirical capillary assay (Dahlquist, *et al.,* 1975).

2.7.   Selection of Nonchemotactic Mutants

Mutants that have lost the ability to sense a gradient, but are motile, are defective in the sensing system. One type of mutant that will be discussed is defective in a specific receptor, but a second type, called generally *non-chemotactic,* is defective in the more general transmission system. Two methods for selection of such mutants have been devised.

Armstrong *et al.* (1967) have used bacterial swimming on soft agar plates to select generally nonchemotactic mutants, and the technique has been called the *swarm plate method*. This involves spotting a colony on the center of a semisolid tryptone agar plate and incubating for 16–18 hr at 35°C in a water-saturated incubator. After this time the wild-type bacteria have swarmed out in several rings. Samples of the cells remaining at the origin are removed and suspended in 1 ml tryptone broth. After sedimentation at 4600*g*, a loopful of the pellet is deposited on the center of a second plate and the procedure is repeated. After ten such transfers, swarming is considerably reduced and the origin has become dense with nonmotile and nonchemotactic bacteria. A final plug is removed from the origin. This plug contains a very few residual wild-type bacteria, some nonchemotactic mutants, and various paralyzed mutants with abnormal flagella. To get rid of the large number of nonmotile cells, the culture is suspended in suitably diluted antiflagella antibody serum, which had been previously preabsorbed with nonflagellated mutant to eliminate nonspecific antibodies. Only colonies that do not swarm normally but react fully with antiserum are further characterized. This procedure has been used both for observing natural mutations and specifically mutagenized cultures. Parkinson (1975) has applied similar methods to isolate mutants adding the use of an *F′* episome.

Aswad and Koshland (1975) have used a "preformed" liquid gradient technique as shown in Figure 10. The bacteria are placed in the center of the column and the wild-type bacteria are attracted to the top of the column by an attractant in that region. The nonmotile mutants stay at the center position at which the bacteria were introduced, but the motile nonchemotactic mutants diffuse away from the center and, therefore, occupy the bottom of the tube essentially free of wild-type and nonmotile bacteria. After 60 min of incubation, all but the bottom 1 ml of the tube contents are aspirated off and the walls of the tube are rubbed with ethanol-soaked cotton to kill bacteria that might be adhering. A 2.0-ml volume of sterile VBC is added to the tube, which is then capped and incubated for 24 hr. Appearance of colonies on a plate allows final separation of nonchemotactic from nonmotile or wild-type bacteria. The method has the advantage of separating nonmotile as well as wild-type bacteria from the motile but nonchemotactic mutants. It is furthermore amenable to the selection of temperature-sensitive mutants.

3. THE SENSING MECHANISM

3.1. Nature of the Sensing Mechanism

At first glance it might seem impossible that an organism as small as a bacterium measuring 2–5 μm in length could have a sensing system. Even

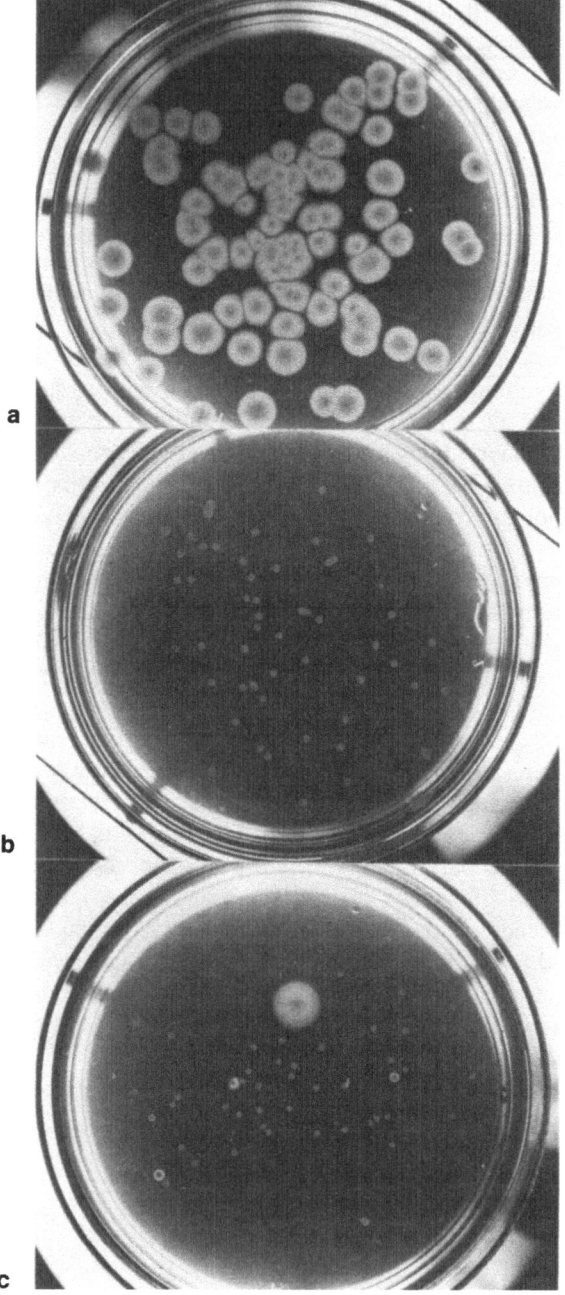

FIGURE 10A. Swarm plate method for isolating mutants. (a) Wild type, (b) a che⁻ smooth mutant, (c) mixture of wild type che⁻ and nonmotile cells.

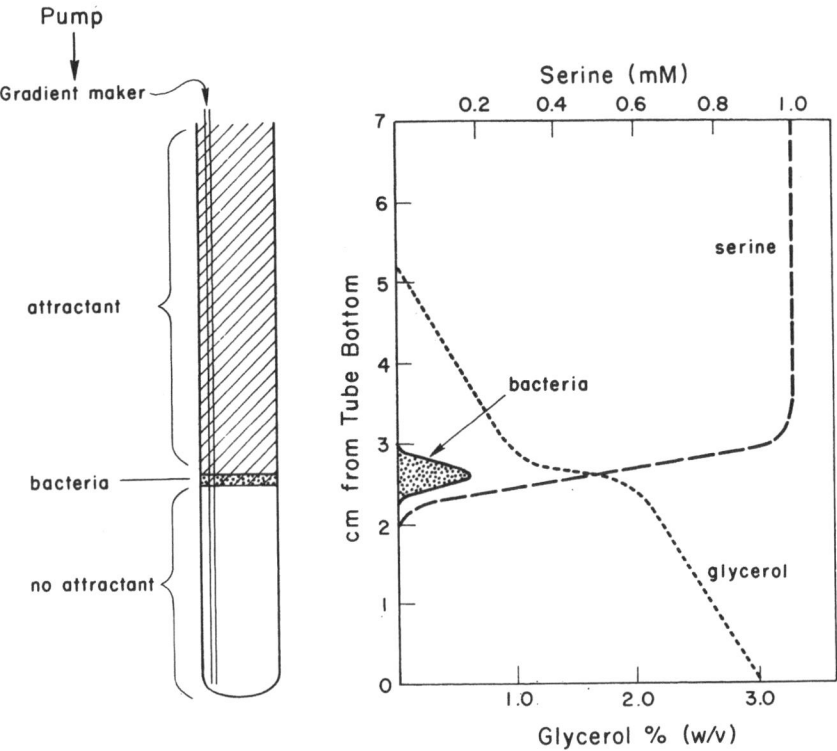

FIGURE 10B. Preformed "liquid" gradient method for isolation of generally non-chemotactic mutants. The drawing on the left shows how the gradient was constructed. The drawing on the right shows the distribution of gradients at the time zero. Wild-type bacteria swim up to attractant. Nonmotiles stay at the origin. Nonchemotactic mutants which cannot sense a gradient but are motile wander into the lower region where they are collected.

more extraordinary is that it can sense a chemical gradient. Measured in the population-migration apparatus, bacteria respond readily to gradients in which the difference in concentration of attractant between the head and the tail of a bacterium is one part in 10^4. A sensing system that can detect the difference between 1.000 and 1.001 needs a quite extraordinary analytical device.

An easy way to avoid this difficulty would be to assume that the bacterium did not compare differences, but simply modified its swimming behavior based on the absolute concentration of attractant. One version of this hypothesis would have the bacteria accumulate at high concentrations of attractant by simply slowing down as a function of the attractant concentration until they all accumulated at the highest concentration. This hypothesis is easily discarded by a variety of evidence. In the first place, the early experiments of Pfeffer indicated that bacteria responded to ratios of concentrations and this was confirmed by the quantitative measurements in the population migration apparatus (Dahlquist *et al.*, 1972). Secondly, the velocity and tumbling frequency and angle of turning are not a function of absolute concentration for *Salmonella* (Macnab and Koshland, 1972) (typical data are shown in Table 1) and are only a minor function of concentrations for *E. coli* (Berg and Brown, 1972). Thus, bacteria sense concentration differences, not absolute concentrations.

Two general types of mechanisms seemed reasonable based on what has been studied on other microorganisms and in higher mammalian species (Clayton, 1953; Clayton, 1959; Rothert, 1904; Beidler, 1971). One alternative could be called *instantaneous spatial sensing* and involves a comparison between sensors at the head and tail of the bacterium. A second, which we call *temporal sensing,* involves a comparison taken over time. The difference in concentration occurs because a gradient exists in space, but the bacterium would detect the gradient because it senses over a time interval and it arrives at one position of the gradient at a later time than the other. The question is how the difference between these two criteria can be tested in bacteria that are constantly swimming through three-dimensional space. It was solved by the temporal gradient apparatus described in Section 2.3.

When the bacteria are subjected to the sudden mixing of the temporal gradient apparatus, the situation illustrated in Figure 4 is obtained. The initial

TABLE 1. Motility of *Salmonella typhimurium* at Constant Attractant Levels[a]

Serine concentration (mM)	Velocity of bacteria in sample (μm sec^{-1})	Overall average velocity (μm sec^{-1})
0	27.4 ± 4.7	
	29.9 ± 6.0	28.8
	29.0 ± 4.6	
0.01	27.6 ± 4.7	
	28.7 ± 3.5	27.2
	25.2 ± 6.0	
1.0	30.2 ± 4.0	
	29.0 ± 2.9	28.8
	27.1 ± 6.0	

[a] Procedure: each sample represents 15 bacterial trajectories taken from a single photograph. The trajectories each contained 10 successive images of a bacterium taken at 1/3-sec intervals. Three samples were examined for each set of conditions.

concentration of the attractant (C_i) is suddenly changed to a final concentration (C_f), and the bacteria are observed on the stage of the microscope only after mixing has become complete. If, therefore, they respond by instantaneously comparing readings between their heads and their tails, they would sense only a uniform distribution of attractant. Since the absolute concentration of attractant has no effect on their motility patterns (Table 1), they would swim normally. If, on the other hand, they have a temporal mechanism and their "memory" is long enough to span the time of mixing, they would "remember" their initial concentration C_i, compare it with their final concentration C_f after mixing, and behave as if they were swimming down a gradient.

The evidence (shown in Figure 5) is clear-cut. When the bacteria are subjected to a decrease in concentration of attractant over time, they tumble more frequently. When they are subjected to an increase in concentration of attractant, they tumble less frequently, and when the concentration does not change, they swim normally. In both cases of a change in concentration, the bacteria revert to a normal swimming pattern over a period of time. When subjected to equivalent changes in the concentrations of repellents, the effects are similar but are inverse to the effects of attractants (Tsang *et al.*, 1973); i.e., increases in repellents increase tumbling, while decreases suppress tumbling.

These experiments established two conclusions. First, the mechanism by which bacteria migrate up gradients is a regulation of the tumbling frequency. When they move in a favorable direction ($\Delta C > 0$ for attractants, $\Delta C < 0$ for repellents) tumbling is suppressed, so they swim more than an average run length. When they swim in an unfavorable direction, tumbling is increased, which means that the bacteria travel only a short distance before tumbling. If, as a result of the tumble, they again proceed in the wrong direction, they soon tumble again. Eventually, a random tumble leads them to start swimming in the favorable direction and tumbling is suppressed so that they swim for a long distance before tumbling. The net effect of suppression of tumbling in the favorable direction and an increase of tumbling in the unfavorable direction leads to migration in a favorable direction. The tracking experiments of Berg and Brown independently led to the same conclusion.

Second, the bacterium uses a temporal sensing mechanism. It senses the gradient in space, not by comparing the concentrations at its head and tail but, rather, by traveling at moderate velocities through space and comparing its observations of the environment over time. Macnab and Koshland called this temporal gradient mechanism a "memory" mechanism because the bacteria were comparing past with present experience. We will return to the term *memory* later because its use may mean different things to different people. However, there is no doubt that a temporal sensing device is used by the bacteria to govern their behavior patterns. Confirmation of these conclusions was achieved by Brown and Berg (1974) in experiments in which the concentration of an attractant was increased by enzymatic generation to achieve a temporal gradient. Similar decreases in tumbling were observed.

3.2. Nature of the Biased Random Walk

Watching a bacterium in a microscope, one can see it traveling in a series of runs and tumbles, but this behavior can be observed far more elegantly and quantitatively using tracking devices. Berg and Brown (1972) followed the swimming of 35 wild-type bacteria selected from three different cultures. They found that the mean speed of *E. coli* was about 14 μm/s, with a mean run length of 0.86 s. The average change in direction from one run to the next occurred at an angle of about 62°. Figure 11 shows the distribution of the changes in direction. A gradient was provided by allowing attractant to diffuse into the observation chamber as described in Section 2.2. The average run length increased from 0.83 s in the control to 1.67 s in a serine gradient and to 0.90 s in an aspartic acid gradient. When these data were analyzed further, it was found that the mean run length moving up a gradient of attractant was extended appreciably, but the mean run length moving down a gradient was essentially the same as the normal control. The distribution of angles between runs was unchanged, i.e., the tumble pattern of a bacterium does not cause it to orient its tumble in the favorable direction.

These results agreed with the temporal gradient experiments and indicated that *E. coli* and *Salmonella* operate by a biased random walk mechanism and that a bias is provided by suppression of tumbling in the favorable direction. However, they appeared to disagree with the finding that tumbling is increased in the unfavorable direction, since the tracking experiments showed no decrease in run length in an unfavorable gradient direction as compared to the isotropic control. This apparent discrepancy was resolved by quantitation. The time of relaxation in the temporal gradient apparatus, i.e., the time to return to normal after the rapid mixing, was longer in the case of a positive gradient than in the case of a negative one. Moreover, the experimental error in tracking experiments was sufficient to make it difficult to detect small changes in run length. The answer, therefore, appears to be that there is some increase in tumbling frequency and, hence, some decrease in path length in traveling in an unfavorable direction; but the effect is far less than the increase in path length on going in the favorable direction.

These results established that the bacteria do not sense a gradient by

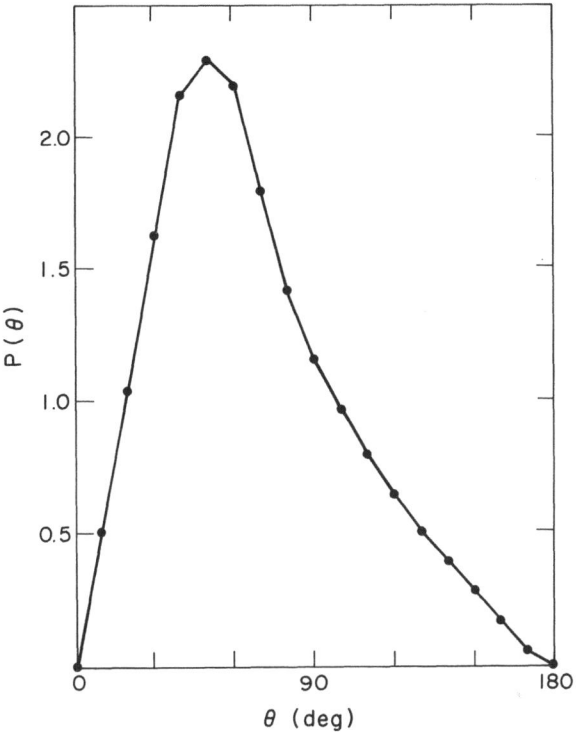

FIGURE 11. Probability distribution function for the angle between successive vectors, based on the data of Berg and Brown (1972). Average angle $\theta = 62°$. The probabilities are normalized so that the average probability of the angle falling within a 10° interval is 1 (cf. Macnab and Koshland, 1973).

simply utilizing an *avoidance response*. There is some avoidance, but the more important contribution to their movement is a positive response to a favorable situation.

3.3. Quantitation of the Sensory Response

The bacterial response just described has characteristics of a neuron in the sense that a chemical (attractant, repellent, or neurotransmitter, respectively) induces a response in the cell that leads to an all-or-none event (a tumble or a depolarization, respectively). It seemed desirable, therefore, to quantitate the response in a manner similar to an action potential and this was done by the quantitative assay procedure described in Section 3.4. (Spudich and Koshland, 1975). This quantitative *tumble frequency assay* could be used to answer several questions about the sensory response. In the first place, it could be asked whether the rate of delivery of the stimulus affected the recovery time, i.e., Did the response depend on dc or dc/dt, where c is the concentration of chemoeffector?

The experiment to resolve this is shown in Figure 12. At time equal to 0, the bacteria were subjected to a sudden increase in serine concentration delivered by steady mixing. In Experiment 1, the two solutions were mixed over a period of 1 s; in Experiment 2, over 5 s; and in Experiment 3, over 15 s. The recovery times were found to be identical (Spudich and Koshland, 1975). In other words, when the total change in chemoattractant concentration was kept constant, it did not matter whether it was delivered over a 1-s, 5-s, or 15-s interval.

At first glance this would seem anomalous since it has just been established that bacteria sense gradients over time, and these results indicate that tumble suppression is not dependent on rate of delivery of the stimulus. In fact, there is no discrepancy. The sensory system affects tumbling frequency by comparing concentrations over a time interval. The quantitative temporal gradient assay, however, is not measuring net migration through the medium nor the number of tumbles per unit of time. It is simply analyzing the total number of tumbles suppressed as a result of the

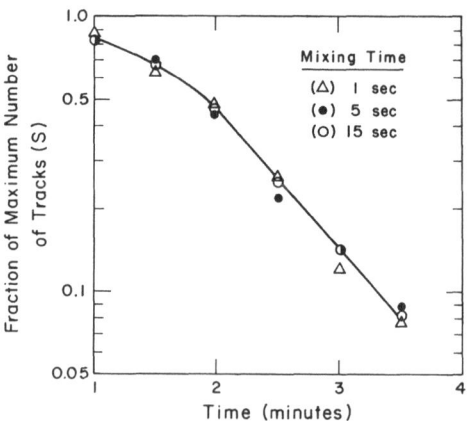

FIGURE 12. Effect of mixing time on recovery curves. Serine (0.1 ml of 5 mM; gradient $0 \rightarrow$ 0.5 mM) was delivered into 0.9 ml of a constant-tumbling mutant suspension with a micropipette over the period of time indicated, while the bacterial suspension was gently shaken. At 0.4 min after initiation of the stimulus, the bacteria were placed in the chamber and data were recorded by the tumble frequency assay. Points are an average of duplicates. The symbol ⊙ represents coincidence of the points ○ and •.

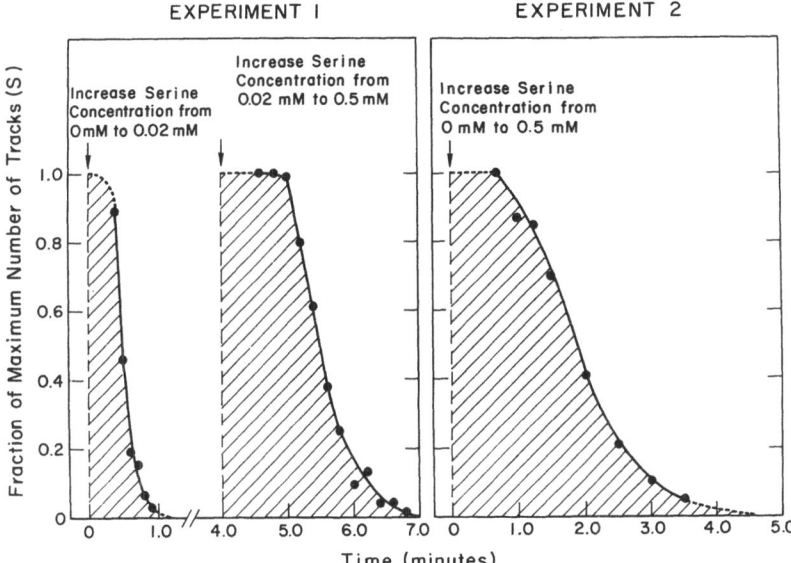

FIGURE 13. Additivity experiment. In Experiment 1, the response of constant tumbling mutant to a 0.02 mM serine stimulus was assayed as described in the text. After 4 min of incubation on a gyratory shaker in the 0.02 mM serine, the bacteria were rapidly mixed with a concentration of serine that yielded a final concentration of 0.5 mM, and the assay was initiated. In Experiment 2, the response to a single stimulus of 0.5 mM was assayed. The mean recovery time, t_R, for each stimulus is the area under the recovery curve. The areas in experiment 1 are 0.5 min and 1.5 min. In experiment 2, the area is 2.0 min.

addition of a chemoattractant. The results say that the total tumbles suppressed are a function of the total amount of chemoattractant added and not of the rate of its addition. If, therefore, a hypothetical bacterium swims up a gradient from C_i to C_f, it may suppress Z tumbles over time t. If it swims up the same concentration gradient in half the time, it will suppress exactly the same *total number* of tumbles but twice as many tumbles will be suppressed per unit of time.

A second question relates to the additivity of stimuli and the results of one such experiment are shown in Figure 13. A given stimulus, from 0 to 0.5 mM L-serine, was broken up into two smaller stimuli, from 0 to 0.02 mM and from 0.02 to 0.5 mM serine. What effect would this have on the response? As seen in Figure 13, the results are additive. The areas under the curves of the two small stimuli add up to the area under the curve of the one large one. Since the areas under the curves represent the average recovery times in the experiment, it means that the $\tau_{13} = \tau_{12} + \tau_{23}$, where τ_{13} is the average time of recovery to the overall stimulus, and τ_{12} and τ_{23} are the average times of recovery to the fractional stimuli. This same experiment was performed with the attractant aspartate as well as serine and with a variety of fractional increments. In all cases, the results were the same: additivity in the tumbles was suppressed. This is quite consistent with the previous evidence in regard to the time independence of the delivery of chemoeffector.

The quantitative method allowed a comparison between responses of the whole organism and the physical parameters of a purified receptor. Figure 14 shows the data obtained for the responses to a D-ribose stimulus of a living bacterial cell of *S. typhimurium*. The purified receptor utilized was the purified ribose-binding protein isolated in milligram quantities from the *Salmonella* (Aksamit and Koshland, 1972; Aksamit and Koshland, 1974). The theoretical line is calculated from the dissociation constants (K_D) of sugar from the ribose-binding protein which had a K_D for ribose of 3×10^{-7} and for allose of 3×10^{-4}. The change in the receptor–chemoeffector complex can be deduced from the mass action law [Eq. (1)] and the change in receptor occupancy [Eq. (2)].

$$K_D = \frac{(R)(C)}{(RC)} \tag{1}$$

where R is the protein receptor, C is the chemoeffector, and RC is the receptor chemoeffector complex. The solid lines are based on the calculation of the change in receptor occupancy, i.e., $\Delta(RC)$ of Eq. 2, using the K_D's from the purified protein.

$$\Delta(RC) = \frac{R(C_f - C_i)}{K_D} \tag{2}$$

The points are the recovery times obtained from the tumble frequency assay on the whole living bacterium. The agreement is excellent for ribose and almost as good for allose (Figure 14).

These experiments, together with the previous ones, show that the tumbling suppression is proportional to $\Delta(RC)$, the change in receptor occupancy, and is not proportional to $d(\Delta RC)/dt$. Mesibov *et al.* (1973) have shown that the sensitivity curves based on the capillary response are consistent with the change in receptor occupancy.

3.4. Additivity Between Different Types of Stimuli

Repellents have been known to exist from early days (Pfeffer, 1883, 1884, 1888), and this phenomenon was called *negative chemotaxis*. It has been studied by a number of investigators (Weibull, 1960; Lederberg, 1956; Smith and Doetsch, 1969; Doetsch and Seymour, 1970; Tso and Adler, 1974). Repellents were assumed to be compounds that would be harmful to

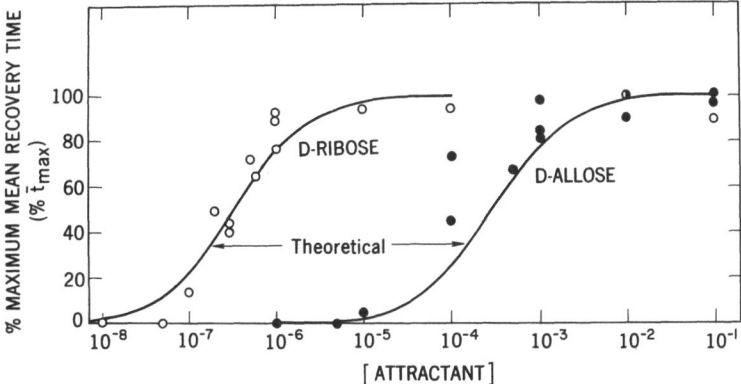

FIGURE 14. Comparison of response with receptor occupancy. Points are percent maximum mean recovery times (t_R) for D-ribose, and D-allose concentration increases from 0 to the concentration shown on the abscissa. Each point represents the average of three consecutive assays. Theoretical curves were calculated assuming (i) noncooperative chemoreceptor binding constants of 3.3×10^{-7} for ribose and 3.0×10^{-4} for allose (as determined *in vitro*), and (ii) the response t_R is proportional to the change in fraction of binding protein occupied. The symbol ⊘ represents coincidence of the points ○ and ●.

the bacteria. Hence, it was interesting to trace their mode of action. One possible mode would have attractants acting through receptors, but repellents acting by damaging the machinery of the transmission process. The first evidence of repellents acting as the opposite of attractants was obtained in experiments carried out by Pfeffer (1883, 1884, 1888). Tso and Adler (1974) showed that the specificity of repellents and their competitive behavior fell in categories characteristic of receptors and, hence, similar to attractants. Experiments by Tsang *et al.* (1973), utilizing the temporal gradient apparatus, showed that repellents caused decreases or increases in tumbling in an exactly inverse relationship to attractants. When phenol concentrations were utilized in the same way as serine concentrations in Figure 5, smooth swimming was caused by a decrease in phenol concentrations; tumbling, by an increase; and normal swimming, by no change. Thus repellents generate the same behavior as attractants except for the difference in sign.

This result could then be exploited to determine whether bacteria had "decision-making" powers. If a mixture of attractants and repellents were presented to the bacteria at the same interval, could it integrate these two signals or would the combination simply jam the machinery? The results are unequivocal. The bacterium integrates the signals in an algebraic manner. In Figure 15, the results of such decision-making experiments in

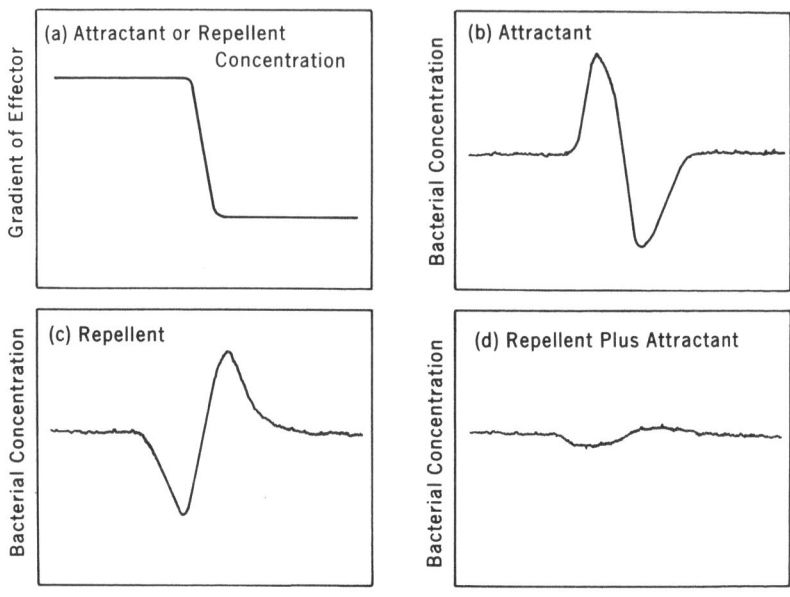

FIGURE 15. Additivity of attractants and repellents. (a) Step gradient established in population migration apparatus (cf. Figures 8 and 9, and Section 2.6). (b) Bacterial migration after 15 min exposure to the attractant, serine. Bacteria migrate to high plateau concentration and are depleted at low concentrations. (c) Bacterial migration after 15 min exposure to the repellent, phenol. (d) Bacterial migration to combination of serine and phenol. Effects cancel because bacteria integrate signals.

the population migration apparatus are shown. Experiments in the temporal gradient apparatus led to the same conclusion (Tsang *et al.*, 1973). Tso and Adler (1974) utilized the capillary assay to test the decision-making powers of *E. coli* and also found the same result.

The same type of additivity seems to be true in the case of a light response and the chemoattractant response. At high serine gradients, the chemoattractant response can swamp out the light-induced response (Taylor and Koshland, 1975). With lesser serine stimuli, a light response can be observed. Hence an additive relationship (with algebraically opposite signs) is observed between the light and attractant effects. These results, together with those of the additivity experiments described in Section 3.3 and similar experiments with combinations of attractants and repellents (Spudich and Koshland, unpublished data; Rubik and Koshland, unpublished data) indicate that all such effectors add to the same transmission system with appropriate algebraic signs. The neuron's capacity to integrate the effect of inhibitory and excitatory stimuli seems quite obviously analogous to this integrative ability of the bacteria.

3.5. Weber–Fechner Law

In the 1880s Pfeffer performed ingenious experiments with meat extracts and showed that to a first approximation the bacteria followed Weber's law. Weber's law defines a response in terms of a just noticeable stimulus, leading to the conclusion that $\Delta\Psi/\Psi$ is a constant, where Ψ is the stimulus and $\Delta\Psi$ is the just noticeable change in the stimulus (Stevens, 1970). The advent of a quantitative method for measuring bacterial population migration made it possible for this phenomenon to be examined more closely. Using the population-migration apparatus previously mentioned, Dahlquist et al. (1972) investigated the movement of bacteria under the influence of various gradients of attractant. In a linear gradient of attractant (ΔC per unit distance equals a constant with distance) the bacteria showed a peculiar pattern that was difficult to analyze quantitatively. However, when the same experiment was performed with a logarithmic gradient ($\Delta C/C$ per unit distance equals a constant), the bacteria gave a uniform migrational velocity over the entire range of the logarithmic gradient as shown in Figure 8. This result can only be explained by the assumption that the bacteria are migrating with a velocity following $d(\ln c)$. From this result alone, one would conclude that bacterial migration obeys the Weber–Fechner law. However, these data were put to a further test because the apparatus allowed study of a wide range of concentrations. A horizontal straight line would be observed if the Weber–Fechner law were obeyed at all concentrations. The results in Figure 16 clearly show they do not.

Similar conclusions were reached by Mesibov et al. (1973) by studying capillary tests. By ingenious methods of correction, Mesibov et al. were able to factor out a roughly quantitative Weber–Fechner law relationship over a narrow range and found that deviations occurred over a wide range. A theoretical curve which gave rather good agreement was obtained for α-methyl-L-aspartate, an extremely strong attractant. But agreement was not nearly as good in the case of poor attractants, such as D-galactose. In the process of this analysis, they found that the needed threshold for the bacteria to detect a gradient was a 12% change in the occupancy of the receptor for galactose but only a 0.4% change for α-methyl-L-aspartate.

If we now assess these conclusions in terms of the excellent correspondence between change in receptor occupancy and the dissociation constant of the pure protein (Figure 14), it is clear why the capillary assay and Weber's law are correct only over a narrow range. The plot of percentage saturation vs. the log of the concentration of the free ligand is as shown in

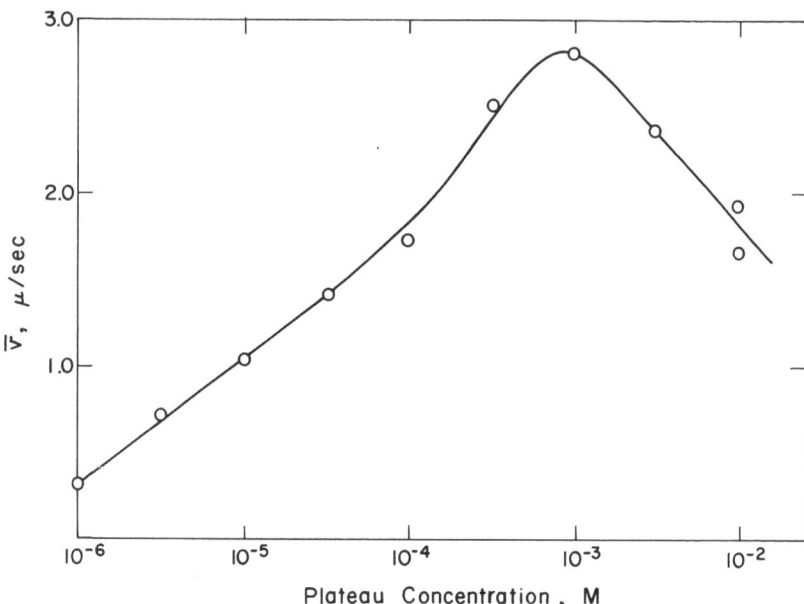

FIGURE 16. Plot of the average velocity of the response of the bacterial population versus plateau concentration at constant relative gradient of decay distance 6.4 mm. These data were collected at 25°C in VBC medium.

the theoretical curves of Figure 14 (cf. also Beidler, 1954, 1962; Stevens, 1970). This is the curve obtained if one plots a biological response against stimulus, provided the stimulus obeys a simple receptor-occupancy law. It is quite clear though that this curve is not linear over its total concentration range, but is approximately linear only over small regions, in particular, over its middle portion. Complex behavioral phenomena subjected to numerous inputs will certainly not, *a priori,* be proportional to receptor occupancy alone. However, it is quite possible that the initial trigger, whether it is a chemical occupying a receptor or a photon activating a photon receptor, can induce a chain of very complex events, which produces a response that is proportional to the initial stimulus over a limited range. Therefore, if the biological response in a complex assay, such as the capillary assay, or a direct assay, such as the tumble frequency assay, is studied over a limited range, it may obey the Weber–Fechner law. This is probably also the case in more complex phenomena in higher species (Stevens, 1970). As the range is increased, however, both assays should deviate and that seems to be the case.

The results of Section 4.3 also throw light on statements about "just noticeable response" and "thresholds" of stimuli. According to the receptor occupancy curves of Figure 14, there is complete correspondence with a theoretical curve of binding to a pure isolated receptor. This means that there is no threshold as far as the stimulus is concerned; the smallest amount of attractant changes the probability of tumbling immediately. Whatever threshold seems to be present exists because the analytical techniques for measuring the behavioral responses have different sensitivities. Thus when capillary assays assign thresholds above those of the tumble frequency assay, it simply means that one assay is more sensitive than another. Thresholds, therefore, are a practical matter but not a theoretical one.

3.6. Speed of the Response

In the temporal gradient experiments, the bacteria are mixed with a new concentration of attractant in less than 200 ms and all of them show modified behavior (Macnab and Koshland, 1972). In experiments where tumbling is induced with light, the response appears immediately in *Salmonella* and is slightly more delayed in *E. coli* (Taylor and Koshland, 1975). It is difficult to quantitate the precise length of time in either of these effects, but certainly it is less than a fraction of a second. Measurements of diffusion in membranes give a diffusion constant of 10^{-8} cm²/s for a phospholipid molecule in a membrane (Scandella *et al.* 1972). From this value it can be calculated that a phospholipid molecule will take about a second to travel from one end of the bacterium to the other. If an integrating system is to explain the additivity experiments just described, it seems difficult for the effect to depend on equilibration of a molecule the size of a phospholipid. Three alternatives seem possible to explain this phenomenon. The first is that the integrating phenomenon occurs through a membrane depolarization which spreads rapidly and generates a signal at distant positions in a very short interval of time. The second possibility is that the molecule, which is the integrating device, is appreciably smaller and more mobile than a phospholipid molecule. Molecules such as benzene might travel through membranes ten times more rapidly than a phospholipid. Recent reevaluations by McConnell (personal communication) suggest that rate constants appreciably greater than 10^{-8} might be obtained. The third possibility is that a few of each type of receptor are located in the region of each flagellum. In that case, the signal generated would have a very small distance to travel. Such a clustering of receptors would require that representative numbers of the various receptors, of which 30 have been identified so far, would have to be located in proximity to each flagellum. Although selection between these alternatives is not possible at this stage, they indicate limits on each type of mechanism. The membrane depolarization hypothesis seems most reasonable.

3.7. Relationship of Artificial Gradients to Normal Bacterial Behavior

It might well be asked whether the temporal gradient studies that subject the bacterium to larger stimuli than they normally confront in swimming through solutions are not a measure of an artifactual condition and whether a totally different mechanism might operate under true physiological conditions. To test this possibility, experiments have been done with shallow gradients using both the enzyme degradation procedure (Brown and Berg, 1974) or the temporal gradient method (Macnab and Koshland, unpublished data). The results provide a smooth extrapolation up to the point at which no detectable change can be observed. There does not appear to be any discontinuity between the shallow gradients in which a small difference in concentration is experienced and the steeper gradients where the responses are easier to measure. It appears, therefore, that a single memory device is operating over the entire concentration range.

This result can be confirmed in a second way. Dahlquist *et al.* (1976) have taken the results from the population-migration apparatus and analyzed them in terms of the prevailing theory making a number of ad hoc, but reasonable, assumptions. They assume, for example, that the chemotactic potential is caused by suppression of tumbling when the bacteria are moving in a favorable direction and involve essentially no alteration in mean free paths when the bacteria are moving in the unfavorable direction. This is in rough agreement with the observation described in which the quantitative contribution to the favorable direction to chemotaxis is certainly far greater. With these and other assumptions, the calculated movement of the bacteria in three-dimensional space agrees with observations. This agreement means that the temporal gradient and tracking studies are, indeed, studying the real mechanism by which bacterial behavior is controlled in its real world.

3.8. Useful Memory

The bacteria sense their environment by some kind of memory device. *A priori,* this sounds extremely reasonable in terms of solving two difficult problems. The first of these is a statistical problem and involves the difficulty of detecting a gradient at very low concentrations of attractant for a very small organism. This difficulty is caused by the statistical fluctuations of molecules in the environment of the bacteria. Thus, a gradient that changes by a factor of e for every 6.7 mm of pathway, which is the type utilized in the population-migration apparatus, leads to a change of only 1 part in 10^4 over the body of the bacterium. At 10^{-6} M concentration, it can be readily calculated that there are approximately 60 molecules within a 0.1 μm × 0.1 μm × 1 μm volume. If this is the effective volume in contact with the receptors near a flagellum or the head or the tail of a bacterium, its Poissonian standard error is ±8 molecules, or greater than 10%. Yet the apparatus must detect differences much less than 10%. The statistical fluctuation problem obviously becomes smaller if the concentration is higher. The second problem is analytical and involves the detection of 1 part in 10^4 over the length of a bacterium even in the absence of statistical fluctuations. Moreover, this small change in concentration must be some-how amplified to generate an all-or-none response, i.e., tumbling vs. nontumbling.

A time-dependent response would be attractive to solve each of these problems. An integration over time of the molecules appearing in the environment tends to smooth out statistical fluctuations. Some type of collection area, which could be the 10^4 receptors in the membrane of the bacteria or the mucous area of the sensilla of insects, can help in the statistical problem. These would all tend to produce time-average type responses. It should be remembered, however, that the greater the length of the time averaging, the less sensitive is the system to the gradient. However, the bacteria must detect a gradient in order to migrate properly, and therefore a temporal sensing system and a moderately large number of receptors are utilized so that bacteria can detect even very small gradients. As the concentration gets higher, the need for statistical averaging disappears, but the excess receptors are not detrimental.

A memory device also helps the analytical problem. If this is so, it might be asked from a theoretical point of view what type of memory would be desirable. Consultation with Figure 17 indicates that there will be some

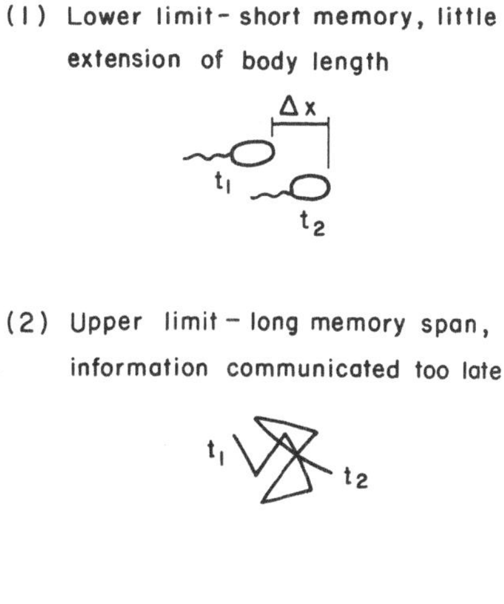

(1) Lower limit- short memory, little extension of body length

(2) Upper limit - long memory span, information communicated too late

FIGURE 17. Useful memory. Too short a memory will give bacteria little advantage over having receptors at head and tail. Too long a memory will give bacteria signals at a time when direction of motion may have changed to opposite direction. Optimal memory is a compromise between extremes.

limitations on the time span of the memory if an optimal sensing of a gradient is desired. If the memory is very short, a bacterium traveling through space at a finite velocity could not travel many body lengths before its information from the past would be forgotten. Then one of the major advantages of a temporal sensing system, i.e., the ability to integrate over several body lengths, would be eliminated. Thus, the memory should be as long as possible to allow a comparison over the greatest number of body lengths and, hence, reduce the analytical problem.

This need for a long memory, however, is extremely disadvantageous in regard to the correlation of the information with the direction of motion. A bacterium traveling in the absence of a gradient proceeds for short runs followed by tumbles; tumbling occurs in a random manner. If the memory spans too long an interval, the bacterium will have tumbled and will proceed in a different direction when it receives the signal from its memory. Thus, a signal that indicates that the bacterium is going in a favorable direction and should suppress tumbling might be received at a time when the bacterium is actually moving in an unfavorable direction. Clearly, then, the need for increasing the sensitivity of analytical devices (optimized by long memory) must be weighed against the need for a one-to-one correlation between the type of signal and the direction of motion (optimized by a short memory).

It is appropriate under these circumstances to describe a "useful memory" that would optimize these two quantities by a compromise between them. This problem has been approached in several ways. One of the approaches has been to examine the bacterium by means of the tracker in a stable, uniform gradient, utilizing the concept of persistence (Macnab and Koshland, 1973).

The important factor in bacterial motion is its motion relative to the gradient of chemoeffector. As shown in Figure 2, many of the turns, tumbles, or twiddles of individual bacteria may still leave it traveling with a positive component in the direction of the gradient even though this component is not the same as the previous run. If we now define a bacterium moving in a favorable direction of a gradient such that the z component of its velocity is positive, we say that the bacterium has persisted in the $+z$ direction if a turn continues this positive component. Persistence number, then, is the number of successive segments or runs with the same sign. Persistence time is the time during which the velocity component of a bacterium maintains the same sign. The components of persistence are illustrated in Figure 2 by the heavy lines ($+z$) and light lines ($-z$).

The advantage of such a procedure is clear in relation of the actual movements of bacteria. It is very difficult to define a twiddle, tumble, or turn. Bacteria do not swim in an absolutely straight line and an individual run frequently proceeds in a gentle arc. If the bacteria turn by 10°, should this be called a twiddle, a tumble, or simply a midcourse correction? The line between such minor adjustments of direction to prolonged end-over-end tumbling is continuous and dividing lines are arbitrary. Persistence, on the other hand, requires only that we measure the sign of the z axis projection of the velocity and this quantitatively can be well defined, particularly in the tracking apparatus which involves a unidirectional gradient. A second advantage is that the tracking apparatus utilized for obtaining data in regard to persistence time was designed so that the tracking of bacteria could easily register changes from $+z$ to $-z$, but was less useful in delineating details in the angle of individual turns. The third advantage to this type of analysis relates to the useful memory concept previously described. The actual information fed to the memory system of the bacterium presumably integrates over the entire time the bacterium is proceeding in a favorable direction. An alteration in the course of its trajectory, which finds it proceeding at a new angle in the gradient, may actually decrease its rate of movement up a gradient, but if it continues in a positive direction, it will still continue to receive some stimuli informing it to suppress tumbling. Hence it will acquire an extended spatial comparison, which was not eliminated by an adjustment of course correction at some point in the trajectory. One can calculate a potential trajectory in a simple way by considering a random walk around a circle.

A bacterium in a unidirectional gradient in the z direction starts up the gradient, as shown in Figure 2, and makes successive turns, all random in orientation. (By random we mean that the direction of the new trajectory is not based on its previous direction and is distributed in a uniform manner around an average angle.) In the case shown in Figure 2, it makes one turn which leaves it traveling in a $+z$ direction, but the second turn causes it to move in the $-z$ direction. One can ask, What is the average number of turns that will cause a reversal in direction? The analysis of this problem involves fairly complex mathematics and statistical theory, but the result is quite simple. It states that the persistence number $\overline{N}$ is equal to $180°/\theta_k$, where θ is the average turn angle (Macnab and Koshland, Jr., 1973). This means that an angle of 90° gives an $\overline{N}$ of 2 and 60° gives an $\overline{N}$ of 3. Of course, no one trajectory can be predicted absolutely, but an $\overline{N}$ of 3 means that on the average a bacterium will make three tumbles before altering the z component of its direction in two-dimensional space.

Extending this logic to three-dimensional space, the calculations of the average trajectories were performed by Monte Carlo methods, and the relationship between the average persistence number $\overline{N}$ and the angle between successive vectors was found to be the same as in the two-dimensional case $\overline{N} = 180°/\theta_k$ (Macnab and Koshland, 1973). Lovely and Dahlquist (1976) confirmed this calculation analytically and further showed that the average trajectory and the randomly distributed trajectories gave the same average value. For simplicity, we will use the average value since this gives the same results as the true case in which there is a skewed distribution of angles around this average value. Utilizing these results and actual measurements in the tracking microscope, Macnab and Koshland observed an average value of 12 s for the persistence of *S. typhimurium* in the absence of a gradient with a mean persistence number of 3. This gives an average time between turns of 3.7 s. Similar results by Dahlquist and co-workers have given results that are very much in agreement, providing a value of the order of 6.7 s for the mean persistence time. Since the bacterium travels at 20 μm/s and is 2 μm long, this means that the bacterium in an average movement up the gradient obtains information over a distance of 20–100 body lengths.

The question arises whether there is a carry-over between tumbles. Does the bacterium that is traveling up a gradient and tumbles proceed down a gradient for a longer interval than one already heading down a gradient that tumbles? Attempts to decide whether there is carry-over between runs was studied further, but the results were on the borderline of experimental accuracy (Macnab and Koshland, unpublished data). Some carry-over was indicated, but it was small. Studies of Dahlquist and co-workers and of Brown and Berg suggest there is none. The experimental accuracy is insufficient to resolve the question precisely, but it suggests that the effect is slight if it does exist. An effective comparison over 50 body lengths would reduce the analytical problem from 1 part in 10^4 to 5 parts in 10^3, which is a help, but still a formidable analytical problem remains.

Whatever the detailed conclusion, it seems quite clear that the memory is finite and that its usefulness has been optimized as a compromise between the extremes outlined above.

4. THE RECEPTORS

4.1. Specificity

Bacteria have a number of receptors, each of which is rather specific. A list of the receptors, which have been identified so far for the organism *E. coli,* are shown in Table 2 (Adler, 1969; Tso and Adler, 1971; Adler *et al.* 1973; Mesibov and Adler, 1972). *S. typhimurium* has not been studied nearly as intensively, but studies on it seem to indicate a similar distribution and similar specificity for the receptors that have been studied (Dahlquist *et al.,* 1972; Tsang *et al.,* 1973; Aksamit and Koshland, 1972; Mesibov and

TABLE 2. List of Chemoeffectors for Chemotaxis in *Escherichia coli*

Compound	Threshold activity M	Effect
N-Acetyl-D-glucosamine	1×10^{-5}	Attractant
Benzoate	1×10^{-4}	Repellent
D-Fructose	1×10^{-5}	Attractant
Fatty acid (acetate)	3×10^{-4}	Repellent
D-Galactose	1×10^{-6}	Attractant
D-Glucose	1×10^{-6}	Attractant
Indole	$\sim 1 \times 10^{-6}$	Repellent
Maltose	3×10^{-6}	Attractant
D-Mannitol	7×10^{-6}	Attractant
Isopropanol	6×10^{-4}	Repellent
D-Ribose	7×10^{-6}	Attractant
D-Sorbitol	1×10^{-5}	Attractant
Trehalose	6×10^{-6}	Attractant
Tryptophan	$\sim 1 \times 10^{-3}$	Repellent
L-Aspartate	6×10^{-8}	Attractant
L-Serine	3×10^{-7}	Attractant
H^+	pH 6.5	Repellent
OH^-	pH 7.5	Repellent
O_2	—	Attractant

Adler, 1972). A study of a variety of other microorganisms indicates a similar distribution, but different compounds are found to serve as attractants or repellents (Seymour and Doetsch, 1973). For example, cyclic AMP is an attractant for slime mold, but does not seem to serve as an attractant for *E. coli* or *S. typhimurium* (Gerisch and Hess, 1974; Bonker *et al.*, 1969).

Receptors listed in Table 2 are not absolutely specific for the compounds indicated (Adler, 1969). For example, the receptor for galactose also has affinity for glucose and fructose. The receptor for serine binds cysteine, L-alanine, and glycine. The receptor for L-aspartate also binds L-glutamate. However, the receptors are not of wide specificity such as those characterized as bitter or sweet in the tastebuds of man. Rather, they appear to be designed for specific molecules as are the pheromone receptors, and the number of compounds that are attracted to them, to a significant extent, is rather small.

Evidence for this number of specific receptors was obtained by competition studies (Adler, 1969). An individual receptor, such as the one for galactose, is tested in the following manner. Galactose is placed inside the capillary at concentrations that produce a high measurable response of bacteria swimming into the capillary. A high concentration of the sugar to be tested, e.g., glucose, is then added to both the capillary and the outside solution. If it prevents the normal accumulation of bacteria in the capillary, presumptive evidence has been obtained that the glucose acts by binding to the galactose receptor. If, on the other hand, a high concentration of aspartate does not inhibit galactose taxis, it is presumed that aspartate acts via some other receptor when it is acting as an attractant. In those cases in which the search for a receptor has succeeded in isolating a pure protein, the results have confirmed the capillary analyses.

It is of some interest to invoke the teleological rationale using the particular list of compounds for a particular organism. While it is likely that the total list of receptors contained in bacteria is not yet complete, and from the known total content of protein in the membrane and the variety of compounds screened thus far, it also appears unlikely that the list of Table 2 is representative and comprises even half the total receptors. Hence, it is significant that almost all the known attractants are also nutrients for the bacteria, and many of the repellents are toxic compounds when added to the bacteria at high concentrations. An easy generalization would appear to be that bacteria swim toward compounds that are needed for their metabolism and away from compounds that are detrimental to their survival.

This easy generalization, however, is not possible to maintain rigorously. A variety of compounds needed for the survival of the organism are not attractants, including a number of the amino acids; also, some of the repellents, e.g., short-chained fatty acids, do not appear to be toxic, even at relatively high concentrations. An explanation for these anomalies could be as follows: A bacterium that needs amino acids will migrate toward protein sources in which these amino acids are being broken down. Since most proteins contain all 20 amino acids, attraction to a few of these amino acids should be sufficient; hence, it is not necessary for the bacterium to synthesize receptors for all of them. In the case of repellents, one can argue that some of the repellents, such as pH and short-chain fatty acids, might be an indication of crowded conditions because they are excretion products of metabolizing cells. Again, it may not be necessary for the bacteria to have receptors for every excretion product, since one or two of the typical ones might be sufficient. Finally, it should be emphasized that some of these repellents are very weak at the concentrations that one would find them under physiological conditions. For example, tryptophan is a very weak repellent and would certainly never overcome the attraction of serine, which is a very powerful attractant, if both were present in the same medium. Possibly the tryptophan repellent action arises as a result of the indole side chain and does not indicate that an organism is programmed to leave an environment containing tryptophan.

4.2. Metabolism and Transport

Although bacteria swim toward nutrients, the chemotactic response does not depend on the metabolism of the compound. This has been demonstrated for *E. coli* in a variety of ways (Adler, 1969). First of all, some compounds that are extensively metabolized are not attractants. Histidine, for example, is a nutrient and is not an attractant. Secondly, some chemicals that are not metabolized at all attract bacteria. Specifically, these can include inhibitors of the receptors listed in Table 2. For example, D-fructose, a 6-deoxy analogue of D-galactose, inhibits D-galactose chemitaxis, undoubtedly by binding to the D-galactose receptor; α-methylaspartate acts in a similar manner to the aspartate receptor. Thus, these compounds, in analogy to saccharine in higher species, can exert their effect without being metabolized themselves. Furthermore, it is possible to block normal metabolism so that a compound that is a normal nutrient is not metabolized. Thus, mutants of *S. typhimurium* were made that could not metabolize ribose and yet the microorganism is attracted to ribose as in the wild type (Aksamit and Koshland, 1974).

In the case of repellents it is clear that many of these compounds are not metabolized. Sometimes they are compounds that are toxic, such as phenol (Lederberg, 1956), but the chemotactic response is observed well below the levels that are toxic for the organism. This makes sense teleologically since it would induce the organism to swim out of the area long before the concentration became high enough to kill it. Some repellents, such as the amino acids in Table 2, are themselves not harmful to the bacteria and yet they still cause negative chemotaxis.

The chemoreceptors that have been studied most so far are periplasmic proteins (Hazelbauer and Adler, 1971; Aksamit and Koshland, 1972). That is, they are easily released by osmotic shock and, hence, are assumed to be present in the space between the inner and outer membrane. Each has been shown to be a binding protein for transport as well as chemotaxis. It might be asked whether transport is essential for the chemotactic response. It is not. In the case of the galactose receptor, it has been shown by mutation studies that transport can be disengaged from chemotaxis (Adler, 1969; Hazelbauer and Adler, 1971; Ordal and Adler, 1974). Some mutants cannot carry out transport, but this does not affect chemotaxis. In the case of ribose (Aksamit and Koshland, 1974), similar experiments have been performed using mutants that demonstrated quite clearly that transport is not necessary for chemotaxis. Thus, in the case of the "shockable" proteins, the binding protein serves both as a receptor for chemotaxis and the initial protein involved in transport.

Other proteins involved in chemotaxis cannot be released by osmotic shock. Adler and Epstein (1974) have shown that the enzyme II of the phosphotransferase system (Kundig and Roseman, 1971) serves as a binding protein for the chemotactic process and it is, of course, part of the transport system delineated by Kundig and Roseman (1971). Enzyme II of the glucose phosphotransferase system serves as a chemoreceptor for glucose, and enzyme II of the mannose system serves as the chemoreceptor for mannose (Adler and Epstein, 1974). In these cases, the other proteins of the phosphotransferase system are needed for chemotaxis. Whether this means that the entire system must be intact, or whether there is some enhancing effect by having these additional enzymes present, is not yet clear.

4.3. Purified Receptors

Two receptors identified with chemotaxis have been purified and characterized: the galactose receptor of *E. coli* and the ribose receptor of *S. typhimurium*. Both are periplasmic proteins. In the case of the galactose receptor, the galactose-binding protein was originally isolated by Anraku (1968) and shown to be involved in transport. Further study on this protein by Boos (1969, 1972; Boos and Gordon, 1971) has revealed some of its properties, and an *E. coli* mutant was obtained by Boos and Gordon (1971) that was lacking this protein, which was deficient in transport. This mutant was tested by Hazelbauer and Adler (1971), who showed that it was also defective in chemotaxis, providing the first evidence that the bacterium lacking the galactose-binding protein was incapable of migrating specifically toward galactose. Further, they showed that mutants defective in other compounds of galactose metabolism responded to galactose. Chemotaxis toward other attractants was found to be unaffected (see Figure 18). Further evidence was obtained by adding the shock fluid from a wild type to the mutants and observing restoration of the chemotactic ability. The experiment was encouraging because loss of a function in a mutant could be the result of several causes. These reconstitution experiments seemed to provide strong evidence that the protein was indeed the chemotactic protein, but, unfortunately, it has not been possible to repeat them. In fact, no reproducible reconstitution of either transport or chemotaxis in the whole bacterium has been achieved so far. Nevertheless, it appears quite certain that the galactose-binding protein is the receptor for chemotaxis.

The ribose-binding protein was isolated from *S. typhimurium* by osmotic shock methods and shown to be the receptor for chemotaxis by mutation and characterization studies (Aksamit and Koshland, 1972, 1974). About 10^4 molecules of the protein per bacterium were isolated. By preparing antibodies to the purified isolated ribose-binding protein, it was possible

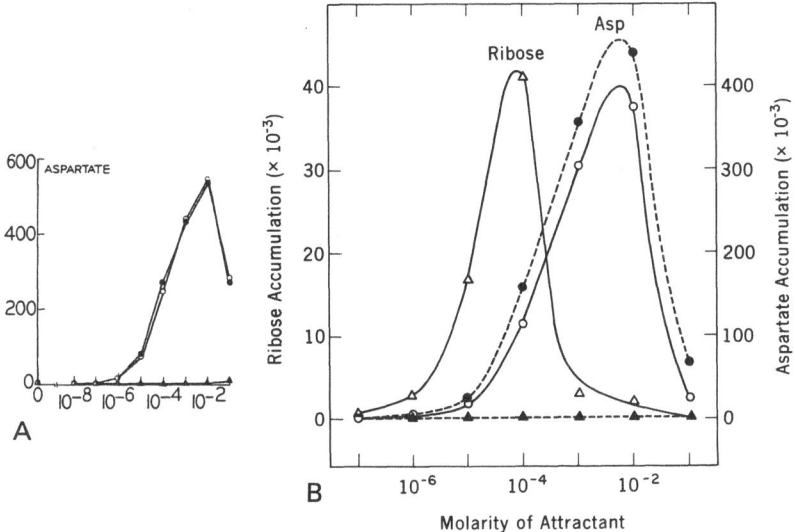

FIGURE 18. Response of bacteria with mutations in a specific receptor protein. (A) *E. coli* wild type and a mutant lacking response to galactose. (B) *Salmonella typhimurium* wild type and mutant lacking the ribose-binding protein. Solid lines show the response of wild type; dashed lines show the response of a mutant lacking receptor to the sugar but normal to other attractants.

to detect minute quantities of the protein present in the bacterium as a whole or in the membrane of the bacterium. A mutant was obtained that had no ribose-binding activity and it lacked the ribose-binding protein. This mutant did not respond to ribose but showed chemotaxis to all other attractants of which aspartate is shown as a control in Figure 18. Revertants were obtained that restored the ability to migrate toward ribose and at the same time restored the presence of the ribose-binding protein reacting with antibody. Finally, the specificity pattern of the protein was shown to be identical to the specificity for chemotaxis in the whole organism. As shown in Table 3, modification of ribose in any position except carbon 5 produced compounds that could not interact with the ribose-binding protein in the purified state, and none of these compounds were attractants. Modification of position 5 led to a compound, allose, which was an attractant and was bound to the purified protein, although it required 10^3 higher concentrations than ribose in order to produce similar effects. When assayed in the quantitative tumble frequency assay (see Figure 14), allose was shown to give a chemotactic response; also, the quantitation was that to be expected from the binding affinity to the pure protein. Further studies

TABLE 3

D-ribose   D-allose

Position	Ribose modification	Binding to ribose-binding protein
C1	–OH replaced by:	
	–H	No
	$\begin{array}{c} OH \\ \| \\ O{-}P{-}O \\ \| \\ O{-} \end{array}$	No
C2	epimer	No
	–OH replaced by:	
	–H	No
C3	epimer	No
	–OH replaced by:	
	–H	No
	–OCH₃	No
C4	epimer	No
C5	–H replaced by:	
	–CH₂OH(allose)	Yes

showed a lack of correlation between transport of ribose and chemotaxis. Evidence appears to be overwhelming that the ribose-binding protein is, indeed, the true receptor for chemotaxis (Aksamit and Koshland, 1974).

There are some conspicuous differences between the two isolated receptor proteins. It has been reported that the galactose-binding protein has two binding sites (Boos and Gordon, 1971), whereas the ribose-binding protein clearly has only one over the entire range tested (Aksamit and Koshland, 1974). A maltose-binding protein has been isolated (Schwartz and Kellerman, 1972) and a galactose receptor for *S. typhimurium* has also been isolated (Strange and Koshland, 1976).

These proteins are the binding proteins identified with transport, are in a 30,000-molecular-weight range, and are periplasmic proteins. A protein of this sort, which has been examined crystallographically, is the sulfate-binding protein initially isolated by Pardee (1968). It has the dimensions of a long cigar-shaped molecule that could easily span the entire membrane (Langridge *et al.*, 1970). Whether this is important in its function is not known at present. The tumble frequency assay showed a precise agreement between the binding constant of the purified ribose-binding protein and the ribose chemotactic response in the whole bacterium. At first, this seems surprising because the protein is presumably buried in a lipid membrane *in vivo*. However, periplasmic proteins may not be immersed in the lipid layer, or the part of the molecule that binds the ribose may be projecting out into the aqueous solution while the second part is immersed in lipid. The quantitative agreement between these numbers is intriguing and provides evidence for an aqueous environment of the active site of the protein, the rest of which may be partially in the membrane.

5. THE TRANSMISSION SYSTEM

5.1. Mutants

In the bacterial system, the use of genetic mutation is a classical tool, and it has been applied in a helpful way in regard to bacterial chemotaxis. In initial studies, Armstrong *et al.* (1967) used the swarm plate method to select generally nonchemotactic mutants from other types of mutations. By "generally nonchemotactic," Adler meant mutants that were defective in the sensing system and not in motility. These mutants could not respond to gradients of any attractant and, hence, appeared defective in the transmission system and not in a specific receptor. Forty independent mutants were isolated, which fell into three complementation groups called *cheA, cheB,* and *cheC*. Armstrong and Adler (1969*a,b*) subsequently mapped these mutants in three groups, which they placed in between the *his* and *trp* regions in the *E. coli* chromosomes. Using the same techniques with the addition of an *F'* episome, Parkinson (1975) has added a fourth gene, which he called *cheD*. This *cheD* mutant was different from the others in that it was nonchemotactic to serine and repellents but did respond to sugars and other attractants.

Aswad and Koshland (1976) used the preformed liquid gradient technique for selection of mutants from *S. typhimurium*. This technique, which separated nonmotile from nonchemotactic mutants as well as from wild types, produced a total of 72 mutants, of which 58 were identified and separated into six complementation groups. This work was then carried further by Warrick *et al.* (1976*a,b*), who separated 110 additional generally nonchemotactic mutants and proceeded to characterize them by abortive transduction, complementation, episome, and deletion techniques. A total of nine different complementation groups were found, of which seven were mapped in specific regions of the chromosome of the *Salmonella* and two of them in separate locations as yet not mapped. These were named *che* genes *P, Q, R, S, T, U, V, W, X,* and *Y.* In the future there will probably be a close correspondence between the *E. coli* and *Salmonella* genes, but at the moment the identities are unknown. In Figure 19 are shown the potential alignment of the *Salmonella* genes together with those of *E. coli.*

At the moment, only the beginnings of information on the function of these genes are available. Some of the mutations eliminate individual receptor responses. For example, the light effect is not observed in genes *cheP* and *cheQ.* Other genes seem to be essential for the machinery for transmission of all receptors because their absence blocks every type of signal from being received at the flagella. Two of the nonchemotactic genes, *cheU* and *cheV*, map precisely in regions previously identified with *fla* genes (Koshland *et al.,* 1976). Silverman and Simon (1974*a*) also observed overlap between a chemotactic gene in *E. coli,* and similar results have been obtained by Colins and Stocker (personal communication). *Fla* genes are identified on the basis that the flagella fail to assemble. The nonchemotactic mutants are motile but cannot receive signals. The most simple explanation of these facts is that these genes produce a protein that is part of the flagellar machinery but is defective in that portion which receives signals from the sensory apparatus. The *cheD* gene of Parkinson would appear to be located at the other end of the transmission pathway, since mutants in it block transmission of signals from serine and leucine but do not alter signals from other chemoeffectors. Hence at the moment the transmission system appears to involve some gene products that may receive signals from some groups of effectors and not others, some gene products that act at the interface where the signals are received at the flagella, and others that are common to all transmission of messages. A possible scheme is shown in Figure 20.

Caution must be expressed in concluding that the transmission machinery is composed of nine gene products. An exhaustive search for mutants has not been completed, and therefore more mutants may well show up with future study. Moreover, one or more of the genes identified may involve a single enzyme, e.g., a regulatory gene and a structural gene. However, the virtue of bacterial study is that billions of bacteria can be evaluated by relatively simple selection techniques, and it seems possible that a good fraction of the transmission system has been identified. At least the genetic technique indicates the magnitude of the problem and limits the search. Several of the proteins have been isolated already (Springer and Koshland, unpublished data) and a study of their function will reveal much about the general transmission system.

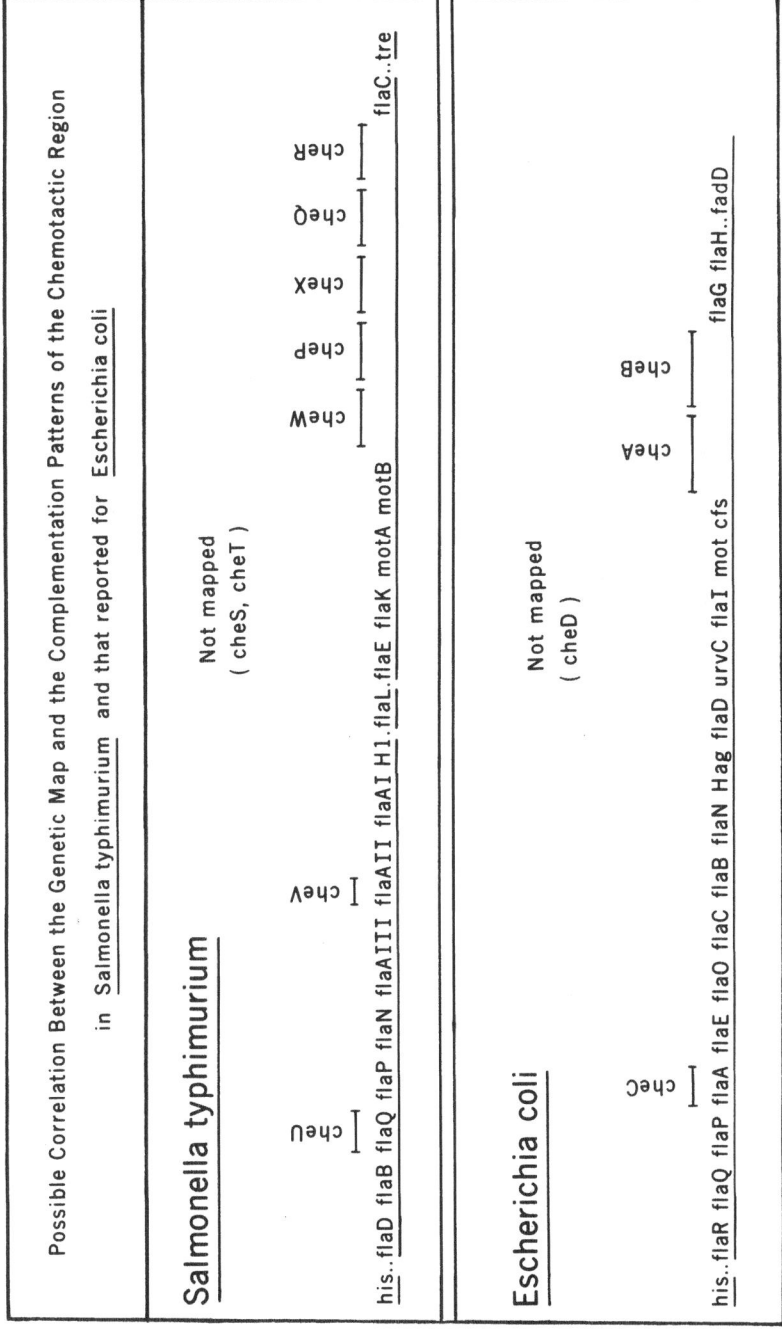

FIGURE 19. Genetic map of generally nonchemotactic mutants in *E. coli* and *Salmonella typhimurium*.

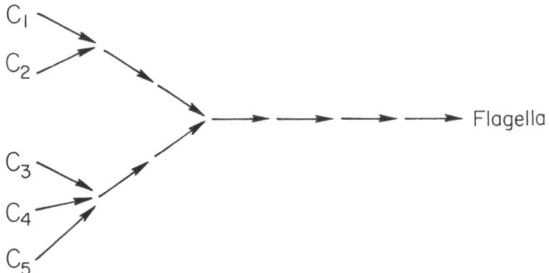

FIGURE 20. Overall scheme for transmission of information in bacterial sensing of gradients.

5.2. Methionine

Armstrong and Adler (1969a) showed that a methionine auxotroph of *E. coli* was nonchemotactic toward all attractants tested when starved for methionine. The effect of methionine starvation was not caused by a lack of protein synthesis because starvation for other amino acids, such as threonine or leucine, had no effect on chemotaxis. Moreover, methionine starvation does not lead to loss of motility but, rather, an alteration in the type of motility displayed. The same effect of methionine was also shown in *Salmonella,* and the dramatic effect of small changes in methionine concentration is shown in Figure 21 (Aswad and Koshland, Jr., 1974). It is seen that 10^{-7} M methionine shows little or no chemotaxis whereas 10^{-6} M produces an optimal response. When the bacteria lacking methionine were examined in the microscope, it was found that they swim smoothly without any tumbles. The absence of methionine therefore inhibits the ability to tumble.

This observation raised the question of whether methionine was involved simply in the mechanical ability of the flagella to tumble or in the sensory apparatus. This was solved by transducing the block in methionine metabolism to a constantly tumbling mutant *ST4* (Aswad and Koshland, 1974). The bacteria continued to tumble in the absence of methionine. Hence the methionine is not needed specifically for tumbling. The mutant was then treated with a serine gradient, increasing the concentration of serine suddenly. Smooth swimming results and then the bacterium returns to the normal tumbling pattern of the mutant. The rate of return to the constantly tumbling state was appreciably longer in the methionine-depleted bacterium than in the bacterium to which methionine had been added. These results indicated that methionine was involved in maintaining the relative levels of the tumble regulator and the threshold for detection of the tumble regulator and not in the mechanics of the tumbling process (Aswad and Koshland, 1974).

It then became of some interest to find whether or not methionine was itself involved or whether it was a precursor for some other compound. Experiments by Armstrong (1972) on *E. coli* and by Aswad and Koshland 1975) on *S. typhimurium* indicated that S-adenosylmethionine had to be synthesized in order to obtain the chemotactic response. These results are obtained in several ways. Methionine analogues that replaced methionine as substrates in the SAM synthetase reaction replaced methionine in restoring chemotaxis, while those that were nonsubstrates showed no replacement ability. Cycloleucine, a methionine analogue that inhibits SAM synthesis, produces the same behavioral changes as methionine starvation. SAM is only one of two metabolites derived from methionine, which is formed rapidly and present in high concentrations during incubation in the chemotaxis medium. Concentration of the other metabolite, spermidine, does not correlate with chemotactic ability. When a methionine-starved, chemotactically wild-type cell is given an exogenous supply of methionine, the SAM pool is replaced at the same time that spontaneous tumbling reappears. Likewise, the rapid turnover of SAM correlates well with the rapid loss of tumbling that occurs when the exogenous supply of methionine is removed. These results provide strong presumptive evidence that SAM is involved in the signaling mechanism of the chemotactic machinery. It does not prove, however, that SAM itself is involved. It could be a precursor for a further step in methionine metabolism, e.g.,

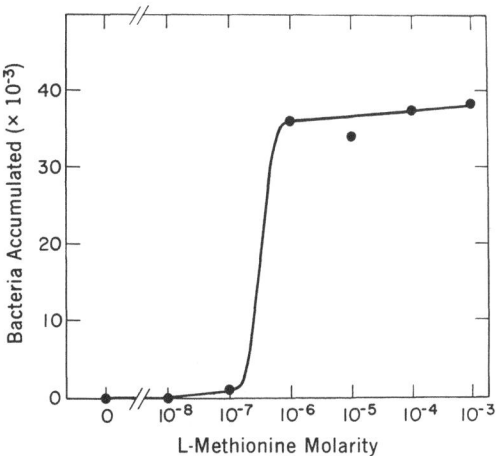

FIGURE 21. Methionine dependence of chemotaxis in a methionine auxotroph. A culture of a methionine auxotroph of *Salmonella* strain *ST2* was washed in methionine-free VBC-glycerol, and the washed suspension was then diluted into tubes containing 3.0 ml VBC-glycerol and L-methionine at the concentration indicated on the abscissa. Chemotaxis was measured at 30°C for 30 min by the capillary assay using 1 m*M* L-serine as an attractant.

methylation of some crucial compound in order to achieve the result. At any rate, it is clear that SAM or some derivative of it is essential for the sensing apparatus to be operating at its fully functional state.

5.3. Light Effect

Exposure of free-swimming *S. typhimurium* to a high-intensity light caused tumbling (Macnab and Koshland, 1973) that was found to be reversible, provided short light pulses were used. The optimal wavelength for inducing tumbling was found to be between 360 nm and 450 nm. It had previously been reported that photoinactivation of bacteria was sensitized by photoinducible dyes and molecular oxygen, a phenomenon that came to be known as the *photodynamic effect* (Raab, 1900). However, the light-induced tumbling effect is quite clearly different from the photodynamic effect and involves a direct perturbation of the transmission machinery. It was subsequently found that there were three distinct light effects (Taylor and Koshland, 1975). One of these was a light-induced tumbling that occurred immediately. A second involved a smooth response, generated on somewhat more prolonged exposures to light. And, finally, a killing or paralyzing effect was found due to very long exposures to light. The paralyzing effect, the so-called photodynamic effect observed earlier, is of relatively little importance to this particular analysis, although it may indeed involve destruction of the chemotactic machinery. However, such extensive changes occur that it would be difficult to disentangle the specific causes.

In the case of the light-induced tumbling, the effect was found to be involved in the chemotactic signaling in several ways (Taylor and Koshland, 1975). First, light-induced tumbling was additive with chemical-induced tumbling as described in Section 3.4. Hence, there is an additive interrelationship between the light-induced tumbling and the normal response to chemoattractants. Second, the mutants that had defective transmission machinery (*cheU* and *cheV*) failed to show the light effect. These mutants were motile, but were swimming without tumbling and the light could not generate tumbling in them. Moreover, methionine auxotrophs did not exhibit the light effect unless methionine was added to the culture.

The action spectrum of the light effect was studied and appeared to be that of a flavin (Macnab and Koshland, 1973). This strongly suggests that the initial absorption of the light occurs in some flavin-like molecule. The light-induced tumbling effect could be generated by externally added dyes (called the *extrinsic effect*) as well as by internal compounds (the *intrinsic effect*). Oxygen was necessary to operate the electron transport machinery in order to observe either light effect, thus suggesting that the light effect causes some interruption in the electron transport chain or modifies an intermediate of the electron transport chain, which in turn induces the tumbling. The light effect was also found in *E. coli,* but seemed to be more sluggish in response. The observation of such a light effect provides a handle in probing the transmission apparatus more deeply, and possibly the flavin action spectrum will help in identifying individual proteins involved in the chemotactic sensing process.

5.4. ATP and Oxidative Phosphorylation

In order to have chemotaxis, bacteria must be motile and both biological processes require energy. It does not follow that both require the same energy source and, indeed, studies have shown that they do not (Larsen *et al.,* 1974).

Studies on transport in bacteria have indicated that some biological processes are driven by the intermediate of oxidative phosphorylation and do not require ATP *per se* (Ernster and Lee, 1967; Kaback, 1972; Hong and Kaback, 1972). Frequently, a high-energy compound (or gradient) of oxidative phosphorylation is the energy source directly or can be produced indirectly by driving the system backward through ATP and an adenosine triphosphatase. Means are available to test the alternative of an ATP or an oxidative phosphorylation energy source (Ernster and Lee, 1967; Kaback, 1972; Hong and Kaback, 1972).

In the case of *E. coli,* Larsen *et al.* (1974a) applied these tests to the problems of motility and chemotaxis. It was found that mutants, blocked in the conversion of ATP to the intermediate of oxidative phosphorylation, because of damage to the ATPase, were motile in the presence of oxygen, showing that their flagella machinery was intact. Conversely, wild-type cells treated with carbonylcyanide *M*-chlorophenyl hydrazone (CCCP), which uncouples oxidative phosphorylation, can inhibit motility even in the presence of ATP. Finally, arsenate, which reduces ATP levels in the bacterium but does not significantly reduce the level of the intermediate in oxidative phosphorylation, did not affect motility. Thus, it is clear that the energy source for motility is the intermediate of oxidative phosphorylation and is not ATP.

On the other hand, ATP is required for chemotaxis (Larsen *et al.,* 1974a). The addition of arsenate to motile cells does not inhibit their motility, but prevents their chemotaxis. This loss of chemotactic ability is coincident with a massive decrease in the amount of ATP present. Addition of lactate and oxygen to these ATP-deprived cells maintains their motility by providing the high-energy intermediate of oxidative phosphorylation but is incapable of allowing them to sense a gradient. These results not only establish the energy bases for the two processes in bacteria, but confirm the conclusion reached from mutant studies that motility and chemotaxis are interrelated but separate and can be controlled by separate genes and by separate enzymes.

6. MOTOR RESPONSES

6.1. Rotation

The construction of the bacterial flagellum and its attachment to the body of the microorganism is now known in considerable detail. There is an elaborate structure involving a helical filament, a hook, and several rings (DePamphilis and Adler, 1971). The outer ring is attached to the outer membrane and the inner ring is attached to the cytoplasmic membrane. At least 20 genes are required for the assembly and function of a flagellum in *E. coli* (Silverman and Simon, 1974*a*) and in *S. typhimurium* (Collins and Stocker, personal communication). In the peritrichous bacteria, such as *E. coli* and *Salmonella,* these flagella are attached at five or six places, apparently at random around the surface of the bacteria. When the bacterium swims, the flagella assemble together to form a bundle whose movement can be seen in the microscope but whose individual flagella are difficult to visualize. As a result, a number of theories were proposed as to the manner of propulsion of bacteria by these flagella. Two theories which gained most support were that: (a) a wave of structural conformational change proceeded from the bases to the tips of the flagella and that this undulating wave pushes the bacteria through the solution, and (b) the flagella are rigid and operate by rotation in analogy to a propellor (Berg, 1974). The answer was recently decided by the elegant experiments of Silverman and Simon (1974*b*), who tethered flagella of bacteria to the wall of a glass slide and observed rotation of the body of the bacterium. Using appropriate tests, they were able to show convincingly that the rotation theory was correct. Analysis of this process by Berg (1974) provided added understanding of the physical forces involved in the processes. Larsen *et al.* (1974*b*) then took such tethered bacteria and added attractants and repellents and found that the reversal of rotation occurred on generation of gradients in the presence of such bacteria. The reversals of rotation correlated precisely with the type of information that had been obtained previously for the tumbling frequency, thus leading to a very simple relationship. Bacteria swimming normally and smoothly utilize counterclockwise rotation of their flagella. When exposed to conditions that generate tumbling, these tethered bacteria showed clockwise rotation of flagella. Hence, the generation of tumbling involved reversal of flagella rotation.

6.2. Tumbling

The correlation of tumbling with the reversal of the flagella rotation was supported by studies on *Pseudomonas citronellolis* (Taylor and Koshland, 1974). This organism contained only a single flagellum and additions of attractants and repellents caused reversal of its rotation. In the case of the monoflagellate, reversal causes the bacterium to back up, i.e., to reverse direction. Since Macnab and Koshland had previously observed that the flagella appeared to fly apart when bacteria were subjected to high-intensity light these experiments led to the logical hypothesis shown in Figure 22 of the tumbling process (Macnab and Koshland, 1973; Taylor and Koshland, 1974). Smooth-swimming bacteria are proceeding with all flagella moving synchronously in a bundle and rotating in a counterclockwise manner. When the tumble regulator signals that tumbling should occur, one or more of the flagella reverse their rotation. In the case of a monoflagellated organism, this involves simple reversal of direction. In the case of a multiflagellated organism, this causes a bundle of flagella to fly apart. Because they are attached to surfaces of the bacterium at random positions and because they continue rotation, this causes a random turning of the bacterium. As the stimulus fades, the flagella return to a counterclockwise rotation and swim smoothly through the solution. Thus tumbling, which is the means by which the bacteria control a directional movement, is caused by a reversal of flagella rotation. Reversal of rotation is itself a simple switching device. So the all-or-none event at the end of the sensory process is caused by a simple switching in the mechanical process of rotation.

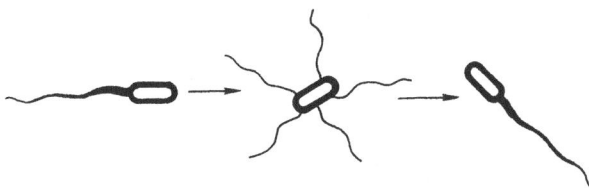

FIGURE 22. Schematic illustration of tumbling in multiflagellated bacteria. Bacteria swim with flagella in a bundle. Flagella fly apart when rotation is reversed. Flagella resume bundle as smooth swimming resumes.

7. CONCLUSION AND SUMMARY

7.1. Nature of the Bacterial System

These studies, although far from complete, do give a rather good overall picture of the bacterial sensing system. On the surface of the bacterium are located 20 or more receptors that report to the bacterium in regard to environmental conditions. Some of the receptors are designed to direct bacterial motion toward attractants, which are usually nutrients, and some are designed to direct bacteria away from toxic or crowded conditions. Most of the receptors are highly specific, interacting with only one or two physiological compounds, but a few are only moderately specific; e.g., one receptor reacts with a variety of fatty acids. The repellent receptors do

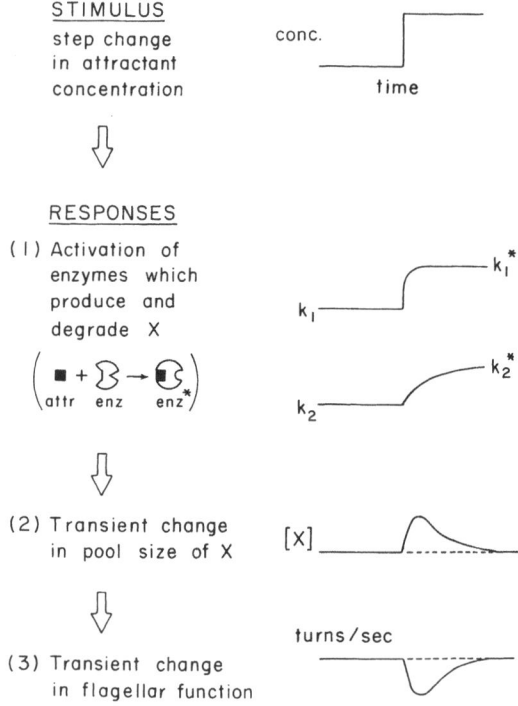

FIGURE 23. Schematic illustration of changes in levels of tumble regulator under influence of attractants and repellents. Increase in attractant increases rate k_1 rapidly to new value k_1^* and increases rate of k_2 more sluggishly to rate k_2^*. As a result, level of X rises temporarily above normal level, thus suppressing tumbling. (See text for more detailed description.)

not record all toxic conditions, nor is there an attractant receptor for all nutrients that are essential to the cell. This indicates that the receptors are monitoring the environment on a sampling basis rather than on a comprehensive basis.

The receptor transmits information through a generalized transmission system that is composed of at least nine gene components in *Salmonella* and at least four gene components in *E. coli*. This analytical system compares changes in concentration based on changes in the fraction of receptor occupancy. It is unaffected by an isotropic uniform concentration of attractant or repellent. The signal generated by a gradient is transmitted to the flagella and results in suppression of tumbling when the bacteria move in a favorable direction (up an attractant gradient or down a repellent gradient) and increases tumbling when moving in an unfavorable direction (down an attractant gradient or up a repellent gradient). The quantitative contribution of these two effects, however, is not equal; the response in the positive direction is the more powerful one. The system is able to integrate the responses from different signals, and to add them with proper algebraic signs, i.e., to integrate the effects of inhibition and activation.

The signal is ultimately transmitted to the flagella in a way that generates reversal of flagella or rotation. Reversal of flagella rotation causes the flagella to fly apart in a case of the multiflagellated bacteria, thereby generating a tumble. The frequency of the tumbling then represents the behavioral response by which the bacterium biases its random walk so that it can migrate toward favorable environmental situations and away from unfavorable ones.

A mechanism that can rationalize the behavior of the bacteria is symbolized as follows and illustrated in Figures 23 and 24. A *tumble regulator,* the concentration of which is varying in a Poissonian manner, generates tumbling when it falls below certain threshold values and suppresses tumbling when it rises above them. Increasing receptor occupancy stimulates the formation of a compound X, which we will use as the symbol for the parameter called *the tumble regulator* [Eq. (3)]:

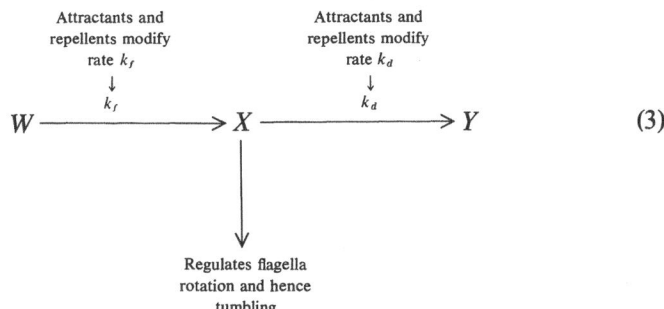

$$W \xrightarrow{\ \ k_f\ \ } X \xrightarrow{\ \ k_d\ \ } Y \tag{3}$$

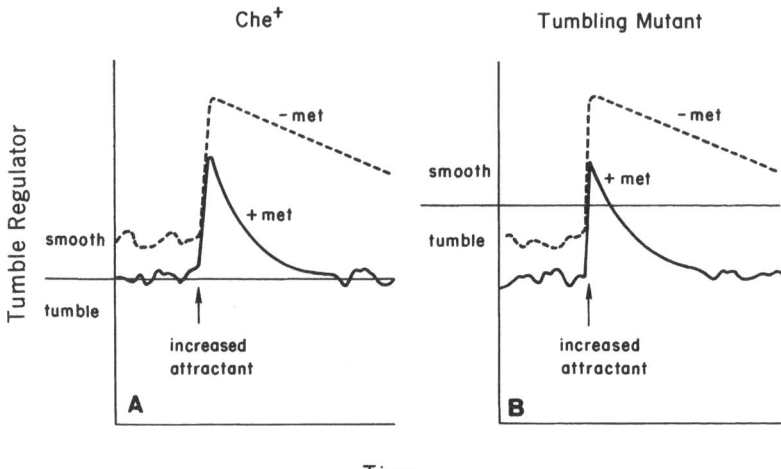

Time

FIGURE 24. A model for the regulation of tumbling frequency by gradients. (A) When the tumble regulator (wavy line) rises above the threshold value (horizontal solid line), tumbling is suppressed; when it falls below, tumbling is increased. In the absence of a gradient, the regulator level fluctuates randomly near the threshold level, but a rapid temporal increase in attractant makes it rise far above the threshold level, suppressing tumbling completely until it returns to normal. The regulator is maintained at a steady-state level by a delicate balance between its synthesis and degradation. S-adenosylmethionine (SAM) may be necessary for the degradation reaction. Thus, during methionine starvation, the level of regulator is higher than normal, so tumbling is suppressed, and when its level is further perturbed by a temporal gradient, it takes longer than normal to return to its steady-state level. By assuming that the uncoordinated mutant has an abnormally high threshold, one can easily explain its behavior in both the presence and the absence of methionine. (B) A model for the way in which the attractant can alter the rate of formation and degradation of the tumble regulator so that its level rises and falls on ascending or descending gradients.

If this stimulus leads to an increased rate in the formation of X relative to its rate of decomposition, X will accumulate as the bacteria go up a gradient of attractant and will therefore suppress tumbling. Its depletion when bacteria swim down a positive gradient leads to a decreased concentration of X and, hence, an increased probability of tumbling. A mutant that tumbles constantly has a level of tumble regulator below the threshold, a situation that could be produced either by an insufficient concentration of tumble regulator or by a change in threshold level. Conversely, a mutant that never tumbles produces tumble regulator above the threshold level.

In Figure 24 this approach is applied to some of the situations described. In the wild-type che^+ strain or a methionine auxotroph-fed methionine, the tumble regulator fluctuates in a Poissonian manner in an isotropic environment (the heavy line). A sudden increase in attractant increases the level of X and suppresses tumbling for a period of time before returning to normal. In an auxotroph deprived of methionine, the level of tumble regulator is always above threshold levels, so tumbling is always inhibited.

In the constantly tumbling mutant, the tumble regulator is always below threshold levels, and the bacteria are constantly receiving signals to tumble. If this mutant is given a sudden increase in attractant artificially, as was done in Figure 6, the level of X rises above threshold briefly. Traveling up a gradient could achieve this for the bacteria, but it never travels long enough in a straight line to straighten itself out. (The analogy to human psychopathology is obvious.)

In Figure 23, a way in which Eq. 3 could explain the responses to attractants and repellents is illustrated. Since an isotropic environment does not change the level of tumble regulator, chemoeffectors (attractants and repellents) must have equal ultimate effects on k_f and k_d, the formation and decomposition of the regulator X. However, it is not necessary that they respond equally rapidly. If k_1 and k_2 represent the initial rates, k_f and k_d in the isotropic case, we assume that an attractant activates both enzyme 1 and enzyme 2, but k_1 responds rapidly, whereas k_2 responds slowly. This could be caused by a slow conformational change in enzyme 2 versus a fast one in enzyme 1 or by the presence of enzyme 2 inside a membrane, whereas enzyme 1 is outside, or by a variety of other alternatives.

The effects of these differences would be shown in Figure 23 for an increase in attractant. Any increase in attractant will change k_1 more rapidly than k_2 and, for a short interval in time, it will increase the rate of formation of X above its rate of decomposition ($\Delta k_1 > \Delta k_2$). The level of X will rise, thus suppressing tumbling. Eventually the slow response of enzyme 2 will allow it to catch up with k_1 and then the rate of formation and decomposition of X will be equal once again. Some device for returning X to normal levels must also be built into the kinetics and that could be achieved in a variety of ways. At this point the return to normal levels is shown by a decay period. The area under the curve, therefore, represents the functioning memory time of the bacterium.

It can be seen that the illustration in Figure 23 applies equally well to increases in repellents or to decreases in attractants. In those cases k_1 decreases more rapidly than k_2, the supply of X is depleted, and tumbling is generated. Moreover, it applies equally well for changes causes by bacteria swimming in a gradient to artificial changes in temporal gradient situations. The length of the response may be longer, the initial rate of stimulus may be more rapid, but the basic process is similar.

If a methionine deficiency is created in the bacterium, as described in Section 5.2, the level of X rises, but not enough to produce a wild type. However, an attractant stimulus now raises the level of X higher above the threshold and increases the time to return to tumbling, as is observed. Additivity between attractants and repellents is also easily accommodated on a similar basis.

Other illustrations could be used. However, these suffice to provide a relatively simple picture of how bacterial memory could work. It is worthy of note that the memory is integrative and begins at the instant of a change. Thus, a bacterium swimming up a gradient of attractant immediately raises the level of X, and hence immediately decreases its probability of tumbling. As it continues to move up the gradient, it further suppresses tumbling, the amount depending on the steepness of the gradient and its velocity. In going down gradients the converse occurs. Thus the optimization of useful memory is easily accommodated by adjustment of the rates k_f and k_d. By their values and their rates of adjustment to the chemoeffectors, the length of the bacterial memory is controlled.

7.2.   Relationship to More Complex Systems

The relationship of the bacterial system to higher systems can be seen at several levels. In the first place, there is the formal analogy described in the introduction. Receptors are located in the periphery as in higher species, there is a distinct transmission system common to many receptors as in higher species, and there is a motor response as in higher species. Moreover, this sensory system is designed to provide survival value for the organism.

There is also an historical analogy. Nature tends to repeat herself in biochemical mechanisms, but each new phenomenon must prove this postulate over again to a skeptical world. In the early days of energy metabolism it was assumed that mammalian species could not possibly use the primitive energy sources of bacteria, but, indeed, ATP proved to be a universal energy source. In the early days of the genetic code, it was assumed that mammalian cells with a nucleus would use a genetic code quite different from the bacterial code, but they have turned out to be the same. These analogies in no way prove that sensory systems are identical, but they give comfort to the assumption that basic molecular mechanisms will be similar over the entire living system. A basic problem in neurochemistry is the amplification of the signal that involves a small fractional change in stimulus and generates an all-or-nothing response. Bacteria, obviously, have developed this capacity and their unit analytical device for detecting small changes appears in every way as extraordinary as the analytical devices of higher species. The manner in which a small change in environmental stimulus can lead to the firing of a neuron appears to have more than casual similarities to the way in which a bacterial receptor ultimately leads to the reversal of flagella rotation.

This analogy in no way implies that the precise chemical involved will be the same in all species. Since different neurotransmitters produce very similar effects, a precise identity between the chemical involved in the bacterial sensing system and that of higher species is not suggested. What is probably similar is the manner in which the change in receptor occupancy can be interpreted by the flagellum and the way the change in receptor occupancy in a higher neuron can be interpreted by that neuron.

Many other features of the bacterial system have analogies to the higher systems. The bacteria have been shown to have constantly tumbling mutants, which from the above example appear to be getting an excessive bombardment of stimuli to their sensing system. These bacteria cannot function because they never make sufficient translational progress in any direction to detect a gradient. This may be analogous to some problems in schizophrenia (Snyder, 1974). On the other hand, other types of mutations can damage the sensory apparatus in just the opposite way. Some of the mutants in structural genes of the flagella make it impossible for the flagella to respond even though the rest of the sensory apparatus is working perfectly as a kind of motor defect. Moreover, a chemical such as S-adenosylmethionine (which also has close relations to higher neural systems) appears to be important in the transmission of the sensory information. It is particularly impressive that the temporal gradient experiments, in which a sudden increase in concentration of attractant generates a time-dependent response, have such a close analogy to the action potential of neurons. It is also of interest that the light-generated response can produce perturbations of the transmission machinery not unlike electric shock on the neural system. Thus, bacteria, like higher species, have a sensing system that can be perturbed by chemicals, genetic defects, and physical stimuli.

7.3. Components of Higher Neural Processes

The study of the bacterial system focuses interest on what we mean by higher sensory processes. If we take words normally identified with complex mental functions, such as perception, choice, discrimination, memory, and judgment, the bacteria can be said to contain each of these functions in more than a formal or semantic sense. The bacteria can perceive changes in its environment. It can tell the direction of a gradient; it can choose whether to travel up or down a gradient. It can show discrimination since it responds to certain compounds positively, others negatively, and, to a third category, not at all. Moreover, its discrimination system is imperfect just as in higher species; it is attracted to some compounds that are not good for it and repelled by compounds that are not harmful.

The bacterium has a memory that allows it to compare events in the past with events in the present. It might well be asked whether this memory is simply a semantic relationship to higher memory processes. Certainly there is no evidence of long-term memory of the type identified with humans, but in this case as in all the others, it is important to draw a distinction between a simple process and a more complex manifestation of the same process. In this case the ability of the bacterium to compare its past with its present is not simply a fortuitous accident. It is an essential component of a biological process, which is essential for survival of the organism. Mutations that destroy this memory destroy the ability of the bacterium to sense gradients and can easily be selected against in biological situations.

Moreover, the analogy to short-term memory in man may be very real. As previously discussed, the bacterium has a memory span that is useful to it. Too long a memory would actually be nonfunctional because it would decrease the ability of the bacterium to sense gradients and confuse it with memories of events that were no longer relevant to its current situation. Short-term memory in man may operate the same way. The brain must operate with devices to allow it to forget certain observations as well as remember others. Otherwise, we would be cluttered with nonessential facts and could not possibly select the information needed for survival. Bacteria have no need for a long-term memory and their actual memory span is optimized for the specific functions they must perform.

The molecular features of bacterial memory have components to be expected in the sensory response of higher species. The receptor generates a change in level of a tumble regulator, a parameter X that might be a chemical pool, an electrical gradient, or a conformational change. Whatever it is, the formation and degradation of this parameter can serve as a time-dependent process, which stores the bacterial memory. In higher species, it is again likely that some kind of chemical or potential signal of a similar sort is involved in short-term memory. This chemical signal can possibly induce protein synthesis, creating new dendrites or adding enzymes to existing dendrites in such a way as to encode long-term memory. It seems unlikely, however, that protein synthesis can, *per se,* operate rapidly enough for the initial signal in bacteria or in the short-term memory of man. If the initial memory is, indeed, some transient pool, it can trigger more permanent devices for reinforcing or decreasing this memory and then generate protein synthesis leading to a permanent recording.

These considerations, then, lead to the last property: judgment. Bacteria can integrate the information from repellents and attractants in an algebraic manner to assess what is best for the organism. This pooling of excitatory and inhibitory events has a formal analogy to excitatory and inhibitory neurons. The firing of a particular neuron must have some kind of integrative machinery to assess incoming signals in the same way the bacteria can assess the variety of signals from its many receptors. It would seem probable that these two processes have biochemical similarities.

Once it is conceded that the biochemical mechanisms are similar at rudimentary and higher levels, words such as *choice, discrimination, judgment,* and the like are applicable to the continuum, and there is no precise point at which they are inappropriate. As one goes from the simplest species to the most complex, one is proceeding along the path from the simplest computer to the most complex computer. No new principles are developed, but the complexities of the inter-connections and the amount of information processed becomes more numerous and more complex. Thus, the molecular biology of monocellular species such as bacteria may well tell a great deal about the mechanisms by which the most complex regulatory device of all, the brain of man, operates.

8. REFERENCES

Adler, J., 1969, Chemoreceptors in bacteria, *Science* **166**:1588–1597.

Adler, J., 1973, A method for measuring chemotaxis and use of the method to determine optimum conditions for chemotaxis by *E. coli, J. Gen. Microbiol.* **74**:77–91.

Adler, J., 1975, Chemotaxis in bacteria, *Annu. Rev. Biochem.* **44**:341–356.

Adler, J., and Epstein, W., 1974, Phosphotransferase system enzymes as chemoreceptors for certain sugars in *E. coli* chemotaxis, *Proc. Natl. Acad. Sci. U.S.A.* **71**:2895–2899.

Adler, J., Hazelbauer, G. L., and Dahl, M. M., 1973, Chemotaxis towards sugars in *Escherichia coli, J. Bacteriol.* **115**: 824–847.

Aksamit, R., and Koshland, D. E., Jr., 1972, A ribose binding protein in *Salmonella typhimurium, Biochem. Biophys. Res. Commun.* **48**:1348–1353.

Aksamit, R., and Koshland, D. E., Jr., 1974, Identification of the ribose binding protein as the receptor for ribose chemotaxis in *Salmonella typhimurium, Biochemistry* **13**:4473–4478.

Anraku, Y., 1968, Transport of sugars and amino acids in bacteria, *J. Biol. Chem.* **243**: 3116–3122.

Armstrong, J. B., 1972, Chemotaxis and methionine metabolism in *E. coli, Can. J. Microbiol.* **18**:591–596.

Armstrong, J. B., and Adler, J., 1969a, Complementation of nonchemotactic mutants of *E. coli, Genetics* **61**:61–66.

Armstrong, J. B., and Adler, J., 1969b, Location of genes for motility and chemotaxis on the *Escherichia coli* genetic map,

Armstrong, J. B., Adler, J., and Dahl, M. M., 1967, Nonchemotactic mutants of *Escherichia coli, J. Bacteriol.* **93**:390–398.

Aswad, D., and Koshland, D. E., Jr., 1974, Role of methionine in bacterial chemotaxis, *J. Bacteriol.* **118**:640–645.

Aswad, D., and Koshland, D. E., Jr., 1975*a*, Isolation, characterization and complementation of *Salmonella typhimurium* chemotaxis mutants, *J. Mol. Biol.,* in press.

Aswad, D., and Koshland, D. E., Jr., 1975*b*, Evidence for an *S*-adenosyl methionine requirement in the chemotactic behavior of *Salmonella typhimonium, J. Mol. Biol.* **97**:207–223.

Beidler, L. M., 1954, A theory of taste stimulation, *J. Gen. Physiol.* **38**(2):133–139.

Beidler, L. M., 1962, Taste receptor stimulation, *Prog. Biophys. Biophys. Chem.* **12**:107–151.

Beidler, L. M., 1971, Taste receptor stimulation with salts and acids, in: *Handbook of Sensory Physiology* (L. M. Beidler, ed.), Vol. 4, Part 2, pp. 200–220, Springer-Verlag, New York.

Berg, H. C., 1971, How to track bacteria, *Rev. Sci. Instrum.* **42**:868–871.

Berg, H. C., 1974, Dynamic properties of bacterial flagellar motors, *Nature (London)* **249**:77–79.

Berg. H. C., and Anderson, R. A., 1973, Bacteria swim by rotating their flagellar filaments, *Nature* **239**:500–504.

Berg, H. C., and Brown, D. A., 1972, Chemotaxis in *Escherichia coli* analyzed by three dimensional tracking, *Nature (London)* **239**:500–504.

Bonker, J. T., Barkey, D. S., Hall, E. M., Konirn, T. M., Mason, J. W., O'Keefe, G., and Wolfe, P. B., 1969, Acrasin acrasinase and the sensitivity to acrasin in *Dictyostelium discoideum, Dev. Biol.* **20**:72–87.

Boos, W., 1969, The galactose binding protein and its relationship to the β-methylgalactoside permease from *Escherichia coli, Eur. J. Biochem.* **10**:66–73.

Boos, W., 1972, Structurally defective galactose-binding protein isolated from a mutant negative in the β-methylgalactoside transport system of *Escherichia coli, J. Biol. Chem.* **247**: 5414–5424.

Boos, W., and Gordon, A. S., 1971, Transport properties of the galactose binding protein of *Escherichia coli, J. Biol. Chem.* **246**:621–628,

Brown, D. A., and Berg, H. C., 1974, Temporal stimulation of chemotaxis in *Escherichia coli, Proc. Natl. Acad. Sci. U.S.A.* **71**:1388–1392.

Clayton, R. C., 1953, Studies in the photaxis of *Rhodospirullum rubrem, Microbiology* **19**:141–195.

Clayton, R. K., 1959, Phototaxis of purple bacteria, in: *Encyclopedia of Plant Physiologie* (W. Ruhland, ed.), Vol. XVII, pp. 371–387, Springer-Verlag, Berlin.

Dahlquist, F. W., Lovely, P., and Koshland, D. E., Jr., 1972, Quantitative analysis of bacterial migration in chemotaxis, *Nature (London)* **236**:120–123.

Dahlquist, F. W., Elwell, R. A., and Lovely, P., 1976, Studies of bacterial chemotaxis in defined concentration gradients: a model for chemotaxis toward L-serine, *J. Supramol. Struct.* **4**:354–362.

DePamphilis, M. L., and Adler, J., 1971, Fine structure and isolation of the Hook-Basel body complex of flagella from *Escherichia coli* and *Bacillus subtilis, J. Bacteriol.* **105**:394–395.

Dimmitt, K., and Simon, M., 1971, Purification and thermal stability of intact *Bacillus subtilis* flagella, *J. Bacteriol.* **105**:369–375.

Doetsch, R. N., and Seymour, W. F. K., Negative chemotaxis in bacteria, *Life Sci.* **9**:1029–1032.

Dryl, S., 1958, Photographic registration of movement of protozoa, *Bull. Acad. Polonaise Sci. Ser. Sci. Biol.* **6**:429–430.

Engelmann, T. W., 1881, Neue Methode zur Untersuchung der Sauerstoffausscheidung pflanzlicher und thierischer Organismen, *Pflügers Arch. Ges. Physiol.* **25**:285–292.

Ernster, L., and Lee, C. P., 1967, Energy-linked reduction of NAD$^+$ by succinate, in: *Methods in Enzymology,* Vol. 10 (R. W. Estabrook, ed.), pp. 729–744, Academic Press, New York.

Futrelle, R. P., and Berg, H. C., 1972, Specification of gradients used for studies of chemotaxis, *Nature (London)* **239**:517.

Gerisch, G.; and Hess, B., 1974, Cyclic AMP controlled oscillations in suspended dictyostelium cells their relation to morphogenetic cell interactions, *Proc. Natl. Acad. Sci. U.S.A.* 71:2118–2122.

Harold, F. M., 1972, Conservation and transformation of energy by bacterial membranes, *Bacteriol. Rev.* 36:172–230.

Hazelbauer, G. L., and Adler, J., 1971, Role of the galactose binding protein in chemotaxis of *Escherichia coli* towards galactose, *Nature (London) New Biol.* 30(12):101–104.

Hazelbauer, G. L., Mesibov, R. E., Adler, J., 1969, *Escherichia coli* mutants defective in chemotaxis toward specific chemicals, *Proc. Natl. Acad. Sci. U.S.A.* 64:1300–1307.

Heppel, L. A., 1967, Selective release of enzymes from bacteria, *Science* 156:1451–1455.

Hilman, M., Silverman, M., and Simon, M., 1976, Conference on assembly mechanisms, Squaw Valley, California, March 3–8, 1974, *J. Supramol. Struct.* 2, in press.

Hong, J. S., and Kaback, H. R., 1972, Mutants of *Salmonella typhimurium* and *Escherichia coli* pleiotropically defective in active transport, *Proc. Natl. Acad. Sci. U.S.A.* 69: 3336–3340.

Jennings, H. S., 1906, *Behavior of the Lower Organisms* (republished by Indiana University Press, Bloomington, 1962).

Kaback, H. R., 1972, Transport across isolated bacterial cytoplasmic membranes, *Biochim. Biophys. Acta* 265:367–416.

Kaissling, K. F., 1969, Kinetics of olfactory receptor potentials, in: *Olfaction and Taste* (Pfaffman, ed.) Vol. 2, pp. 52–87, Rockefeller University Press, New York.

Kalckar, H. M., 1971, The periplasmic galactose binding protein of *Escherichia coli, Science* 174:557–565.

Koshland, D. E., Jr., 1974, Chemotaxis as a model for sensory system, in: *FEBS Lett.* 40 (Supplement), North-Holland Publishing Co., Amsterdam, pp. S3–S9.

Koshland, D. E., Warnick, H., Taylor, B., and Spudich, J., 1976, in: *Cell Motility,* Book A (R. Colman, T. Polland, and J. Rosenbaum, eds.), Cold Spring Harbor Laboratory, Cold Spring Harbor, New York.

Kundig, W., and Roseman, S., 1971, Sugar transport. I. Isolation of a phosphotransferase system from *Escherichia coli, J. Biol. Chem.* 246:1393–1406.

Langridge, R. H., Shinagawa, X. X., and Pardee, A., 1970, Sulfate-binding protein from *Salmonella typhimurium:* physical properties, *Science* 169;59–61.

Larsen, S. H., Adler, J., Gargus, J. J., and Hogg, R. W., 1974a, Chemomechanical coupling without ATP. The source of energy for motility and chemotaxis in bacteria, *Proc. Natl. Acad. Sci. U.S.A.* 71:1239–1243.

Larsen, S. H., Reader, R. W., Kort, E. N., Tso, W.-W., Adler, J., 1974b, Change in direction of flagellar rotation is the basis of the chemotactic response, *Nature (London)* 249:74–77.

Lederberg, J., 1956, Linear inheritance in transduction clones, *Genetics* 41:845–871.

Lovely, P., and Dahlquist, F., 1975, Statistical measures of bacterial motility and chemotaxis, *J. Theoret. Biol.* 50:477–496.

Lovely, P., Dahlquist, F. W., Macnab, R., and Koshland, D. E., Jr., 1974, An instrument for recording the motions of microorganisms in chemical gradients, *Rev. Sci. Instrum.* 45: 683–686.

Lofgren, K. F., and Fox, C. F., 1974, Attractant-directed motility in *Escherichia coli* requirement for a fluid lipid phase, *J. Bacteriol.* 118:1181–1182.

Macnab, R. M., and Koshland, D. E., Jr., 1972, The gradient sensing mechanism in bacterial chemotaxis, *Proc. Natl. Acad. Sci. U.S.A.* 69:2509–2562.

Macnab, R. M., and Koshland, D. E., Jr., 1973, Persistence as a concept in the motility of chemotactic bacteria, *J. Mechanochem. Cell Motil.* 2:141–148.

Mesibov, R., and Adler, J., 1972, Chemotaxis towards amino acids in *Escherichia coli, J. Bacteriol.* 112:315–326.

Mesibov, R., Ordal, G. W., and Adler, J., 1973, The range of attractant concentrations for bacterial chemotaxis and the threshold and size of response over this range, *J. Gen. Physiol.* **62**:203–223.

Moncrieff, R. W., 1967, *The Chemical Senses,* Leonard Hall, London.

Ordal, G. W., and Adler, J., 1974, Properties of mutants in galactose taxis and transport, *J. Bacteriol.* **117**:517–526.

Pardee, A. B., 1968, Membrane transport proteins, *Science* **162**:632–637.

Parkinson, J. F., 1974, Data processing by the chemotaxis machinery of *Escherichia coli, Nature (London)* **252**:317–319.

Pfeffer, W., 1883, Locomotorische Richtungsbewegungen durch chemische Reize, *Ber. Dtsch. Bot. Ges.* **1**:524–533.

Pfeffer, W., 1884, Locomotorische Richtungsbewegunge durch chemische Reize, *Untersuch Bot. Inst. Tübingen* **1**:363–482.

Pfeffer, W., 1888, Über chemotaktische Bewegungen von Bacterien, Flagellaten und Volvocineen, *Untersuch Bot. Inst. Tübingen.* **2**:582–661.

Raab, O., 1900, Über die Wirkung fluorescirender Stoffe auf Infusioren. *Z. Biol.* **39**:524–546.

Rothert, W., 1904, Über die Wirkung des Äthers und des Chloroforms auf die Reizbewegungen der Mikroorganismen, *Jahrb. Wiss. Bot.* **39**:1–70.

Scandella, C. J., Devaus, P., and McConnell, H. M., 1972, Rapid lateral diffusion of phospholipids in rabbit sarcoplasmic reticulum, *Proc. Natl. Acad. Sci. U.S.A.* **69**:2056–2060.

Schwartz, M., and Kellerman, O., 1972, The maltose binding protein, in: *Diplome d'Etudies Approfondies,* University of Paris VII.

Seymour, F. W. K., and Doetsch, R. N., 1973, Chemotactic response by motile bacteria, *J. Gen. Microbiol.* **78**:287–296.

Silverman, M., and Simon, M., 1974a, Positioning flagellar genes in *Escherichia coli* by deletion analysis, *J. Bacteriol.* **117**:73–79.

Silverman, M. R., and Simon, M. I., 1974b, Flagellar rotation and the mechanism of bacterial motility, *Nature (London)* **249**:73–74.

Smith, L. M., and Doetsch, R. N., 1969, Studies on negative chemotaxis and the survival value of motility in *Pseudomonas fluorescens, J. Gen. Microbiol.* **55**:379–390.

Snyder, S., 1974, *Madness and the Brain,* McGraw-Hill, New York.

Spudich, J. L., and Koshland, D. E., Jr., 1975, Quantitation of the sensory response in bacterial chemotaxis, *Proc. Natl. Acad. Sci. U.S.A.* **72**:710–713.

Stevens, S. S., 1970, Neural events and the psychophysical law, *Science* **170**:1043–1050.

Strange, P., and Koshland, D. E., Jr., 1976, Receptor interactions in a signaling system: Competition between ribbon receptor and galactose receptor in the chemotactic response, *Proc. Natl. Acad. Sci. U.S.A.* **73**:762–766.

Taylor, B. L., and Koshland, D. E., Jr., 1974, Reversal of flagellar rotation in monotrichous and peritrichous bacteria: generation and changes in direction, *J. Bacteriol.* **119**:640–642.

Taylor, B., and Koshland, D. E., Jr., 1976, The intrinsic and extrinsic light responses of *Salmonella typhimurium* and *Escherichia coli, J. Bacteriol.* **123**:557–569.

Tsang, N., Macnab, R. M., and Koshland, D. E., Jr., 1973, Common mechanism for repellents and attractants in bacterial chemotaxis, *Science* **181**:60–63.

Tso, W.-W., and Adler, J., 1974a, Negative chemotaxis in *Escherichia coli, J. Bacteriol.* **118**:560–576.

Tso, W.-W., and Adler, J., 1974b, "Decision" making in bacteria: chemotactic response of *Escherichia coli* to conflicting stimuli, *Science* **184**:1292.

Vaituzis, A., and Doetsch, R. N., 1969, Motility tracks: techniques for quantitation study of bacterial movement, *Appl. Microbiol.* **17**:588.

Weibull, C., 1960, Movement, in: *The Bacteria* (I. C. Gunsalus and R. Y. Stanier, eds.), Vol. 1, pp. 153–205, Academic Press, New York.

Yamaguchi, Y., Iino, T., Horiguchi, T., and Ohta, K., 1972, Genetic analysis of *fla* and *mot* cistrons closely linked to H1 in *Salmonella abortusequi* and its derivatives, *J. Gen. Microbiol.* **70**:59–75.

Ziegler, H., 1962, Chemotaxis, in: *Handbuch der Pflanzenphysiologie*, Vol. 17, Part II, pp. 484–532.

INDEX

Glucose, in sleep, 229-235
Glucose-6-phosphatase, and sleep, 225-231, 235, 254
L-Glutamate
in PKU, 110, 115
as sensory transmitter, 195, 201, 203, 204, 205-206
Glutaminase, and sleep, 223, 248-249
Glutamine, in PKU, 90-91
Glutamine synthetase, and sleep, 223, 248-249
Glycogen, and sleep, 235-237
Glycolysis, and sleep, 237-240

Hartnup disease, and tryptophan, 178
Heparin, 143
Hepatic coma, and tryptophan, 174
Homogentisic acid, 7, 160
Hydrocortisone, and tryptophan, 139
o-Hydroxyphenylacetate, in PKU, 4, 5, 50, 52
p-Hydroxyphenyllactic acid, in PKU, 4, 81
p-Hydroxyphenylpyruvate hydroxylase, 52, 82
p-Hydroxyphenylpyruvic acid, in PKU, 4, 81, 82-83
5-Hydroxytryptamine (5-HT), 133, 135, 137, 139-140, 141, 145, 154, 155-158, 159, 160, 164, 167, 168-170, 174-178, 255
metabolites of, 158-159
(see also Serotonin)
Hyperphenylalaninemia, 25, 29-65
heterozygotes for, 65-69
transient, 61-62
(see also Phenylketonuria)

Imbecilitas phenylpyruvica
(see Phenylketonuria)
Indican, 4, 81, 82
Indoleacetic acid, 4, 82
Indomethacin, 142
Instantaneous spatial sensing, 298
Isonicotinic acid hydrazide, 111
Isoprenaline, 143

Kynurenine pathway, 138-141

Lactate, and sleep, 232-235
Lioresal, 204-205, 206, 207

Lithium treatment and tryptophan, 174
Luteinizing-hormone-releasing-hormone (LHRH), 155
Lysolecithin, 17, 43, 63, 67, 114

Mandelate, in PKU, 4
Maple syrup urine disease, 94, 107-108
Melatonin, 158, 160-161
and sleep, 258, 265
Mental retardation, in PKU, 2, 3, 70, 74, 75-113
Methotrexate, 64
α-Methyl paratyrosine, 257
DL-α-methyltryptophan (αMeTrp), 139, 141, 156
Metrazol, 246
Monoamine oxidase (MAO), 156, 158, 159, 163
Myelinization, in PKU, 3, 91-94, 104-106, 113

Negative interallelic complementation, 67-68
Nicotinic acid, 143
Non-albumin-bound tryptophan, 142-144, 163, 165
and electroconvulsive therapy, 172-173
and hepatic coma, 176
and sleep, 164
Nonchemotactic mutant, 294-295, 296, 297, 321
Norepinephrine
in PKU, 41, 54, 83, 88-90
and sleep, 222, 253, 256-257

Paradoxical sleep (PS), 213, 214, 236, 237
and activity, 219-220
and brain temperature, 220
and cerebral blood flow, 216-219
and intracranial pressure, 220-221
and neurotransmission, 254-258
and protein metabolism, 250-254
Paradoxical sleep deprivation
and amino acids, 243-250
and ammonia metabolism, 248-249
and cerebral enzymes, 221-231
and glycogen, 235-237
and glycolysis, 237-240
method, 214, 237
Pedestal method, 214, 237